Earth and Environmental Sciences Library

Earth and Environmental Sciences Library (EESL) is a multidisciplinary book series focusing on innovative approaches and solid reviews to strengthen the role of the Earth and Environmental Sciences communities, while also providing sound guidance for stakeholders, decision-makers, policymakers, international organizations, and NGOs.

Topics of interest include oceanography, the marine environment, atmospheric sciences, hydrology and soil sciences, geophysics and geology, agriculture, environmental pollution, remote sensing, climate change, water resources, and natural resources management. In pursuit of these topics, the Earth Sciences and Environmental Sciences communities are invited to share their knowledge and expertise in the form of edited books, monographs, and conference proceedings.

Shalini Yadav • Abdelazim M. Negm •
Ram Narayan Yadava
Editors

Water Quality, Assessment and Management in India

Springer

Editors
Shalini Yadav
Rabindranath Tagore University
Bhopal, Madhya Pradesh, India

Abdelazim M. Negm
Faculty of Engineering
Zagazig University
Zagazig, Egypt

Ram Narayan Yadava
Madhyanchal Professional University
Bhopal, India

ISSN 2730-6674 ISSN 2730-6682 (electronic)
Earth and Environmental Sciences Library
ISBN 978-3-030-95686-8 ISBN 978-3-030-95687-5 (eBook)
https://doi.org/10.1007/978-3-030-95687-5

This Springer imprint is published by the registered company Springer Nature Switzerland AG
The registered company address is: Gewerbestrasse 11, 6330 Cham, Switzerland

Preface

Water, environment, and energy sustainability is fundamental to sustainable economic development, functioning of healthy ecosystems, reliable agricultural productivity, maintenance of desirable environmental quality, enjoyment of quality lifestyle, management of surface and groundwater quality, and assessment, which are considered as priority requirement by the scientists, academicians, researchers, practicing engineers, consultants, planners, policy makers, economists and social scientists, managers, and leaders from around the world to share their knowledge, skills, experiences, and expertise through research findings and the case studies under different climatic environment.

The motivation of this volume titled *Water Quality, Assessment and Management in India* is to support India and the developing countries by providing scientific, technological, and skilled information on the past, present, and future scenario of the management and assessment of quality of surface as well as groundwater in India; with the growing population, demands for water for domestic, agricultural, and industrial uses are skyrocketing including major concerns over decreasing freshwater resources per capita in terms of quantity and quality.

The nineteen chapters of the book are divided into seven parts dealing with various methodologies and technologies on water quality, assessment, and management of surface as well as groundwater. Part I deals with *Introduction* on overview of water resources, quality, and management in India. Part II covers the *Status of Water Resources in India*, which deals with the interlinking of rivers and overview on water resources management, whereas Part III—*Water Harvesting and Water Qualities*—consists of seven chapters dealing with drought characterization, rainwater harvesting, assessment of groundwater, forest water interaction, water quality, and groundwater exploration. Part IV—*Water Quality Assessment and Management*—comprises **four** chapters and discusses the assessment of river water quality and multi-industrial-effluent outfalls. Part V deals with the *Groundwater Quality Assessment* and has **three** chapters, whereas identification of heavy metals and radionuclides in soil is discussed in Part VI—*Control of Water Quality*. Part VII is devoted to conclusions and recommendations of the book. The novelty of each chapter under all the above-cited parts is summarized as follows:

The first chapter titled "Overview of Water Resources, Quality, and Management in India" discusses the important surface water and groundwater resources in India including several strategies regarding national water policy, water resources planning and management, and water treatment and purification that have been addressed by the government, policymakers, and scientists. The content of this chapter would assist local people, stakeholders, government officials, and industrialists, dealing with water issues in India.

The second chapter titled "The Interlinking of Indian Rivers for Sustainable Use of Water Resources in India: An Overview" focuses on the policy and the benefits for the people of India and provides certain alternatives to solve water issues in India. Also, it suggests the different sustainable options keeping in mind the economic, technological, and environmental feasibility analysis.

The third chapter titled "An Overview: Water Resources Management Aspect in India" gives an overview of the various aspects associated with water resources management in India. Also, it elaborates on how these aspects are important from water resources management point of view in the country and developments made so far in those areas toward achieving the goal and further scope of the work. Further various core issues of the water resources sector are discussed such as total water availability on a global scale and in India, climatic regions, rainfall patterns, and rainfall availability including land and water resources details of India.

The fourth chapter, titled "Drought Characterization During Monsoon Months Based on Standardized Precipitation Index (SPI) in Nuapada District, Odisha, India," discusses the interpretation of drought using drought indices to mitigate the drought by adopting suitable cropping pattern, growing drought-resistant crop varieties, harvesting the runoff water, and water-saving techniques. The interpretation of drought utilizing drought indices such as SPI can guide the water resource planners and experts to formulate the strategies for coping with the drought in case of drought-prone areas.

The fifth chapter titled "Forest-Water Interactions at Catchment Scales for Pollution Management" focuses on forest hydrology, storage of water, and physics of handling rainfall via vegetative stands of forests. Also, a proper understanding of the hydrological cycle is obligatory for the forest–water interactions including determinants of forest–water connection, forest water quality, and climate change issues for forested catchments. Further, the maintenance of good or high ecological status of water bodies of forest catchments could be made easier by applying strategic forest development plans for preserving high-quality drainage waters with lowered nutrient/pesticide/sediments.

The sixth chapter, titled "Assessment of Groundwater Quality In And Around Nemawar, Madhya Pradesh, India," deals with the hydrochemistry of groundwater and derives cause and effect relationship between agriculture practices and the generation of nitrogenous, phosphatic, and potassic compounds leaving the agriculture system and entering into the domain of subsurface water. Also, subsurface water having a high concentration of nitrate, potash, and phosphate is recommended for irrigation as a substitute for the fertilizers. This recycling may lead to an improvement in the quality of subsurface water.

The seventh chapter titled "Hydrological Response of Factors Affecting Rainfall Water Discharge and Water Balance: A Case Study of Tons Watershed" focuses on the understanding of the hydrological processes such as discharge and sediment flux of a watershed and sustainable exploitation of the water resource. Further, it recommends that the outcomes will be worthwhile in planning to reduce the inflow of silt into the reservoir, conserve soil, and minimize runoff in the extreme rainfall years and suggests the formulation of strategies to planners and engineers to control erosion and landslides for watershed management on a precursory basis.

The eighth chapter entitled "Water Quality Status of the Narmada River Across Its Basin: Predicting Water Quality Using Artificial Neural Network" discusses the use of artificial neural network (ANN) model for predicting river water quality and advantages and suggests that the model can be used to predict the chemical oxygen demand (COD) category, i.e., high COD level and low COD level with 85.71% accuracy. The study showcases the ability of the ANN to perform the prediction modeling of a river.

The ninth chapter titled "Rain Water Harvesting Methods in Rajasthan" deals with the rainwater potentials to increase the groundwater level. To meet the demands of water, it is suggested that the surface and groundwater resources should be recharged with the help of rainwater. The water has a neutral pH and it is free from other dissolved impurities. Rainwater harvesting is one of the environmental sound techniques that could be carried out at an individual level. Cities with low annual rainfall should develop a neat setup of their drains as they provide a large catchment area.

The tenth chapter, titled "Groundwater Exploration Using Remote Sensing and GIS Techniques Coupled With Vertical Electrical Soundings From Hard Rock Terrain: A Case Study In Salem District, Southern India," discusses the application of remote sensing (RS) and geographical information system (GIS) techniques coupled with geophysical techniques in the exploration and assessment of groundwater especially in hard rock terrain as it is crucial for successful implementation. Also, it is suggested that the use of high-resolution remotely sensed data, identification of features pertinent to hydrological, geological, and geo-morphological parameters, preparation of corresponding thematic maps, assigning appropriate weights, and integration in GIS environment ensure and enhance the prediction accuracy of location of promising zones of groundwater potential.

The eleventh chapter titled "Management of Coal Fly Ash Leachates Generated From Disposal Sites Near Thermal Power Plants" focuses on the fly ash disposal method by the Thermal Power Authorities. This method should be chosen on the basis of climate and geological strata, groundwater table, and position of surface water bodies from the disposal site, and regular monitoring of water quality near the disposal site must be carried out in and around the Thermal Power disposal sites. Further, regular monitoring of water quality near the disposal site and planning of remediation measures should be taken accordingly.

The twelfth chapter, titled "Assessment of Water Quality by Evaluating Water Quality Index Over Kolkata Using Statistical Approach—A Step Towards Water Pollution Management," describes the assessment of water quality by evaluating water quality index. For this purpose, the analysis of different physicochemical parameters is considered. From the analysis of the parameters, a distinct seasonal variability has been observed. In addition to that, water quality has been monitored through the calculation of water quality index, and the stakeholders can take necessary actions accordingly to deal with.

The thirteenth chapter titled "Assessment of Water Quality of Ramganga River in Moradabad Region" discusses the assessment of the water quality of the Ramganga River in the province of Uttar Pradesh; the water quality of the river has deteriorated severely due to the discharge of domestic wastewater and effluents from metal, food, paper, and sugar industries in and around the region. The water quality has been assessed in terms of the dissolved oxygen, fecal contamination, BOD, pH, and TDS, on the samples taken during the summer, pre-monsoon, post-monsoon, and winter season. The water quality has been assessed and major determinants responsible for water pollution have been identified. Also, a comprehensive remedial method is proposed with great emphasis on the monitoring of the river flow through satellite imagery and assessment of its water quality on a real-time basis. The assessment results should be put in the public domain to make people aware and responsible.

The fourteenth chapter titled "Assessment of Surface Water Quality of Indian Rivers in Terms of Water Quality Index (WQI)" focuses on a comprehensive review of water quality of Indian rivers and can provide an effective approach for the assessment and management of water quality. Further, the water quality index (WQI) as a mathematical tool is suggested to access the river water quality which is useful for the decision-makers and stakeholders to analyze the water quality easily. Water quality assessment has been performed on many Indian rivers in terms of WQI, and the suitability of water resources for human consumption has been addressed.

The fifteenth chapter titled "Groundwater Quality Assessment in a Semi-Arid Block of Rajasthan, India: A Combined Approach of Fuzzy Aggregation Technique with GIS" deals with the overall assessment of the status of groundwater quantity, quality, and its sustainability at the regional level. Further, it discusses a fuzzy-based GW status assessment framework that can be used to quantify the severity level and identify the spatial regions of critical importance in terms of sustainability. The proposed GIS-integrated fuzzy index model has been found to be an efficient tool in preparing priority maps of the chosen region (irrespective of the size) that can be used by decision-makers directly to get a broad perspective of a given GW quality and quantity problem.

The sixteenth chapter titled "Assessment of Groundwater Quality Parameters for Drinking Purpose Using IDW, GIS and Statistical Analysis Methods: A Case Study of Basaltic Hard Rock Area in Mahesh River Basin, Akola and Buldhana Districts" focuses on the selected groundwater quality parameters for drinking and irrigation purposes in the basaltic hard rock area and regularly groundwater quality analysis is needed for human, animal, and sustainable agriculture development. It also suggests that modern tools and software are playing a very significant role in the processing of groundwater quality thematic maps and plotting, which are useful in the planning of drinking and irrigation purposes.

The seventeenth chapter, titled "Evaluation of Hydro-Meteorological Conditions and Water Resource Potential for a Coalfield Area of Damodar Valley, India," discusses the assessment of the hydro-meteorological conditions and understanding of the natural groundwater recharge potentials in a coalfield area. Further besides rainfall, the mine water discharge from the local mining areas and existing water bodies, including water logged in abundant mine quarries, also contributes to the groundwater recharge as return flow.

The eighteenth chapter, titled "Distribution and Statistical Source Identification of Heavy Metals and Radionuclides in Soil: A Case Study From a Proposed Uranium Mining Site, Jharkhand, India," describes the assessment of the metal contamination in the soil with the help of indices like enrichment factor and geo-accumulation index. Also, regular monitoring of soil samples in the vicinity of the mining areas is recommended so as to know the post-mining effect with respect to radionuclide and metals, and human health risk assessment is also suggested due to the exposure of the radionuclides and metals in the soil through the ingestion, dermal, and inhalation pathways.

The nineteenth chapter titled "Conclusions and Recommendations for 'Water Quality, Assessment and Management in India'" summarizes the main conclusions and recommendations from all chapters in addition to an update of the significant literature that is connected to the themes of the book.

The editors want to thank and acknowledge the efforts provided by the Springer team to make this book a reality.

The volume editors would be happy to receive any comments to further improve future editions. Comments, feedback, suggestions for further improvement, and proposal for new chapters for next editions are welcome and should be sent directly to the volume editors.

Bhopal, India
Zagazig, Egypt
Bhopal, India
April 2020

Shalini Yadav
Abdelazim M. Negm
Ram Narayan Yadava

Contents

About the Editors

Shalini Yadav is Professor in the Dept. of Civil Engineering and Head of the Centre of Excellence in Advanced Water and Environmental Research, Rabindranath Tagore University, Bhopal, India. Her research interests include solid and hazardous waste management, construction management, environmental quality, and water resources. She has executed a variety of research/consultancy projects in environmental and water science and technology, and has got rich experience in planning, formulating, organizing, executing, and management of R&D programs. She guides M. Tech. and Ph.D. students.

She has published more than 50 journal articles and technical reports. She is a member of the organizing and scientific committee of several conferences and reviewer in several of international journals Also, she has published a number of edited books, namely *Climate Change Impacts*, *Water Resources Management*, *Groundwater, Energy and Environment, Environmental Pollution, Hydrologic and Modeling*, and *Water Quality Management* in the Water Science and Technology Library series, Springer.

Abdelazim M. Negm is professor of hydraulics (and water resources). Currently, he is interested in sustainability studies, sustainable development, and the green environment in addition to water resources management. Currently, he is the head of the Egyptian permanent scientific committee for Water Resources, Supreme Council of Egyptian Universities, cycle 13, and was the vice head in cycle 12. He has published more than 350 papers and more than 50 chapters and edited more than 35 international books in Springer. He is a member of the organizing and scientific committee of several conferences and associate editor of several international journals. He is reviewer for more than 40 international journals. He is a member of the editorial board of HEC series, Springer.

Ram Narayan Yadava holds the position of Director of Research and International Affairs in Madhyanchal Professional University, Madhya Pradesh, India. Also, he has worked as a founding Vice Chancellor of the AISECT University, Jharkhand,

India. In addition, he served as a Director Gr. Scientist at the Natural Resources Development Center of the CSIR-AMPRI, India His research interests include environment and water resources, hydrologic modeling, and R&D planning and management. Dr. Yadava has executed a variety of research/consultancy projects in the area of his research interest. He has adequate experience in establishing national institutions/organizations.

Dr. Yadava has been recognized for three and half decades of leadership in research and service to the hydrologic, environment, and water resources profession. He has published more than 100 journal articles, 4 textbooks, and 14 edited reference books. He is reviewer of scientific journals and member of the scientific committee of international conferences. He also holds the position of Vice President of International Association of Water, Environment, Energy and Society (IAWEES).

Part I
Introduction

Overview of Water Resources, Quality, and Management in India

Mennat Allah Neama, Michael Attia, Abdelazim M. Negm, and Mahmoud Nasr

Abstract Recently, India has faced major challenges in the domain of water resource management due to the rapid increase in residential, agricultural, industrial, and various domestic requirements. This chapter gives an overview of the important surface water and groundwater resources in India. Previous publications, academic journals, institutional affiliation, and sponsoring agencies available in the SCOPUS database covering the water aspects in India are mentioned. Information about the major rivers, lakes, reservoirs, dams, and groundwater regarding the literature studies is presented. Feasible and cost-effective actions required for the long-term management of the water scarcity issues are recommended. The chapter depicts that several strategies regarding national water policy, water resources planning and management, and water treatment and purification have been addressed by the government, policymakers, and scientists. The outputs of this chapter would assist local people, stakeholders, government officials, and industrialists, dealing with water issues in India.

M. A. Neama
Sanitary Engineering Department, Faculty of Engineering, Alexandria University, Alexandria, Egypt

M. Attia
Irrigation Engineering and Hydraulics Department, Faculty of Engineering, Alexandria University, Alexandria, Egypt

A. M. Negm
Water and Water structures Engineering Department, Faculty of Engineering, Zagazig University, Zagazig, Egypt
e-mail: amnegm@zu.edu.eg

M. Nasr (✉)
Sanitary Engineering Department, Faculty of Engineering, Alexandria University, Alexandria, Egypt

Environmental Engineering Department, Egypt-Japan University of Science and Technology (E-JUST), Alexandria, Egypt
e-mail: mahmoud-nasr@alexu.edu.eg; mahmoud.nasr@ejust.edu.eg

S. Yadav et al. (eds.), *Water Quality, Assessment and Management in India*, Earth and Environmental Sciences Library, https://doi.org/10.1007/978-3-030-95687-5_1

Keywords India · SCOPUS database · Water management · Water research · Water resources

1 Introduction

According to the Central Water Commission (CWC) (http://cwc.gov.in/water-info), India has a geographical area of about 3.29 million km^2, representing 2.4% of the world area. The annual rainfall in India was 1208 mm in 2005, and the primary demand for water (over 85%) comes from the agricultural sector. Moreover, India has about 46 and 12 of medium (0.246 million km^2) and major (2.53 million km^2) river basins, respectively. The total annual utilizable water resources reach 1123 billion cubic meters, corresponding to water availability of about 1720 cubic meters per capita (http://cwc.gov.in/water-info). Recently, the water consumption pattern in India has increased due to urbanization expansion, rapid population growth, lifestyle change, climatic condition, and industrial transformation [1, 2]. In addition, the recent rapid rise of industrialization, agricultural practices, and other human activities has resulted in the deterioration of several water bodies [3, 4]. Moreover, the variability of water resources has been adversely impacted by the unsustainable withdrawal of water from surface and subsurface water bodies [5, 6]. Jain [4] reported that the management of water in India is influenced by several factors, including availability, fluctuation, sharing disputes, and consumption of water, climate change, and land-use cover. In this context, further scientific, environmental, and engineering efforts should be progressed to address the recent and upcoming challenges associated with the water resources in India.

This chapter offers an overview of the status of water resources (surface water and groundwater) in India. The number and type of documents that have been recently published to handle the management of water bodies in India have been reported. Moreover, the chapter gives comprehensive information about the countries, institutional affiliations, and funding organizations and sponsors that concern about the water quality in India.

2 Indian's Water Statistics from SCOPUS Database.

Figure 1 represents the number of documents obtained from the SCOPUS database, covering most water issues in India (https://www.scopus.com/search/form.uri?display=basic). By using the search keywords "Water", "Resources", and "India", the cumulative number of documents found in SCOPUS was 2017 during 2001–2010. This number increased to 4911 during 2011–2019 (Fig. 1a). Moreover, the total numbers of documents retrieved using the search keywords "Water", "Management", and "India" were 2226 and 4670 during 2001–2010 and 2011–2019, respectively (Fig. 1b). Furthermore, by using the search keywords "Water", "Quality", and "India" in the SCOPUS database, the total number of

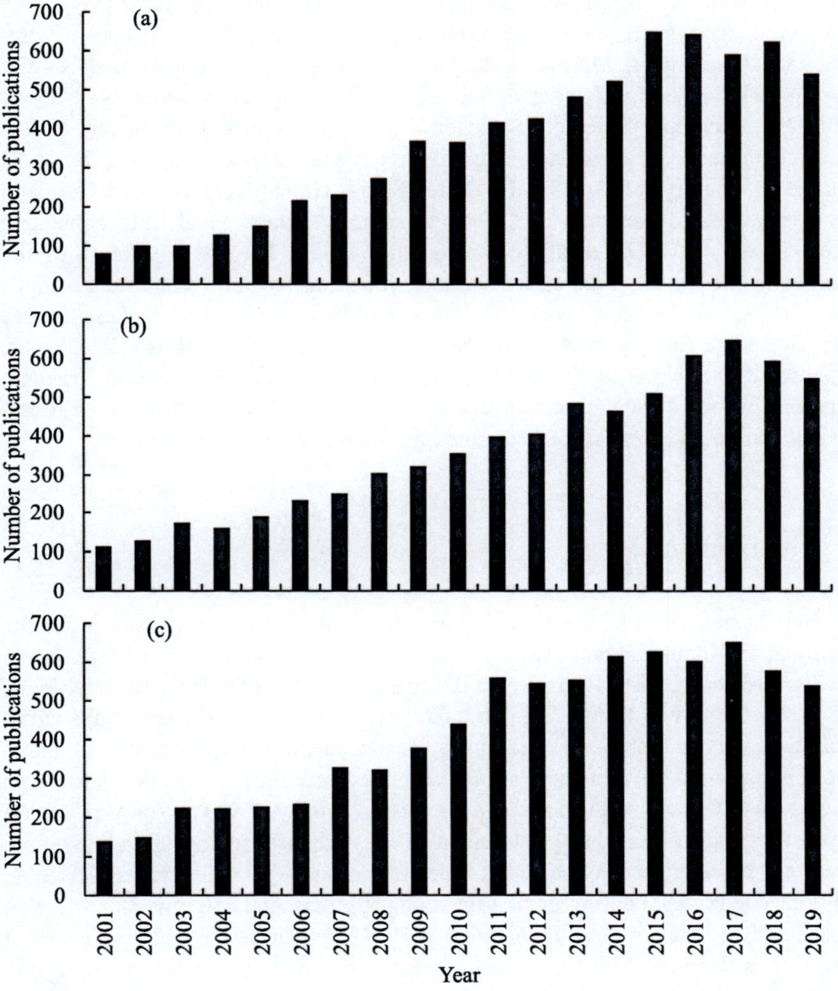

Fig. 1 Increasing pattern of published documents retrieved from SCOPUS database (www.Scopus.com) using research keywords (**a**) "Water", "Resources", and "India", (**b**) "Water", "Management", and "India", and (**c**) "Water", "Quality", and "India"

documents was 2669 during 2001–2010. This number attained about a two-fold increase (i.e., 5281 documents) during 2011–2019 (Fig. 1c). The increasing trend in the number of published documents suggests that the topic of "Water Resources, Quality, and Management in India" has become an essential point of research. The documents were published in various international journals such as Environmental Monitoring and Assessment, Environmental Earth Sciences, Water Resources Management, Journal of Food Science and Technology, International Journal of Applied Engineering Research (IJAER), Agricultural Water Management, International

Journal of Remote Sensing, and Journal of Hydrology. These journals focus on the following subject areas: Geosciences, Development of Monitoring Systems, Pollution Risk Assessment, Marine Sciences, Climate Change, Humans and Natural Resources Interaction, Water and Soil Contamination, and Biosystems. Various publishers, including Elsevier, Springer, Taylor & Francis Group, and Wiley, managed these peer-reviewed and highly reputable international journals. India was the top country followed by the United States, United Kingdom, and Germany that contributed to the topic of "Water Resources, Quality, and Management in India" in the SCOPUS database during 2001–2019. The type of most of the documents (i.e., over 75%) was an article, followed by conference manuscripts, book chapters, and review papers. The important funding sponsors of these articles were the Indian Council of Agricultural Research, University Grants Commission, Department of Science and Technology, Central Mechanical Engineering Research Institute (Council of Scientific and Industrial Research), Indian Council of Agricultural Research, and Council of Scientific and Industrial Research, India.

3 Rivers in India

The river system in India is considered an important source of drinking, domestic, irrigation, and industrial purposes [7]. Jain [4] reported that the rivers of India could be classified into Coastal, Himalayan, Deccan, and Inland Drainage Basin Rivers. For instance, the river Ganga (e.g., Himalayan River) is one of the essential sources of drinking water, outdoor recreation, and various human applications in India, representing about 25% of India's total water resources [8]. It covers a basin area of about 861,404 km^2, with an average population density of 520 persons/km^2. It has played a crucial role in the development of the Indian economy and civilization. However, the river receives different types of wastes (e.g., industrial, agricultural, and domestic sewage) during its passage along approximately 50 towns, 29 cities of class-I, and 23 cities of class-II. A study by Paul [8] reviews the pollution status of river Ganga conducted by the previous researchers, mainly due to anthropogenic sources. Because of this pollution condition, the Ganga Action Plan (GAP) and other River Action Plans (RAPs) have been implemented by the Prime Minister of India to evaluate, monitor, and protect the river water quality in India [9]. The river Yamuna is also an important source of drinking water in India, providing about 70% of Delhi's water requirements [10]. Yamuna is the fifth-longest river in India characterized by a total length of about 1376 km, and it connects to the River Ganga at Allahabad, Uttar Pradesh. However, several parts of the river water are unfit for potable purposes and may cause paucity of the fish organisms due to the discharge of untreated or partially treated wastewater [7]. A previous study by Sharma et al. [11] represented the water quality and pollution status of Yamuna River, India, during 1999–2009. Moreover, Jaiswal et al. [7] reported that Yamuna River is negatively influenced by various natural (e.g., erosion and weathering) and anthropogenic factors. Ghaghara in India is also an essential river of the Ganga Plain, joining the

Ganga River near Maharajganj, Bihar state [12]. Extensive studies have also been conducted to address the quality and management of various Indian rivers such as Teesta [13], Arkansas [14], Brahmaputra [5], Indo-Gangetic [15], Kosi [16], Gandak [1], and Rapti [17]. However, most of these rivers have been recently influenced by intense human activities, urbanization, atmospheric deposition, and anthropogenic sources, leading to the reduction of the self-purifying nature of the water streams [18].

4 Lakes in India

Lakes are defined as large water bodies (fresh- or salt- water) surrounded by land, and they generally compress four distinct zones, namely, Littoral, Limnetic, Profundal (or Hypolimnion), and Benthic zones [19]. India contains several lakes such as Vembanad Lake [20], Moticher Lake [19], Roopkund Lake, Pookot Lake [21], Ennamangalam Lake [22], Prashar Lake [23], and Renuka Lake [24]. For instance, Pookot Lake is a 6.5 m deep lake of rain-fed freshwater, covering an area of about 85,000 m^2 in the Kerala State [21]. Moreover, Ennamangalam Lake is about 1.54 km^2 in Tamil Nadu (southern India), receiving water from the ephemeral streams and a small spring [22]. Prashar Lake covers an area of approximately 0.013 km^2 in Mandi, Himachal Pradesh, and it is used for drinking and residential demands [23]. The catchment area of Renuka Lake is about 2.543 km^2, with a depth of 13.7 m located in the Sirmaur District, Himachal Pradesh [24]. Recently, the lake systems in India have been influenced by urbanization, sewage disposal, uncontrolled industrial development, and massive application of fertilizers. Other natural factors such as evaporation, weathering, rock-water interaction, eutrophication, loss of biodiversity, and particle sedimentation are also affecting the lake ecosystems.

5 Reservoirs and Dams in India

India contains about 91 major reservoirs [25] used for various purposes such as irrigation, water supply, hydroelectric generation, and areas protection against flooding. Among the 91 reservoirs, 18 reservoirs have a surface area greater than 250 km^2, whereas the area of about 26 reservoirs does not exceed 50 km^2 [26]. Some of these reservoirs include Indira Sagar, Nagarjuna Sagar [27], Sardar Sarovar [28], Gobind Sagar [3], and Hirakud Reservoir [29]. The live storage status and variation of water levels for most reservoirs in India are given by the Central Water Commission (http://cwc.gov.in/reservoir-storage). India also contains several dams such as Andhra Pradesh, Telangana, Chhattisgarh, Gujarat, Himachal Pradesh, Jammu & Kashmir, Rajasthan, West Bengal, and Jharkhand, each having a height $\geq$100 m and a storage capacity $\geq$1 km^3. The dams are used for water storage, irrigation, and navigation purposes, as well as disaster control such as floods and droughts.

6 Groundwater in India

India is a water-stressed country that relies heavily on groundwater for drinking and cultivation purposes [6]. The total utilizable groundwater resources reach 433 billion m^3 (http://cwc.gov.in/water-info). Recently, several appropriate management strategies have been performed to avoid the decline of groundwater levels [6]. In recent years, a vast number of studies have been investigated on the groundwater quality and quantity at various parts in India, viz., Delhi [30], Andhra Pradesh [31], Telangana State [32], and Tamil Nadu [33]. For example, Singh et al. [30] found that groundwater samples collected from New Delhi contained high levels of nitrate (NO_3^-) and fluoride (F^-), exceeding World Health Organization (WHO) standards [34]. In another study, Raju et al. [31] revealed the existence of high levels of HCO_3^- and Na^+ in 2014 and HCO_3^- and Ca^{2+} in 2015 in groundwater samples collected from the Chittoor and Nellore districts of Andhra Pradesh. Other ions such as Na^+, K^+, Mg^{2+}, Cl^-, F^-, SO_4^{2-}, and NO_3^- have been considered as indicators for the pollution of groundwater at Telangana State, India [32]. Duraisamy et al. [33] reported that groundwater of about 49% of the Kangayam taluk area had a "Poor Quality" category; hence, appropriate treatment is essential before potable use.

6.1 Modeling and Statistical Tools for Water Quality Assessment in India

Previous researchers have employed various modeling and statistical techniques to determine the anthropogenic and natural factors affecting the water quality in India. For instance, Kumari and Sharma [23] applied different statistical approaches such as principal component analysis, Pearson's correlation, and cluster analysis to analyze the physicochemical parameters of Lake Prashar. Moreover, Lad et al. [2] applied the geographic information system (GIS) technique to assess the water quality of Tapi Basin, Gujarat, India. In addition, Sajil Kumar and James [35] used geostatistical and geochemical models to investigate the deterioration of groundwater quality in Coimbatore district, South India. Furthermore, Bisht et al. [36] applied artificial neural network as a forecasting tool to control the pollution of Ganga River, India. Elkiran et al. [37] used various artificial intelligence-based models to simulate the dissolved oxygen (DO) pattern in Yamuna River, India.

7 Recommendations

Several recommendations should be considered to protect the valuable water resources in India: Advanced mathematical and statistical models should be used to handle large and complex information that can improve water quality.

Periodic spatial and seasonal assessment of various physico-chemical parameters (e.g., temperature, total solids, dissolved oxygen, biological oxygen demand, and nutrients), bacteriological indicators (e.g., E. coli), and heavy metals (e.g., Hg, Pb, Fe, and Cr) of water bodies should be performed.

Steps should be taken to minimize the discharge of wastewater from the industrial sector into water bodies. Moreover, adequate and suitable waste disposal techniques and wastewater treatment methods should be considered to safeguard human health during water utilization. Special care should be given by the government for the topic of rainwater harvesting and wastewater recycling and reuse to meet the water supply challenges in rural areas.

Suitable awareness, strategies, and scientific communication addressing environmental and health concerns should be offered to locals and farmers.

8 Conclusions

This chapter attempts to give an overview of the main water resources in India. It can be concluded that:

India contains various freshwater resources stored in surface water bodies (rivers, lakes, and reservoirs), and within soil systems as groundwater. Recently, India has faced significant challenges in the topic of water resource management due to the rapid population growth, urbanization expansion, and agricultural and industrial activities. The cumulative number of published documents dealing with water condition in India within 2011–2019 was about two-fold that within 2001–2010. The river system (e.g., Ganga, Yamuna, Teesta, Arkansas, Brahmaputra, Indo-Gangetic, Kosi, Gandak, and Rapti) is considered an important source of drinking, domestic, irrigation, and industrial purposes in India. India contains several lakes such as Vembanad Lake, Moticher Lake, Roopkund Lake, Pookot Lake, Ennamangalam Lake, Prashar Lake, and Renuka Lake. India contains about 91 major reservoirs (e.g., Indira Sagar, Nagarjuna Sagar, Sardar Sarovar, Gobind Sagar, and Hirakud Reservoir) used for various purposes such as irrigation, water supply, hydroelectric generation, and downstream regions protection from flooding. India also contains several dams such as Andhra Pradesh, Telangana, Chhattisgarh, Gujarat, Himachal Pradesh, Jammu & Kashmir, Rajasthan, West Bengal, and Jharkhand. The total utilizable groundwater resources reach 433 billion m^3, supporting important drinking and cultivation purposes. The outputs retrieved from this chapter would support environmental organizations dealing with all problems of water bodies in India.

Acknowledgements The last author would like to acknowledge Nasr Academy for Sustainable Environment (NASE).

References

1. Bhardwaj V, Singh DS, Singh AK (2010) Water quality of the Chhoti Gandak River using principal component analysis, Ganga Plain, India. J Earth Syst Sci 119(1):117–127
2. Lad D, Mehta M, Vashi M (2020) Assessment of surface water quality using GIS: case of Tapi Basin, Surat, Gujarat, India. Lect Notes Civil Eng 33:303–311
3. Sarvade S, Gupta B, Singh M (2016) Composition, diversity and distribution of tree species in response to changing soil properties with increasing distance from water source—a case study of Gobind Sagar reservoir in India. J Mt Sci 13(3):522–533
4. Jain SK (2019) Water resources management in India—challenges and the way forward. Curr Sci 117(4):569–576
5. Saikia L, Mahanta C, Mukherjee A, Borah SB (2019) Erosion–deposition and land use/land cover of the Brahmaputra river in Assam, India. J Earth Syst Sci 128(8):211
6. Adimalla N, Taloor AK (2020) Hydrogeochemical investigation of groundwater quality in the hard rock terrain of South India using geographic information system (GIS) and groundwater quality index (GWQI) techniques. Groundw Sustain Dev 10:100288
7. Jaiswal M, Hussain J, Gupta SK, Nasr M, Nema AK (2019) Comprehensive evaluation of water quality status for entire stretch of Yamuna River, India. Environ Monit Assess 191(4):208
8. Paul D (2017) Research on heavy metal pollution of river Ganga: a review. Ann Agrarian Sci 15(2):278–286
9. Tare V, Bose P, Gupta SK (2003) Suggestions for a modified approach towards implementation and assessment of Ganga action plan and other similar river action plans in India. Water Qual Res J Can 38(4):607–626
10. CPCB (2007) Water quality status of River Yamuna (1999–2005). Assessment and Development Study of River Basin Series (ADSORBS) ADSORBS/41, Central Pollution Control Board Delhi
11. Sharma D, Kansal A, Pelletier G (2017) Water quality modeling for urban reach of Yamuna river, India (1999–2009), using QUAL2Kw. Appl Water Sci 7(3):1535–1559
12. Singh DS, Awasthi A (2011) Natural hazards in the Ghaghara River area, Ganga Plain, India. Nat Hazards 57(2):213–225
13. Sharma A, Goyal MK (2020) Assessment of the changes in precipitation and temperature in Teesta River basin in Indian Himalayan Region under climate change. Atmos Res 231:104670
14. Buchkoski JJ (2019) "Being judged by its fruits": transforming Indian land into orchards along the Arkansas River, 1800–1867. Great Plains Quart 39(1):39–58
15. Nath R, Luo Y (2019) Disentangling the influencing factors driving the cooling trend in boreal summer over Indo-Gangetic river basin, India: role of Atlantic multidecadal oscillation (AMO). Theor Appl Climatol 138(1–2):1–12
16. Roy P, Guha A, Vinod Kumar K (2017) Multi-sensor space-borne earth observation data for characterizing fluvial-geomorphic provinces: a case study from Kosi River, India. J Indian Soc Remote Sens 45(5):847–858
17. Sarkar UK, Khan GE, Dabas A, Pathak AK, Mir JI, Rebello SC, Pal A, Singh SP (2013) Length weight relationship and condition factor of selected freshwater fish species found in river Ganga, Gomti and Rapti, India. J Environ Biol 34(5):951–956
18. BIS (2012) Bureau of Indian Standards (BIS) 10500 (2012) specification for drinking water. Indian Standards Institution, New Delhi, 1–5

19. Bid SD, Patel PL, Christian RA, Patra AK (2020) Hydrodynamic modeling of radionuclide effluent in Moticher Lake, Kakrapar atomic Power Station, India. J Hazardous Toxic Radioact Waste 24(1):4019033
20. Suja G, Lijo J, Kripa V, Mohamed KS, Vijayan KK, Sanil NK (2020) A comparison of parasites, pathological conditions and condition index of wild and farmed populations of Magallana bilineata (Roding, 1798) from Vembanad Lake, west coast of India. Aquaculture 515:734548
21. Kizhur S, Shankar R, Warrier AK, Yadava MG, Ramesh R, Jani RA (2019) Late Holocene palaeovegetational and environmental changes inferred from organic geochemical proxies in sediments from Pookot Lake, southern India. Arab J Geosci 12(21):643
22. Mishra PK, Ankit Y, Gautam PK, Lakshmidevi CG, Singh P, Anoop A (2019) Inverse relationship between south-west and north-east monsoon during the late Holocene: geochemical and sedimentological record from Ennamangalam Lake, southern India. Catena 182:104117
23. Kumari R, Sharma RC (2019) Assessment of water quality index and multivariate analysis of high altitude sacred Lake Prashar, Himachal Pradesh, India. Int J Environ Sci Technol 16(10): 6125–6134
24. Kumar P, Meena NK, Diwate P, Mahajan AK, Bhushan R (2019) The heavy metal contamination history during ca 1839–2003 AD from Renuka Lake of Lesser Himalaya, Himachal Pradesh, India. Environ Earth Sci 78(17):549
25. Mishra V (2020) Long-term (1870–2018) drought reconstruction in context of surface water security in India. J Hydrol 580:124228
26. Tiwari AD, Mishra V (2019) Prediction of reservoir storage anomalies in India. J Geophys Res Atmos 124(7):3822–3838
27. Pradeepakumari B, Srinivasu K (2019) An implementation of artificial neural reservoir computing technique for inflow forecasting of nagarjuna sagar dam. Int J Recent Technol Eng 8(1): 860–864
28. Anklesaria Aiyar SS, Kaushal N (2019) Are resettled oustees from the Sardar Sarovar dam project 'better off' today? Econ Polit Wkly 54(12):48–55
29. Gupta KK, Kar AK, Jena J, Jena DR (2017) Variation in rainfall trend at upstream—a threat towards filling schedule of Hirakud Reservoir, India. Int J Water 11(4):395–407
30. Singh CK, Kumar A, Shashtri S, Kumar A, Kumar P, Mallick J (2017) Multivariate statistical analysis and geochemical modeling for geochemical assessment of groundwater of Delhi, India. J Geochem Explor 175:59–71
31. Raju NJ, Patel P, Reddy BCSR, Suresh U, Reddy TVK (2016) Identifying source and evaluation of hydrogeochemical processes in the hard rock aquifer system: geostatistical analysis and geochemical modeling techniques. Environ Earth Sci 75(16):1157
32. Rao NS, Sunitha B, Rambabu R, Nageswara Rao PV, Surya Rao P, Sravanthi B, Marghade M, Deepthi Spandana D (2018) Quality and degree of pollution of groundwater, using PIG from a rural part of Telangana State, India. Appl Water Sci 8:227
33. Duraisamy S, Govindhaswamy V, Duraisamy K, Krishinaraj S, Balasubramanian A, Thirumalaisamy S (2019) Hydrogeochemical characterization and evaluation of groundwater quality in Kangayam taluk, Tirupur district, Tamil Nadu, India, using GIS techniques. Environ Geochem Health 41(2):851–873
34. WHO (2009) World Health Organization (WHO). Guidelines for Drinking Water Quality, World Health Organization, Geneva
35. Sajil Kumar PJ, James EJ (2019) Geostatistical and geochemical model-assisted hydrogeochemical pattern recognition along the groundwater flow paths in Coimbatore district, South India. Environ Dev Sustain 21(1):369–384
36. Bisht AK, Singh R, Bhutiani R, Bhatt A (2019) Artificial neural network based water quality forecasting model for ganga river. Int J Eng Adv Technol 8(6):2778–2785
37. Elkiran G, Nourani V, Abba SI (2019) Multi-step ahead modelling of river water quality parameters using ensemble artificial intelligence-based approach. J Hydrol 577:123962

Part II
Status of Water Resources in India

The Interlinking of Indian Rivers for Sustainable Use of Water Resources in India: An Overview

Ram Karan Singh, Vineet Tirth, Vaishali Sahu, and Mansvee Singh

Abstract In the recent era, due to the rapid increase in India's population, commercialisation and industrialisation the concerns for increasing demands for water are also becoming alarming. The temporal and spatial variability in rainfall over the country has created multiple suffering due to droughts and floods. It is in this context that linking of Indian rivers has been proposed as a solution to the looming water crisis. Interlinking of Indian rivers is a grand project aimed at a number of proposed benefits that would solve a number of water-related problems in the country, by diverting the extra flood water to the areas where the water quantity is limited to meet the demand of various sectors. The project aims at overcoming the regional imbalances in water availability in India. However right since its inception, the project has been shrouded in controversy. The paper analyses the policy, a summary of the benefits and a review of the major concerns. It also questions the viability of such a project to achieve the proposed goals and also provides certain alternatives to solve water issues in India. In this work the effort is made to suggest the different sustainable options keeping in mind the economic, technological and environmental feasibility analysis. It will also suggest the different possible alternatives before the project passes through the final stage of government approvals.

R. K. Singh (✉)
The Tech School, The ICFAI University Dehradun, Uttarakhand, Dehradun, Uttarakhand, India
e-mail: vc@iudehradun.edu.in

V. Tirth
Mechanical Engineering Departments, College of Engineering, King Khalid University, Abha, Kingdom of Saudi Arabia
e-mail: vtirth@kku.edu.sa

V. Sahu
Department of Civil and Environmental Engineering, The NorthCap University, Grogram, India
e-mail: vaishalisahu@ncuindia.edu

M. Singh
Department of Clinical Psychology, The Manipal Academy of Higher Education, Udapi, Manipal, Karnatka, India
e-mail: mansvees@mtu.edu

S. Yadav et al. (eds.), *Water Quality, Assessment and Management in India*, Earth and Environmental Sciences Library, https://doi.org/10.1007/978-3-030-95687-5_2

Keywords Interlinking of Rivers · EIA study · Tran's boundaries water disputes · Sectorial water use, United Nations guidelines

1 Introduction

Rivers are the basic source of water for mankind. Due to urbanization, policy deficit and in recent time due to climate change issue there is major shift in the hydrologic cycle in terms of spatial and temporal distribution of the rainfall this leads to flood and drought situation in the various regions of the India [1–4]. The groundwater table is also moving down fast due to misbalance in the extraction and recharge. In recent time due to melting of the Himalayan glaciers the formation of glacier lakes and increase in the size has been observed also due to burst of one such lake massive disaster in Kedarnath area has happened in the recent past [5–8]. The option desalination is ruled out in the Indian scenario due to very high cost involved in the per unit of the water desalinised and re-mineralised. The per capita potable water availability is decreasing with rapid pace. The global population is about to reach 7.9 billion which is nearly fifty percent more than the population of the year 1990, because of rapid increase in global population, the world may see more than a six-fold increase in the number of people living under conditions of water stress: from 470 million today to 3 billion in 2025 [9–12].In India this problem may be addressed by interlinking of rivers and diversion of the resource which is unevenly distributed and major part of it goes to bay of Bengal in monsoon seasons as a flood water by surface run-off [10–12].

The interlinking of rivers (ILR) project is based on the National Perspective Plan (NPP), formulated by the Ministry of Water Resources for water development in August 1980, ensuring the best possible utilization of water resources. For this purpose the national water development agency was established in 1982. The then president Dr. A.P.J. Abdul Kalam revitalised the proposal in 2002. Later in the same year, as an answer to public interest litigation, the Supreme Court ordered the government to complete the project within 10 years. A Task Force headed by Mr. Suresh Prabhu on interlinking of rivers was constituted in December 2002 to arrive at a speedy consensus amongst states for the sharing and transfer of surplus water to deficit areas, and sustainable development of water resources in all states of the India, to wake up to the authenticities and study the techno- economic feasibility of this project. However serious doubts are being raised about the project mainly because of its size and the considerable financial and environmental costs it involves, after all it's a question of altering the geography of India. Such a project should ideally be preceded by a detailed study of the economic, ecological and technological viability as well as detailed environmental impact assessment [9–13].

According to the theory of planned behavior, the most immediate predictor of an action is behavioral intent (intention to engage in the behavior) which in its turn is shaped by attitudes, subjective norms and perceived behavioral control. Subjective norms are the perception of social support for a perceived behavior control l reflects on the perceived ease of performing the behavior. Specific water-related beliefs have been found to be more predictive of water conversation behaviors compared to more general beliefs about the environment and the place of human beings in it. The challenge for policy makers is to identify the most consequential and salient beliefs associated with water conservation. Behavior is also guided by automatic processes such as habits or routines. There is a need to design interventions that help people break established patterns of water using behavior that result in high water use. Understanding habitual water use may also yield areas in which there is scope for lowering water demand. Personal capabilities such as knowledge and skills can facilitate conservation behaviors and socio demographic variables such as age, education, and income may be proxies for personal capabilities [14].

2 Need for Interlinking Rivers

India is regarded as a better-endowed country in terms of its share in global water resources with an annual precipitation of 4000 cu.km and a per capita water availability of 1820 cu.m (2001). India receives bulk of its rainfall during the monsoon. But there is both a spatial as well as temporal disparity in the precipitation received all over the country. In the time of monsoon season India receives maximum precipitation by rain specially the Himalayan catchments of Ganga-Brahmaputra-Meghna (GBM) basin due to large catchment area. The north-eastern part of the country receives more rainfall than the north-western, western and southern parts. Thus there are often droughts in some areas simultaneously with floods in the other. This has led to acute shortage of domestic water supply in some areas, flooding of farm areas, damage of harvests etc. Every individual in our country is entitled to get his/her share of drinking water and in the national water policy of a country the most important sector should be house hold water supply. However, there may be an overall shortage of water; the primary problem is the improper distribution and vast differential in consumption, including leakages and consumer wastages [6–9].

The proposal to interlink rivers seeks to utilize the available water resources in an equitable and efficient manner. Through this project it is believed that the surplus water, which flows 'waste' into the sea, would be fruitfully utilized. The certain rights may be true about the future water crisis, on a careful examination of the entire proposal to interlink major river basins, the National Commission for Integrated Water Resources Development Plan (NCIWRDP) made following observations: "There seems to be no imperative necessity for massive water transfer. The assessed needs of the basins could be met from full development and efficient utilization of intra-basin resources except in the case of Cauvery and Vaigai basins. Therefore it is felt that limited water transfer from Godavari would take care of the deficit in Cauvery and Vaigai basins" [15].

3 Interbasin Water Transfer Proposals

The concept of transferring water from water surplus areas to water deficit areas got attention by many water resource manager and professionals. The idea of linking rivers of India stems from the views of the stalwart engineer Visvesraya. Later the idea received attention from Sir Arthur Cotton. K.L. Rao and Captain Dastur proposed similar ideas in 1971 and 1977, respectively. K.L. Rao's gave the suggestion of keeping in view of high quantity of water leading to flood and low quantity water with deficit in the environmental flow area. He believed that such river basins could be connected by means of a "National Water Grid". However the Central Water Commission (CWC) declared it to be economically prohibitive. Capt. Dastur proposed the construction of two canals (popularly called "garland canal")—Himalayan canal at the foot of Himalayan slopes connecting Ravi in west to Brahmaputra and beyond and second garland canal covering southern and central parts. However, the proposal, as evaluated by senior engineers from CWC, state governments, professors and scientists, was declared to be technically and economically infeasible.

In the context of inter-basin water transfer, some of the earliest proposals were the Periyar project, Parabikulam-Aliyar, and the Telugu-Ganga project. Of these the Periyar inter-basin transfer project, involving the transfer of water from Periyar basin to Vaigai basin, has been so far the most successful endeavour. The Parabikulam-Aliyar project was built during the second and third five-year plans for delivering water to the drought-prone areas in Coimbatore district of Tamil Nadu and the Chittur district of Kerela. The Telugu-Ganga project has been recently implemented to meet the water demands in the Chennai. The Beas-Satluj link combined with Indira-Gandhi Nahar project, also known as the Rajasthan canal project, exemplifies how large inter-basin transfers have brought about socio-economic development in that region.

4 Proposal to Interlink

The interlinking of rivers (ILR) project is based on the National Perspective Plan (NPP), formulated by the Ministry of Water Resources for water development in August 1980, ensuring the best possible utilization of water resources. For this purpose, the National Water Development Agency (NWDA) was established in 1982. The then president, Dr. A.P.J. Abdul Kalam, revived the proposal on the 14th of August 2002. Later in the same year, as an answer to Public Interest Litigation, the Indian Supreme Court ordered the government to complete this project within 10 years. Soon after, the government to examine the feasibility of such a mammoth project constituted a task force headed by Mr. Suresh Prabhu. The task of preparing the detailed project report was given to NWDA, which is to be completed by June 2008.

4.1 The NPP Comprises of Basic Components

The Himalayan watershed consists of construction of dams for water storage on the major tributaries of the rivers Ganges and Brahmaputra in Nepal, Bhutan and Bangladesh, transferring surplus waters from the eastern tributaries of Ganges to the water deficit west. The other part of this component involves linking of the main Brahmaputra and its tributaries with Ganga and also the linking of river Ganga with Mahanadi. The pre-feasibility studies of all the 14 links have been completed while the complete feasibility studies of only 2 links have been completed. The National civil society committee on interlinking of rivers (NCSCILR) has arranged for a review of the feasibility report for Ken-Betwa link.

4.2 Conceptual Gaps in the Proposed ILR

In India, the interlinking river proposal for the different river basin was based on the surplus, marginally surplus, marginally deficit and deficit water quantity available in the different part of the river with the assumption that these entire rivers were perennial the Fig. 1 below gives the Map of India showing the peninsular component. The world water program defines. 'Water scarcity' as insufficient water of good quality and required quantity to meet the requirement of domestic sector and environmental.

4.3 The Peninsular Component Has Four Major Parts

Interlinking of Mahanadi-Godavari-Krishna-Cauvery Rivers.
Interlinking of west flowing rivers, north of Bombay and south of Tapi.
Interlinking of Ken-Chambal Rivers.
Diversion of other west flowing rivers.

It includes 16 links and feasibility reports of all these links have been completed. This component includes construction of dams and reservoir at various sites to overcome water related problems. The Fig. 2 below shows the Map of India showing the River Basin Watershed.

5 The Proposed Benefits

This ambitious project envisages the following benefits:

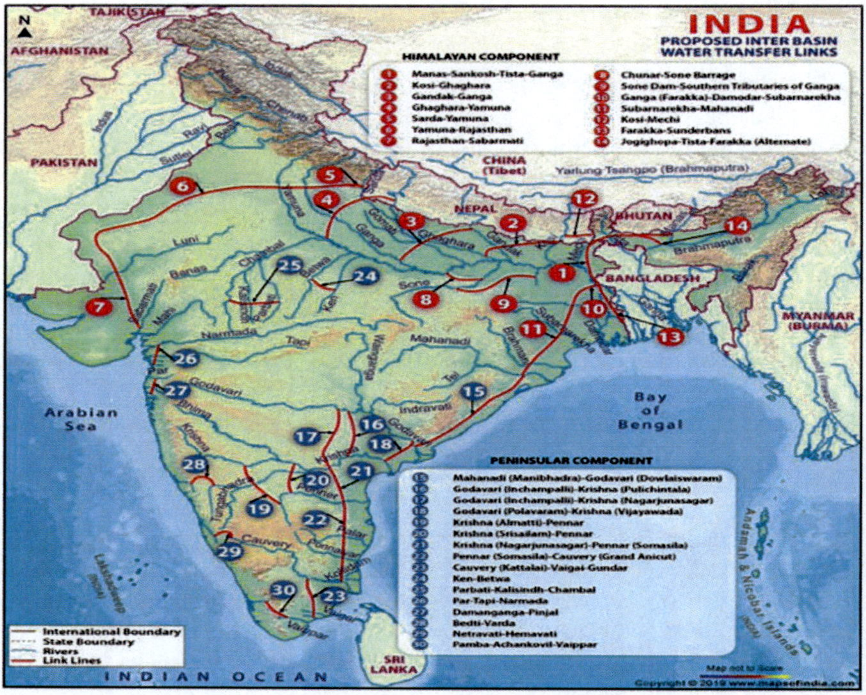

Fig. 1 Map of India showing the peninsular component [16]

Increased irrigation area by 35 million hectares thus augmenting food grain production.
Hydroelectricity.
Drought mitigation.
Employment generation.
Socio-economic development.
Flood control.
Domestic and industrial water supply.
Providing navigational facilities.
Salinity and pollution control.
Rearing fishes.
Recreational facilities.

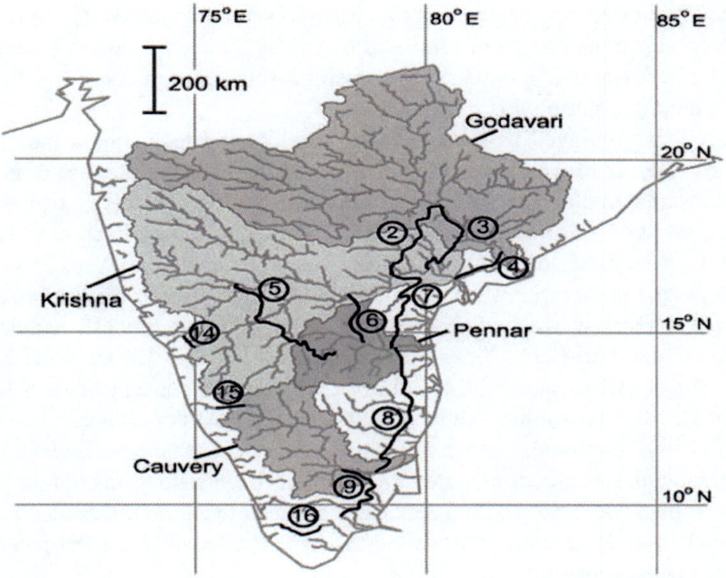

Fig. 2 Map of India showing the River Basin Watershed [16]

6 The Important Factors Link with the ILR

The ILR is believed to be the largest water related project in this world; it should be ideally based on a sound knowledge base. The technical information such as environmental impact assessment report and model results for the watershed at a general and regional scale is not available so the technical and professional analysis of the interlinking project is not possible. This entire project exists on paper as of now, information made available to public domain consist of a series of maps showing the rough positions of dams and reservoirs to be built. There is a lack of communication between the policy planners and executers on one side and the common man on the other, which has further aggravated the conflict. Since environmental assessment report is not available and other detail economic and technical data and information is also not available this may affect the confidence of public on the wisdom behind such a huge investment. At all the level transparency is expected by the government by the citizen of India. The social impact, financial feasibility and sustainability of the current solution and the alternatives need to be studied in detail.

6.1 Political Issues and Domestic Water Supply Priority

After issuing directives to states to interlink rivers, the states were expected to get back on this issue. Tamil Nadu in its affidavit supported the proposal to interlink rivers and requested the construction of a high-powered committee by the Union of India to complete the project in stipulated time. The Supreme Court concluded that since no other state had given any affidavits, it assumed that a consensus has been obtained among states on this issue.

The ILR is envisaged as a project that would serve to integrate the country. Question arises as to whether interlinking would increase the number of conflicts related to water? In the light of the various inter-state disagreements that are likely to arise as a result of the project, national integration seems too far fetch an idea. India is known for multiple inter-state water disputes whether it is between Punjab and Haryana, in the north or between Tamil Nadu and Karnataka in the south. A MOU was reached recently reached between the states of Uttar Pradesh and Madhya Pradesh on the Ken-Betwa link but the states of Kerala, Bihar, West Bengal, Assam, Punjab, Chattisgarh and Goa have criticised this concept of river linking. States of Orissa, Maharashtra, Gujarat, Karnataka, and Andhra Pradesh have given only provisional approval. It is natural that the donors of water are dead against the concept while the recipients of water are in favour. Each state wants to first provide water to its own water stressed and deficit areas. Under these circumstances, a central law to dictate water sharing between the concerned states only has the potential to precipitate new problems.

There have been agreements between riparian states with regard to sharing water, however there are no settled agreements or principles with regard to transfer of water from surplus basins to deficit ones. Non-riparian states are assumed not to have any claims to watercourses. In the process of transferring the water from one basin to other basin for the use of major sector such as domestic, agriculture and industry etc. the water donor state should get some revenue from the water receiving state. It is important that a consensus amongst states be obtained with regard to water sharing, a rather difficult task ahead for the Task force. The project should be implemented with the agreement between states, because it would otherwise challenge the basic fiber of federalism in India. With the Centre having full power with respect to water management, riparian rights of states would be lost. The domestic water supply should get the top priority in the national water policy and ILR must offer the domestic water supply in water scare areas especially in the heart of the country.

6.2 Environmental Social and Economic Concerns

Since the ILR envisages transferring water from Himalayan river basins to other deficit basins, there is a need for studying the impact of such a project on Himalayan ecology. The significant gaps have been identified in the conventional knowledge. Some of which are:

(a) The process to generate the artificial drainage system from Himalayan foothills to floodplains.

(b) The system of generation, movement and deposition of silt and sand particles in the perennial river system originating from the Himalayas.
(c) The risks link with the high intensity earthquakes on mega project dams on the Himalayan Rivers.
(d) The impact of river training works such as embankment walls etc.

The economic feasibility of the above four activities.

The NCIWRDP in the face of the above knowledge gaps has declared that there is an urgent need for a scientific research on Himalayan Rivers and that "the hydrological data from all rivers must be made available on public demand".

The major impacts of this project would be the submergence of forests causing loss of rich biodiversity. One of the proposed benefits of ILR is controlling floods. Floods in the Himalayan foothills and the adjoining areas are a result of complex ecological processes that are not well understood. In the past also engineers by means of some projects have made many such claims of controlling floods in the Himalayan region. However there has been little evidence to prove such a claim. The natural flow in the river system has low and high flow; the low flow should never be less than the environmental flow to maintain the aquatic life at the same groundwater replishment. The high flood water brings the soil which gets deposited in the flood plain and forms the high fertile agricultural land.

Consequently, by eliminating floods completely, the land fertility would decline over years. ILR is also based on the essentially flawed perception that floodwaters flow 'waste' into seas. Proposed damming and diversion to stop the flow of rivers into sea will have a disastrous effect on the formation of deltas. The transfer of flood flows will affect the processes of many ecosystem services that many planners do not comprehend, by not recognizing the importance of the diverse ecosystem services that water performs.

Another cause of great distress is the construction of canals, which will interfere with the natural drainage systems. Diversion of rivers for the proposed links would decrease the water flow, which will increase the pollution levels in rivers. Irrigational benefits from this project can lead to salinization of soil as it has been observed in Punjab where intensive irrigation has rendered 20% of the cultivable land infertile. These and more such probable environmental impacts associated with the project warrant the application of the "precautionary principle".The precautionary principle also gains importance in the context of the large population of people that would be displaced by the project; its estimated that around 5.5 million people would be displaced and nearly 79, 000 ha. Of land would be submerged.

After the recognition of "precautionary principle" beneath the environmental jurisprudence of our country, the "onus of proof" is clearly on the Government to show as to how a project like ILR, oppressed with serious consequences and its potential for irreversible damage, is environmentally feasible [15].

It becomes the duty of the government to foresee and assess the potential risks and to warn the victims of such risks. The government should hence carry out detailed impact assessments and feasibility studies. In the social perspective, it is estimated that around 5.5 million people would be displaced, mostly tribes and farmers due to submergence of thousands of villages. Tribal cultures would also be adversely affected due to their resettlement. India's track record at providing compensation, re-settlement and rehabilitation to the aggrieved is very poor; these costs have not been estimated for the project; thus adding more expenditure to the already enormous amount needed for ILR.

The costs for construction alone of ILR is around US$200 billion, not including specifics like yearly inflation, costs to the environment, wildlife, resettlement and rehabilitation of displaced people. The project is being touted as the largest multipurpose river project. The task force members have admitted that it could exceed Rs. 1,000,000 crores. So how would the government generate such a huge expenditure? It would have to borrow from banks and if it does so, it would pile up on the existing debt and the interest alone could be 30, 000 crores per year.

7 International Conflicts

This should be kept in mind that many rivers origination in different parts of the country specially from the Himalayan region shares number of international boundaries such as Pakistan, China, Nepal, Bhutan, Bangladesh etc. so the treaties related to trans boundaries issues and guidelines related to international boundary disputes related to water sharing should be kept in mind, other than this serious issues are also there within country with state water sharing problems, article 6 of the UN Convention on Law of Non navigational uses of international watercourses, provides 17 considerations to be applied for sharing such resources under such conditions. Almost two dozen dams on the Brahmaputra in the Northeast are already in advanced stage of construction and planning. This would leave little water for diversion to other parts of India. Bangladesh has already made its stand clear on the project and has heavily criticised it declaring it to be technically infeasible and a malpractice from the viewpoint of international water law and practise. The diversions of the Ganga and Brahmaputra waters have raised severe objections from our neighbouring countries which share these waters- Bangladesh, Nepal, etc. Many experts in Bangladesh claim that the ILR flouts the India-Bangladesh Treaty of 1996 on the sharing of Ganga waters.

7.1 Is River Linking the Only Solution?

Tanks were the mainstay of irrigation across many states in India—before the British policies caused great damage. Unfortunately, after independence, the governments continued the neglect and allocated meager resources to develop new tanks and maintain existing ones [3–6] had undertaken water availability studies in as few water scarce areas of India and their study made it clear that if the precipitation available within the concerned watersheds or sub basins is harvested and conserved properly, supply of domestic water needs would not pose a serious a problems in most parts of the country. For promoting domestic water security in drier areas of India, local level water harvesting and conservation has been a proven technology. It is cheap and socially acceptable technological option even today when compared with large storage and long distance diversion facilities, which often carry high financial social and ecological costs. This observation is completely in consonance with the results of numerous community initiatives for water harvesting in India, whether in Maharashtra, Gujarat, Rajasthan, Tamil Nadu, Uttaranchal or anywhere else in the country. Some success stories have been listed below:

8 Conclusion

India has a multiparty system of governance, thus building consensus amongst states is also has the additional hurdle of conflicting political interests. In India disputes between riparian states have gone awry but the interlinking proposal would involve a riparian state to share its water with a non-riparian, an issue of even more fragile nature. It is difficult to expect that riparian states would be generous enough to share. The centre needs to chart out a proposal that would keep all states content. Displacement of a large population and its resettlement and rehabilitation is a major hurdle. With a poor record of rehabilitation of the displaced, the Indian government has to certainly abide by the "Precautionary principle". It should make all information regarding the feasibility of the project public so that an unbiased assessment could be made. So far the government has failed to maintain any transparency, which has further aggravated the suspicion in the minds of the general public, and environmentalists at large who fear adverse environmental impacts due to the project. The enormous cost of the project is another issue of contention. The Government has not disclosed the sources from which it shall raise the money to sponsor the project. Private management of the water is likely considering the huge investment needed. This would be detrimental and would lead to denial of all rights traditionally enjoyed by communities staying on the riverside for centuries. International repercussions of the project are also doing not appear promising. Bangladesh is dead against the project and believes the project to be a huge conspiracy to render its territory a wasteland. Our relations with this state are bound to be strained if we go ahead with the project without keeping Bangladeshi interests in mind. From the viewpoint of International law, article 8 of the UN convention on non-navigational uses of water courses, requires that watercourse states cooperate, thus it becomes India's legal duty to undertake ILR only after it is assured of cooperation from its neighbour states. Water security indeed is a foremost cause of concern in India with a population of more than a billion. But instead of considering the interlinking project as the only way out to the water calamity, warrants more thinking. Interlinking of rivers with all its benefits and impacts should be subjected to a public debate. Planning should be on the basis of ground capability and implemented with greater stress on the socio-political factors, than merely on engineering. Such a massive task requires a detailed

Environmental Impact Assessment (EIA). The impact of climate change on the Himalayan glaciers sustainable and sensible approach to management of water is of utmost importance in this context.

9 Recommendations

The improper distribution and differential consumption of water, not ignoring the aspects of consumer wastages and leakages etc., is one of main culprits behind water shortage in India. Rainwater harvesting at local level has produced many desirable results. When water is allowed to percolate naturally or artificially into the ground, even dried seasonal rivers have been rejuvenated. The IRL is believed to be crucial for food security in India. But improvement in farming technologies and cultivation of high yielding varieties has produced good results. The proponents of IRL need to examine the Indian political system, which is a huge nexus between politicians, beaurocrats, and contractors that doesn't work with public interest in their minds. For them the prime motive is profit; the project is estimated at 5 lakh crores and is a virtual goldmine for them. Even if we assume that the project has zero technical error and execution, political interference may completely defeat the purpose of this project. There is a need to design interventions that help people break established patterns of water using behavior that result in high water use. Understanding habitual water use may also yield areas in which there is scope for lowering water demand. Personal capabilities such as knowledge and skills can facilitate conservation behaviors and socio demographic variables such as age, education, and income may be proxies for personal capabilities.

Acknowledgments The authors gratefully acknowledge resources and facilities given by the Civil and Environmental Engineering Department, The NorthCap University, Gurugram, India.

References

1. Alagh YK, Pangare G, Gujja B (eds) (2006) Interlinking of rivers in India: overview and Ken-Betwa link. Academic Foundation
2. Bandyopadhyay J, Mallik B (2003) Ecology and economics in sustainable water resource development in India. Water Resour Sust Livelihood Ecosyst Serv. Concept Publishers, New Delhi: 55–96
3. Bandyopadhyay J, Perveen S (2003) The interlinking of Indian rivers: some questions on the scientific, economic and environmental dimensions of the proposal. In Seminar on Interlinking Indian Rivers: Bane or Boon
4. Singh RK (2018) Impact of climate change on the retreat of Himalayan glaciers and its impact on Major River hydrology: Himalayan glacier hydrology. In Climate change and environmental concerns: breakthroughs in research and practice (pp 681–694). IGI Global
5. Singh SR, Shrivastava MP (eds) (2006) River interlinking in India: the dream and reality. Deep and Deep Publications

6. Upreti T (2006) International watercourses law and its application in South Asia. Pairavi Prakashan
7. Srivastava A, Gupta V, Agarwal G, Srivastava S, Singh I (2011) Water quality assessment of Ramganga River at Moradabad by Physico-chemical parameters analysis. VSRD Tech Non-Technical J 2(3):119–127
8. India Population 2019, Population of Moradabad (2019). Available on https://indiapopulation2019.com/population-of-moradabad-2019.html. Accessed on 23.01.20
9. Khan MYA, Gani KM, Chakrapani GJ (2016) Assessment of surface water quality and its spatial variation. A case study of Ramganga River, Ganga Basin, India. Arab J Geosci 9(28): 1–9. https://doi.org/10.1007/s12517-015-2134-7
10. Channdra R, Kashyap K, Pandey A (2010) Pollution status of river Ramganga: Physico-chemical characteristics at Bareilly. J Exp Sci 1(5):20–21
11. Nizami G, Rehman S (2018) Assessment of heavy metals and their effects on quality of water of rivers of Uttar Pradesh, India: a review. J Environ Chem Toxicol 2(2):65–71
12. Down to Earth. Available on https://www.downtoearth.org.in/news/water/huge-amounts-of-toxic-heavy-metals-swim-in-indian-rivers-60545. Accessed on 23.01.20
13. Samantray P, Mishra BK, Panda CR, Rout SP (2009) Assessment of water quality index in Mahanadi and Atharabanki Rivers and Taldanda Canal in Paradip area, India. J Hum Ecol 26(3):153–161
14. Russell S, Fielding K (2010) Water demand management research: a psychological perspective. Water Resour Res 46(5). https://doi.org/10.1029/2009wr008408
15. http://ablazeiasacademy.com/river-interlinking-project/ accessed on 21.01.2020
16. Google image (2019) https://www.google.co.in/search?hl=en&authuser=0&tbm=isch&sxsrf=ALeKk026hqTRZINZmHnfRqbu_nRmEUd9Mw%3A1586248223922&source=hp&biw=1366&bih=657&eig...... Accessed on 21.01.20

An Overview: Water Resource Management Aspects in India

R. V. Galkate, Shalini Yadav, R. P. Pandey, Abdelazim M. Negm, and Ram Narayan Yadava

Abstract The present chapter gives an overall understanding of water resource development and management aspects in India after a thorough review of several government publications, research papers, scientific studies, popular articles, and media reports. Water resources management has long been a major concern in India, owing to rising water demand as a result of population growth, agricultural development, industrial growth, worsening water quality, and expansion in other sectors. India has achieved remarkable growth in the water and agriculture sector in the post-independence period however the per capita water availability has been reducing constantly due to increasing demands. Despite having enormous water resources, there has always been a shortage of safe and quality water as per the specified standards for domestic and irrigation supply. Around 70% of India's population lives in rural areas, where agriculture, the most water-demanding sector, is the major source of income. Water demand has increased in both urban and rural areas as a result of the rising population and changing lifestyles. This demands the development of sustainable water resources in the country in order to bridge the rising demand-supply gap and provide enough water supply for domestic, agricultural, industrial, hydropower, and other uses. In India, the southwest monsoon contributes

R. V. Galkate (✉)
National Institute of Hydrology, Central India Hydrology Regional Centre, Bhopal, Madhya Pradesh, India

S. Yadav
Center of Excellent in Water & Environment, and Department of Civil Engineering, Rabindranath Tagore University, Bhopal, Madhya Pradesh, India

R. P. Pandey
National Institute of Hydrology, Jalvigyan Bhawan, Roorkee, Uttarakhand, India

A. M. Negm
Water and Water Structures Engineering Department, Faculty of Engineering, Zagazig University, Zagazig, Egypt
e-mail: amnegm@zu.edu.eg

R. N. Yadava
Madhyanchal Professional University, Bhopal, Madhya Pradesh, India

S. Yadav et al. (eds.), *Water Quality, Assessment and Management in India*, Earth and Environmental Sciences Library, https://doi.org/10.1007/978-3-030-95687-5_3

the majority of the country's rainfall, supplying massive amounts of water to rivers, lakes, reservoirs, and groundwater. However, the country's significant spatial and temporal variation of rainfall are the primary causes of water scarcity due to recurrent droughts.

Although India has a huge network of rivers for the development of surface water resources, the most desired source of water is groundwater. In India, there are still certain regions where surface and groundwater resources may be developed further. Extreme events like floods and droughts are posing threat mainly to water, agriculture, livelihood, economy, and other sectors. More than half of the geographical area can be considered as drought-prone contrarily major part of the country is flood-prone affecting 33 million population. The extent and risk of these extreme events may increase further due to climate change. To address these floods and droughts under climate change scenarios, it is critical to analyze droughts, devise mitigation plans, flood management, policy development, a systematic scientific methodology, and a sustainable water resource development strategy. Water quality is one of the concerns that threaten freshwater availability and poses major health risks owing to surface and groundwater contamination which has to be addressed immediately across the country. Because India is an agricultural-based country with agriculture as the primary profession of the people, the country must prepare to ensure that water is accessible for irrigation to improve food production and satisfy rising food demand. In the 1970s and 1980s, the Green Revolution resulted in remarkable improvements in agricultural and food grain production, which may further be sustained with systematic and scientific techniques for water resource development and management in the country.

Keywords Water resources management · Water availability · Droughts · Floods · Climate change

1 Introduction

Water is one of the most vital and valuable elements on the earth. It is necessary for the survival of human civilization, living creatures, and the natural environment. Water resource development and management are critical to satisfy rising household, agricultural, industrial, and hydroelectric needs. Water resources are equally useful for transportation, recreation, animal husbandry, and commercial uses. Due to its numerous advantages and the difficulties caused by its excesses, shortages, and deterioration in quality, water as a resource needs special attention. Total water availability on a worldwide basis is around 1600 million cubic kilometers. Although there is a vast amount of water available on the planet, 97.5 percent of seawater is saline, which contributes nothing to human usage. Only approximately 0.26 percent of the remaining 2.5 percent fresh water is available in rivers, lakes, and ponds, the majority of which is deep and in a frozen state in Antarctica and Greenland. The present article is an attempt to discuss and elaborate on different aspects of water availability and water resources management in India. It covers land and water resources, current and future water needs, precipitation, river systems, and water availability in terms of surface and groundwater resources. The chapter also covers droughts, floods, climate change, India's action to mitigate climate change, water quality, agriculture, and irrigation development in India. With a total size of 3.28 lakh km^2, India is the world's seventh biggest country. India's water resources have immense potential; the country's total utilizable water supply is estimated to be 1123 billion cubic meters (BCM). The per capita water availability at a national level was 5178 m^3 in 1951 and decreased to an estimated 1820 m^3 in 2001. Given the current pace of population growth, per capita water availability is expected to fall below 1000 m^3 by 2025, potentially indicating a water shortage situation [1]. The monsoon season rainfall pattern in India is highly variable, with 55 percent of precipitation falling on an average in only 15 rainy days and nearly 90% of river flows during just four monsoon months. To cope up with the situation, India has generated a storage capacity of 212.78 BCM through major, medium, and minor irrigation projects [2]. The country's irrigation potential has been assessed to be about 139.9 Mha without inter-basin water sharing, and 175 Mha with inter-basin water sharing [3]. The Central Ground Water Board (CGWB) estimates that taking up rainwater harvesting and artificial recharging of groundwater across the 45 Mha area through excess monsoon runoff may enhance groundwater availability in the country by around 36 BCM. The Indian land and water resources [4] are summarized in Fig. 1.

India has an average annual precipitation of 1170 mm, which equates to roughly 4000 km^3 of water and can provide about per capita 1720 m^3 of freshwater each year. Around 80% area falling in the semi-arid and sub-humid region in the eastern portion of the country receives 750 mm or more of annual rainfall. This rainfall, on the other hand, has an unequal distribution and a lot of variation in space and time. The monsoon seasons rainfall (June to September) account for the majority of India's rainfall, with the north-eastern and northern region receiving considerably more rainfall than the southern and western region. Besides the rainfall, the melting of snow in the Himalayas throughout the winter season feeds the northern rivers to various degrees results in flooding in certain months. The southern rivers, on the other hand, have higher flow fluctuation throughout the year [5]. Despite its large network of the river system, India has a lack of safe, clean potable water and also irrigation water supplies for sustainable agriculture. It has tapped only a tiny portion of its surface water resource which are accessible and recoverable. In 2010, India harnessed 761 km^3 of water, accounting for 20 percent of its total water resources, some of which come from groundwater which is not sustainable [6]. Figure 2 shows the availability of water resources in India [7].

The water sector in India faces numerous challenges, including combating the growing gap between demand and supply, providing adequate water for food production, matching with competing demands, meeting the growing demands of big cities, wastewater treatment, inter-basin transfer, water sharing with neighbouring countries and inter-state disputes, and so on [8]. Figure 3 shows the future water demands of major sectors such as irrigation, drinking water, industry, energy and other purposes in India, as anticipated by the Ministry of Water

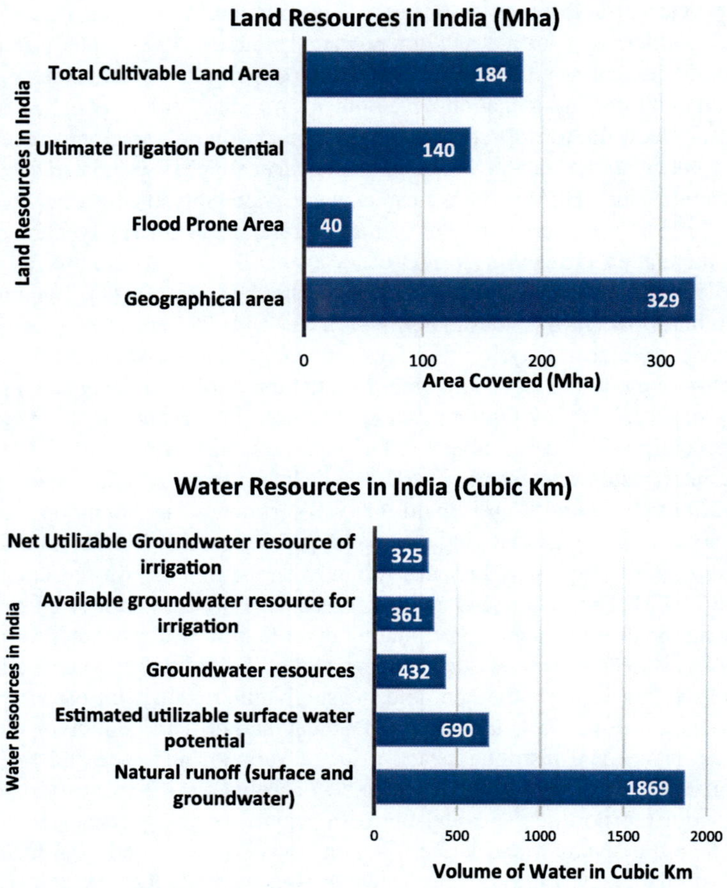

Fig. 1 Land and Water Resources in India

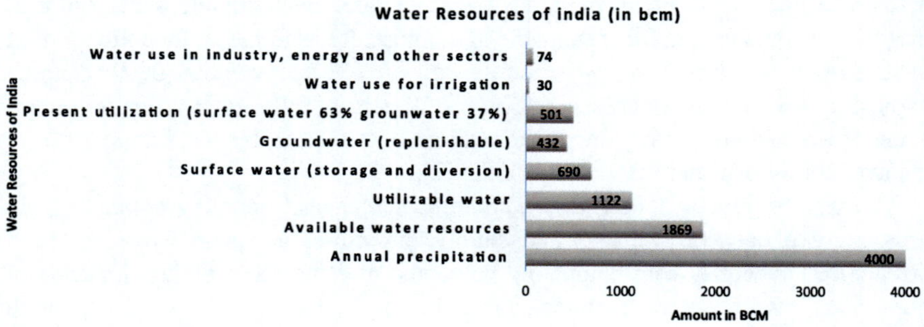

Fig. 2 Water Resources status in India

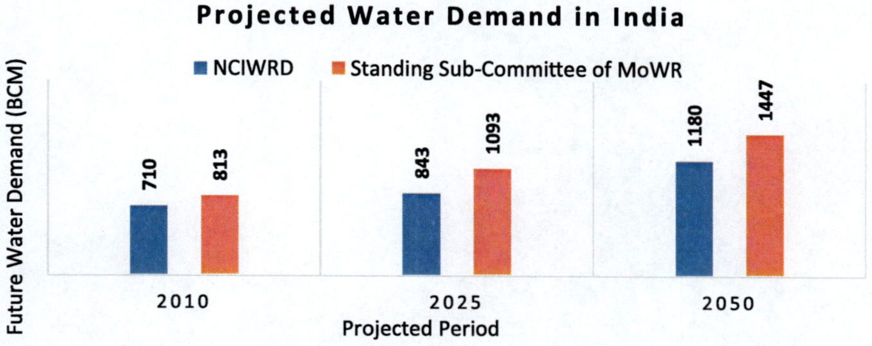

Fig. 3 Water Demand (in BCM) for various Sectors

Resources (MoWR) and the National Commission on Integrated Water Resources Development (NCIWRD) [9].

The NCIWRD and the Ministry of Water Resources have calculated India's future anticipated water consumption with a minimal difference, indicating a large imbalance between supply and demand by 2050. According to the NCIWRD and the Ministry of Water Resources, India's total water consumption would rise to 843 and 1093 BCM by 2025, and 1180 and 1447 BCM by 2050 respectively. As a result, finding new resources, managing existing supplies, and finding innovative ways to meet future water demand will be important concerns.

2 Precipitation in India

India has typical monsoon climatic conditions. Complete reversal of surface winds between January to July, resulting in two types of monsoon systems in India. During the winter, cold and dry air from northern latitudes flows in the southwest direction, creating the northeast monsoon, whereas, during summer, the warm and humid air rises from the sea and flows typically in the opposite direction, causing the southwest monsoon, which accounts for 75 to 90% of yearly rainfall. The southwest monsoon covers a major portion of India from June to September. However, the northeast monsoon influences much of the rainfall in the southern region of the east coast, such as Tamil Nadu and surrounding districts, during October and November [10]. Different climatic regions in India are shown in Fig. 4.

The country's average annual rainfall is about 1170 mm, however, there is a lot of temporal and geographical variance. Rajasthan's northwest desert receives less than 150 millimeters of rain each year on average. The average annual rainfall across the wide region from Central India to the peninsular region is typically less than 600 mm. The Khasi hills in the northeastern area, on the other hand, get more than 10,000 mm of rainfall in a short duration. The average annual rainfall in the west

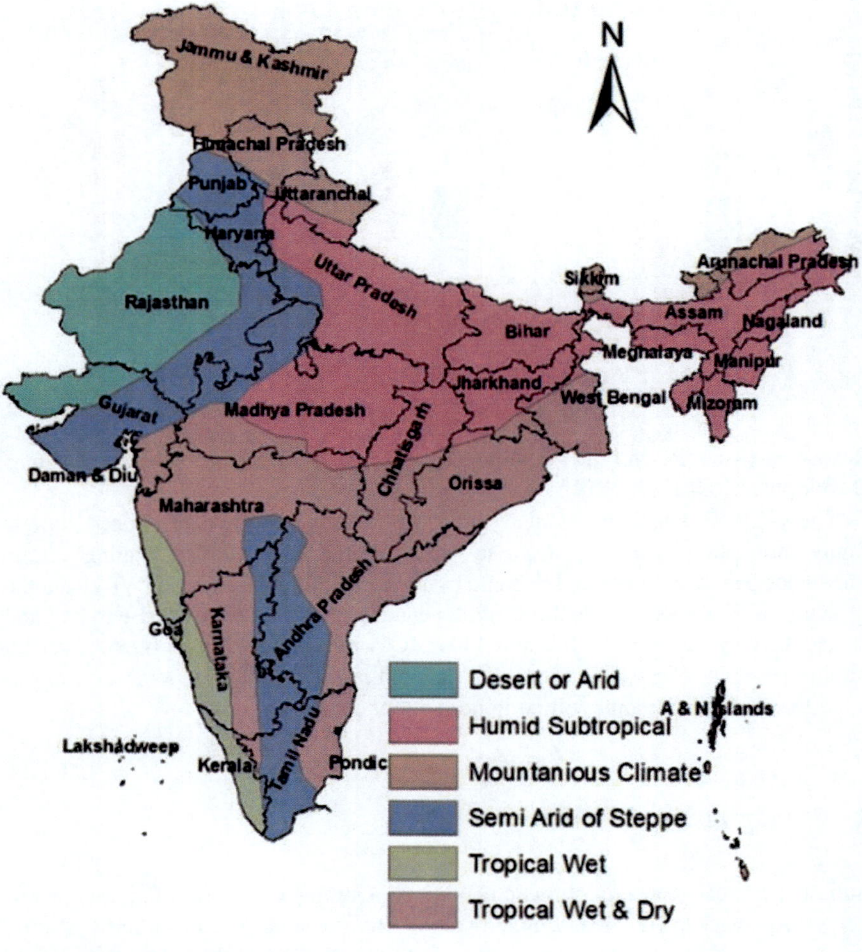

Fig. 4 Different climatic regions in India

coast states of Assam, Meghalaya, Arunachal Pradesh, and sub-Himalayan West Bengal is around 2500 mm.

Despite the fact that the seasonal monsoon dominates the Indian climate, regional variances in microclimate are attributable to different terrain and the impact of the oceans. Because of the Himalayas and the Indian Ocean, the nation has both a tropical and a continental climate. Temperatures in India range from sub-zero at high altitudes in the winter to 48 °C in the summer. The typical maximum temperature in northern India during the coldest months is 21 °C, whereas it ranges from 38 to 43 °C during the summer. The extreme southern part of the nation has less temperature fluctuation than the rest of the country. The Indian Meteorological Department (IMD) categorized four typical climatological seasons: October to February—winter season, April to June—summer or pre-monsoon, July to September—rainy or monsoon season from, October to November—post-monsoon or autumn season, all are based on average weather conditions.

3 River Systems in India

India has a 2.89 million km^2 network of river basin systems, with an average annual water flow output of 1900 BCM. According to the magnitude of their catchments, India's river system may be divided into four categories: major, medium, minor, and desert rivers [11]. All of India's major rivers originate in the Aravalli, Himalayas, Sahyadri, Vindhyan, and Satpura ranges. River systems in India can also be classified by their names, such as the Aravalli range river system, the Ganges, Brahmaputra, Indus river system, and rivers in Peninsular India. The majority of India's major rivers flow east, but the Narmada is the country's largest west-flowing river. Figure 5 depicts India's major river basins.

The Ganga, Yamuna, Brahmaputra, Godavari, Krishna, Kaveri, Narmada, Chambal, and other important rivers are among the major river system. There are around 44 medium river systems, with catchment sizes ranging from 2000 to 20,000 km^2 and a total area of 0.24 million km^2 within basins. These rivers have an annual average flow of 112 BCM, which irrigates just 0.08 million km^2 of cultivated land. Minor river basins, with a drainage area of smaller than 2000 km^2, are comes in the third category. These are mostly small streams that run into the sea from the eastern and western ghats. The catchment area of these rivers is 0.2 million km^2. These rivers are known for their ephemeral flow, steep slopes, and thick silt, which cause devastation from severe floods frequently. These rivers, on the other hand, are critical for coastal irrigation, notably in Kerala and Tamil Nadu. The overall estimated flow of minor rivers is 120 BCM, with the majority of the water going down small rivers that travel towards the west [11]. The fourth category includes rivers that flow for a short distance before disappearing in Rajasthan's desert. Their waters are utilized for irrigation, however, due to the erratic rainfall pattern, their flows are unpredictable in magnitude and timing. Desert rivers have a total basin area of roughly 1 lakh km^2 and an annual average flow of 10 BCM. Thus, in India, all river systems provide 1645 BCM of yearly water volume, with major rivers accounting for 85% of this, while medium and small rivers, as well as desert rivers, contribute 7% and 8%, respectively. The small rivers carry more water than the medium rivers [11].

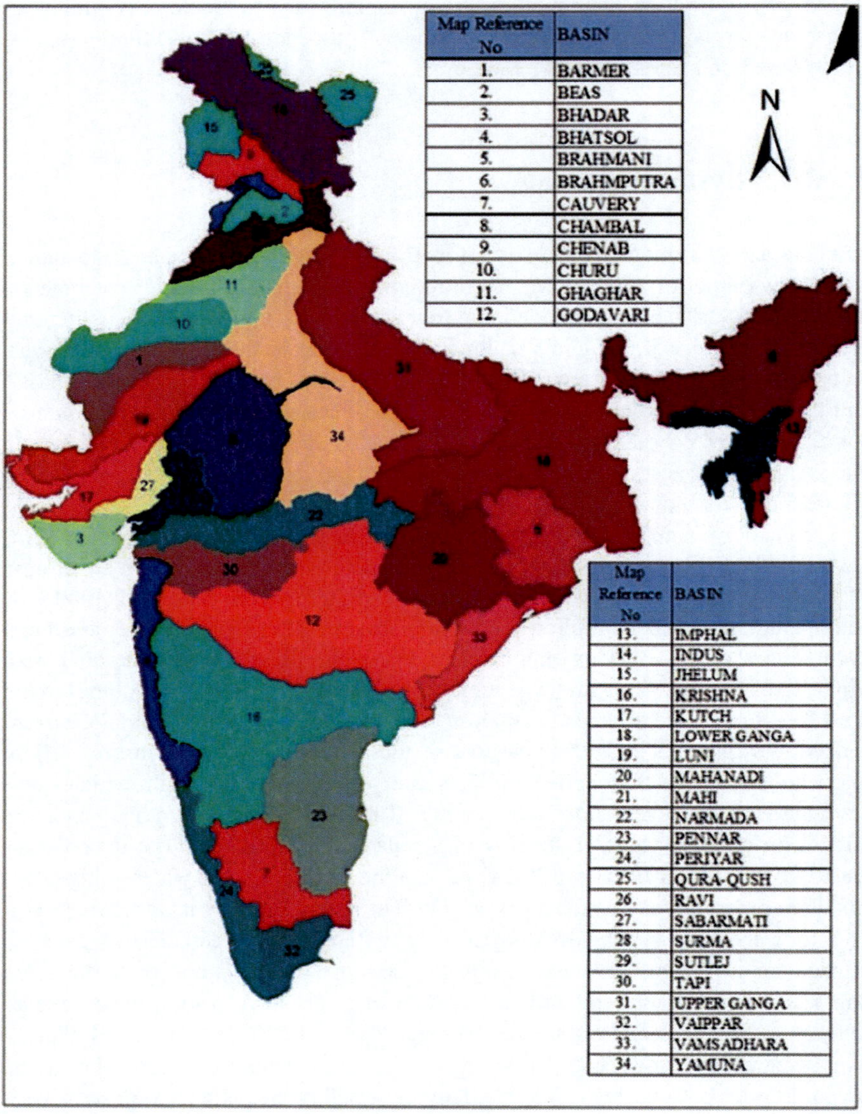

Fig. 5 Major river systems in India

4 Water Availability in India

In India, water availability is very uneven; almost 80% of the annual rainfall occurs in only four monsoon months from June to September. The Brahmaputra, Barak, and Ganga river basins contribute approximately 60% of total surface water resources.

On the other hand, water availability in the western and southern areas is limited. Years after years, this has resulted in the drought—flood—drought syndrome. As a result, the most difficult task is to prepare the plan, develop and manage water resources in order to increase water availability. India obtains 4000 BCM of total water from precipitation, leaving 1869 BCM of water accessible. Out of which there are 609 billion cubic meters of available surface water and 433 billion cubic meters of available groundwater. India covers 3.29 million square kilometers, or 2.4 percent of the world's land area, and is home to over 15 percent of the world's population. India sustains roughly one-sixth of the world's population, with one-fifth of the world's area and one-fifth of the world's water resources. India also has a 500 million-strong cattle population, which accounts for around 20% of the global livestock population. The country's total exploitable and utilizable water resources are estimated to be 1086 BCM [12]. Water resources are classed mostly depending on their available location, such as surface and groundwater resources, which are discussed further below.

4.1 *Surface Water Resources*

Several attempts have been made in the past to estimate water availability in India. Many researchers and government institutions worked in this area in order to get a true picture of the country's water resources. The average annual flow in Indian river systems was estimated to be 1953 km^3 by the National Commission for Integrated Water Resources Development. The utilizable water resource is the amount of water that can be taken away from its natural source. This utilization may be limited by physiographic circumstances and the socio-political context, as well as legal and constitutional limits and current development technology. It can be observed that there is still scope for improving further the water resource utilization from India's numerous river systems.

4.2 *Groundwater Resources*

The near-universal availability, low capital cost, and more reliability make groundwater one of the most desired natural sources in different water user sectors in India. Because of the country's growing reliance on groundwater as a desired source of water, indiscriminate extraction has occurred in many regions of the country, with little consideration for aquifer recharge capabilities or other environmental concerns. On the other hand, despite the availability of sufficient resources, still there are regions of the country where groundwater development is not optimal, including canal command areas that are experiencing waterlogging and soil salinity as a result of the steady rise in groundwater levels [13]. According to the Central Groundwater Board, the yearly potential groundwater recharge occurring naturally from rainfall in

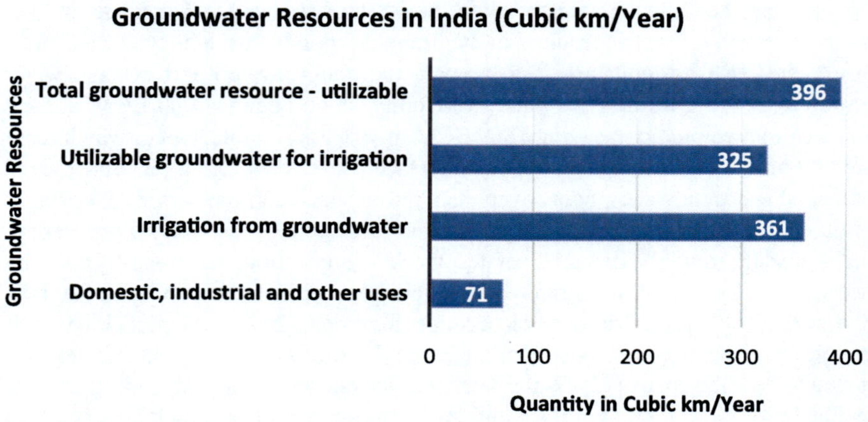

Fig. 6 Utilizable groundwater resources of India

India is around 342.43 km^3 which is around 8.56 percent of the country's total annual rainfall [14]. The canal irrigation system has an annual capacity for groundwater recharge improvement of about 89.46 km^3. As a consequence, the total replenishable groundwater resource for the country is calculated at 431.89 km^3. The rest of this quantity can be utilized for irrigation after 15 percent is set aside for drinking and 6 percent is set aside for industrial usage. Figure 6 shows the Central Groundwater Board's (CGWB, 1995) [14] estimates of total replenishable groundwater resources utilization for residential, industrial, and irrigation purposes, and future utilizable groundwater resources in India.

In the Indo Gangetic alluvial plain, there is a large area where the development of groundwater resources is sub-optimal and there is still scope for future development. Similarly, in water-stressed areas, immediate efforts are necessary to supplement groundwater artificially. However, in order to ensure the sustainable, judicious and appropriate use of groundwater resources, the focus on development activities must now be balanced by scientific planning and management systems [15]. Because groundwater development is mostly driven by demand, it may be boosted by appropriate agricultural, financial, subsidy, and energy policies, as well as the formation of appropriate markets. Furthermore, the flood plains along the country's major river systems provide excellent opportunities for groundwater development. Similarly, there are artesian conditions in some parts of the country that may be identified and appropriate development plans are to be developed. There is a need to create techniques for assessing the development potential of deeper aquifers in alluvial regions with multi-aquifer systems. For developing and implementing appropriate groundwater management plans in the country, there is a pressing need for scientific, integrated, and coordinated efforts by different Central and State Government departments, non-governmental and social service organizations, academic institutions, and stakeholders [13].

5 Droughts

Drought is referred to as the regional occurrence of relatively below-average water availability, whether in the form of precipitation, river flow, or surface and groundwater storages. It's a natural and common recurring phenomenon of climate that may happen everywhere on the earth, in any climate zone. Droughts are caused by a lack of precipitation as a result of natural climatic fluctuation in space and time [16]. Drought may affect each section differently, resulting in varying degrees of severity and impact. Drought has a subtle but long-lasting effect. Droughts may be divided into four categories. Meteorological drought is defined as a period when precipitation averages below the threshold level. Drought in agriculture refers to a shortage of soil moisture for crops, woods, rangelands, and cattle. A shortage of water in streams, tanks, and the aquifer is linked to hydrological drought. The fourth form of drought is socio-economic drought, which occurs when a lack of water impacts a region's social and economic activity. Droughts in India are caused by differences in the amount and distribution of monsoon season rainfall in time and place relative to the average value.

India has endured more than 14 significant droughts in the previous fifty years, out of which 1972, 1979, 1987, and 2002 droughts were the worst and affected almost half the geographical area of the country. Geographically, over 68% of India is vulnerable to drought, primarily covering the arid, semi-arid, and sub-humid climatic regions i.e. peninsular, central, western and north-western parts of India. Thus, in the planning and management of water resources and drought mitigation activities, determining the frequency, size, and persistence of droughts, as well as the susceptibility of an area to drought, is critical. Droughts influence both surface and groundwater resources, resulting in reduced water availability, worsened water quality, crop failure, reduced agricultural productivity, reduced power generation, disrupted riparian ecosystems, and halted recreational activities, among other things [17]. Droughts have an impact on water quality because mild climatic changes modify hydrologic regimes, which have a significant impact on river and lake chemistry [18].

Drought as the recurrent calamity has always threatened the Indian water and agriculture sector over decades and centuries. The detrimental impacts of drought are manifest in the sharp drop in agricultural production and farm income. It causes a reduction in rural employment imposing widespread economic impoverishment among farmers, farm laborers, rural artisans, and small rural businesses. Droughts are infamous for their complexity, thus the problems of preventing, mitigating, and managing this disaster need a scientific understanding of the symptoms, meticulous planning, concerted action, and cooperation amongst many concerned groups.

6 Floods

In India, floods are the most common natural disaster, with floods happening almost every year. Floods have lately wreaked havoc in India, particularly in the eastern states of Bihar, West Bengal, Orissa, and Andhra Pradesh. In India, the major causes of floods include a lack of capacity inside river banks to retain heavy flows, riverbank erosion, and riverbed silting. Landslides obstructing flow and changing the river's path, tidal and backwater effects delaying flow, inadequate natural drainage in flood-prone areas, cyclones and accompanying severe rainstorms or cloud bursts, snowmelt, and glacier outbursts, and dam-break floods are among the other factors. The material and intangible damages caused by floods in India are growing at an alarming rate, according to data provided by several government organizations. According to the Central Water Commission (CWC), which is part of India's Ministry of Jal Shakti, D/o Water Resources, the yearly average area impacted by floods in the country is 7.563 Mha. From 1953 to 2000, floods harmed around 33 million people on average. Due to the fast growth of the population and the growing encroachment of the flood plains for dwelling, farming, and other activities, this figure is certain to rise.

Following the devastating floods of 1954 in India, a nationwide flood management program was established. The Indian government has implemented a variety of flood-prevention measures, formulated policies from time to time, established working groups, task forces and committees to address flood-related issues. This includes Policy statement in 1954, high-level committee on flood in 1957, policy statement in 1958, ministerial committee on flood control in 1964, ministerial committee on flood and flood relief in 1972; working groups on flood control for five-year plans; Rashtriya Barh Ayog, National Water Policy in 1987, National Commission for Integrated Water Resources Management in 1996, Regional Task Force in 1996, and National Water Policy in 2002 and 2012 [19]. The government's committees and commissions have made useful suggestions on a variety of flood management concerns. To limit the damage in flood plains, a variety of structural and non-structural solutions have been used. In certain Indian states, structural interventions such as the construction of levees, embankments, spurs, and other similar structures have been adopted. The entire length of built embankments is 16,800 kilometers, while the total length of drainage canals is 32,500 kilometers. Flood protection is now in place for 1040 cities and 4760 villages. These have given reasonable safety and protection to an area of roughly 15.07 Mha, except for occasional breaches in embankments. A huge number of reservoirs have been built, and the intensity of floods has decreased as a result of these reservoirs. Non-structural flood management strategies, such as flood forecasting, flood classification based severity and flood warning, mass awareness, scientific studies on Dam break analysis and Emergency Action Plans are also being used. Flood forecasts and warnings in Delhi for the Yamuna River began in 1958. It has grown to encompass the majority of India's interstate river basins which are flood-prone and vulnerable. The Central Water Commission of India has established a large flood forecasting system that spans 62 major rivers and has over 157 stations, allowing it to give flood predictions for nearly all flood-prone states in the country. The response from all the state governments to the action on the flood plain zoning has been seen as encouraging. Although some states, such as Manipur and Rajasthan, Bihar, Uttar Pradesh have passed flood legislation on plain zoning, however the other flood-affected states, such as Himachal Pradesh, Goa, Sikkim, and Assam are yet to take action on this.

Flood control methods must be more concentrated and aimed at the predetermined goals within a set time frame. Methods for flood plain zoning must be developed in cooperation with local governments in order for flood plain zoning laws to be approved. As proposed by the Working Group of the tenth five-year plan, any concerns about the difficulty of writing legislation should not be used to dismiss the concept of flood plain zoning. One of the most essential aspects of flood catastrophe preparedness is flood forecasting. Technical progress in a well-planned flood forecasting and warning system can assist provide more lead time for the prompt response. The creation of suitable flood storage in reservoirs is widely acknowledged as a long-term solution to flood concerns. In India, completed projects have a total live storage capacity of around 174 km^3. A good flood management scheme necessitates a substantial flood storage area in reservoirs. Flood management also necessitates community involvement. Flood management is something that farmers, professional bodies, companies, and non-profit organizations must be aware of. Participation of the public in disaster preparedness, flood control, and disaster response is essential. Radio, television, newspapers, and social media platform may all play a part in flood control. Due to the fact that India shares river systems with six bordering nations, namely Pakistan, Bangladesh, Bhutan, Nepal, Myanmar, and China bilateral cooperation for flood control is required. The Indian government has taken some steps in this direction, but more active engagement is necessary. Figure 7 depicts India's major flood zones.

7 Climate Change

Climate change has been well defined by the United Nations Framework Convention on Climate Change (UNFCCC) as a change in climatic conditions caused directly or indirectly by anthropogenic activities that alter the global atmosphere's balance, in addition to natural climate variability seen over long time periods. Climate change is a long-term, continuous shift in average weather conditions that is either rising or decreasing. It may result in more frequent and severe extreme events like floods and drought, wreaking havoc on the region's agriculture, water supplies, livelihood, and economy. Unlike year-to-year fluctuation, climate change is a slow and steady process. Climate change is now broadly acknowledged as one of the most serious problems facing the world in the twenty-first century. Measurements over the past 157 years reveal that temperatures at the surface have risen globally, with considerable regional differences, according to the Intergovernmental Panel on Climate

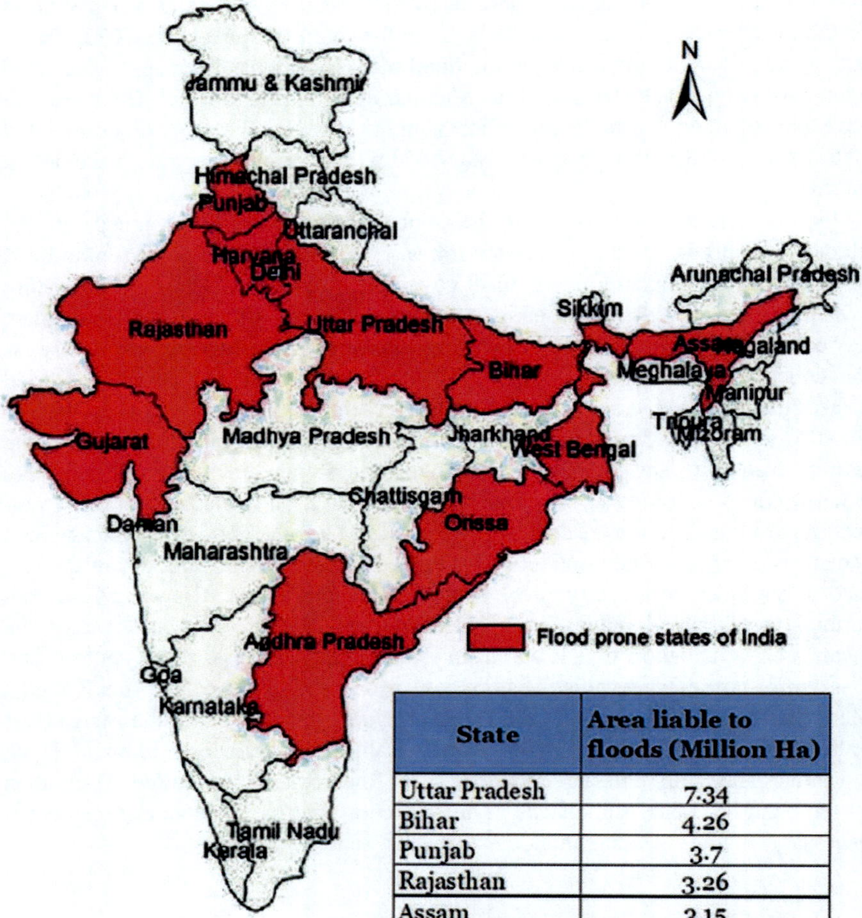

State	Area liable to floods (Million Ha)
Uttar Pradesh	7.34
Bihar	4.26
Punjab	3.7
Rajasthan	3.26
Assam	3.15
West Bengal	2.65
Haryana	2.35
Orissa	1.4
Andhra Pradesh	1.39
Gujarat	1.39

Fig. 7 Major flood zones of India

Change (IPCC) report [20]. Over the previous 25 years, the rate of warming has accelerated, with 11 of the 12 hottest years on record occurring in the last 12 years. In general, this warming exacerbates the global hydrological cycle [21], and it is well documented that the earth's mean surface temperature has been rising since the last glacial maximum 21,000 years ago [22], resulting in increased globally averaged precipitation, evaporation, and runoff.

Biotic activities, variation in solar energy received by the planet, plate tectonics, and volcanic eruptions all contribute to climate change. Temperature and rainfall intensity has increased, indicating that the world and local climates are changing. Human activities have also been recognized as important contributors to climate change in recent times, often known as "global warming," which has the potential to have a wide-ranging, complex, and unpredictable influence on the environment. Climate change will have a significant influence in the future, according to IPCC assessments, which predict a decrease in freshwater availability as a result of climate change. This has also indicated that by the middle of the twenty-first century, yearly average runoff and water availability will perceive decrease by 10–30% [20]. A change in a region's long-term weather patterns, such as temperature, wind, and precipitation, is referred to as global climate change. According to the IPCC's fourth assessment report, rising the mean seasonal temperatures will have an impact on agriculture, water, and the global economy. Due to its agrarian economy, India is one of the most vulnerable countries to climate change, with 70% of its population reliant on agriculture for a livelihood.

Although numerous climatic factors influence water supplies, temperature and precipitation are the two most important. The hydrological cycle, which is a thermally driven system, is projected to accelerate as a result of global warming. Warmer temperatures will cause more precipitation to fall as rain rather than snow. In the long run, this will result in glacier shrinkage and a reduction in snow and glacier melt. There may be variations in the process of run-off production and its timing in areas such as India's northern plains, where significant river flows are generated by Himalayan snow and glacier melt. Increased evapotranspiration from crops, plants, and land surfaces, as well as greater water needs, may result from rising temperatures [23]. Extreme weather events such as floods, droughts, and rainstorms are projected to become more common as a result of global warming. The precipitation pattern is likely to alter in two ways: temporal patterns, such as the monthly distribution of yearly precipitation, and geographical patterns, such as some places getting high precipitation receiving less and vice versa. The three sorts of effects are distinct and need distinct adaptive responses, but they are also intricately linked [8]. Climate change in India may have an impact on long-term rainfall patterns, reducing water supply and raising the risk of droughts and floods. Climate change in India, notably the South-West monsoon, would have a substantial influence on agricultural productivity, water management, the general economy, and the country's livelihood sector. During the non-monsoon months, the great concentration of rainfall in the monsoon months leads to a scarcity of water in this area of the nation. Droughts may become more common when yearly rainfall decreases and wet days increase, despite increased rainfall and decreasing rainy days.

7.1 *Climate Change and India's Actions*

In 2007, India's emissions were expected to be in the range of 1331.6 million tonnes of carbon dioxide equivalent to Green House Gas (GHG) emissions. The emissions have seen increased by 4.2 percent each year since 1994. India's CO_2 emissions, on the other hand, account for just around 4% of total world CO_2 emissions and much less than the historical concentrations taken into consideration. Nonetheless, India has been aware of the worldwide threat of climate change. India has agreed to communicate information about the Convention's implementation in accordance with the obligations imposed on parties to the United Nations Framework Convention on Climate Change (UNFCCC). In 2012, India presented the UNFCCC with its Second National Communication (NATCOM). In 2004, the first National Communication was submitted by India. According to India's Second National Communication to the UNFCCC, the annual mean surface air temperature rises by the end of the century are expected to range from 3.5 to 4.3 degrees Celsius, while sea level along the Indian coast has been increasing at a pace of around 1.3 millimeters per year on average.

In 2006, the Government of India outlined the National Environment Policy as a critical component of India's response to climate change, with the goal of identifying India's key vulnerabilities to climate change, including impacts on water resources, forests, coastal areas, agriculture, and health, as well as assessing the need for climate change adaptation and encouraging Indian industry to participate in the Clean Development Mechanism (CDM). Later, India's Prime Minister established a High-Level Advisory Group on Climate Change, which includes both government and non-government members. The Council organizes national action plans for climate change assessment, adaptation, and mitigation. Since 2008, the Ministry of Environment and Forests has been implementing the National Action Plan on Climate Change (NAPCC) through nodal ministries in key sectors. State governments are also drafting State Action Plans on Climate Change, with the help of the federal government, to build institutional capacity and undertake sectoral actions to combat climate change. So far, 21 states have prepared State Action Plan on Climate Change (SAPCC), including Manipur, Meghalaya, Mizoram, Nagaland, Assam Arunachal Pradesh, Tripura, Sikkim, Jammu & Kashmir, Andaman and Nicobar, Lakshadweep, Delhi, Odisha, Punjab, Rajasthan, Uttarakhand, Andhra Pradesh, Karnataka, Kerala, West Bengal, and Madhya Pradesh.

8 Water Quality

In India, the degradation of water quality due to pollution is a serious environmental problem. The majority of India's rivers, lakes, and surface water are contaminated. The most significant sources of water pollution in India are untreated sewage from cities and industrial effluents. Unregulated small-scale enterprises and agricultural

Fig. 8 Picture showing the severity of pollution in Ganga

runoff are two further causes of pollution. The condition of India's rivers, lakes, and ponds is under growing threat as a result of population growth, urbanization, and industrialization. The quality of water degrades every time it is utilized for something. Unfortunately, the country's reverence for rivers does not guarantee that they are clean.

The severity of water pollution varies by region, based on the density of urban growth, agricultural and industrial activities, and wastewater collection and treatment infrastructure. India's Central Pollution Control Board (CPCB) has identified some of the contaminated river lengths as well as potential pollution sources. The majority of the contaminated regions are found in and around metropolitan cities. Water pollution is a significant problem in India, with biological, toxic, chemical, and inorganic pollutants contaminating almost 70% of its surface water resources and an increasing percentage of its groundwater reserves. The high frequency of severe contamination near urban areas suggests that the contribution of the industrial and residential sectors to water pollution is considerably more than their proportional prominence in the Indian economy suggests. In terms of total influence on water quality, agricultural activities also have a role. Aside from a rapidly falling groundwater table in various regions of the country, the government is also dealing with a severe groundwater pollution problem that has afflicted 19 states, including Delhi. Groundwater has been contaminated by geogenic pollutants such as salt, iron, fluoride, and arsenic in over 200 districts across 19 states. More than 70% of our country's pure water in liquid form has been rendered unsuitable for human use. Other countries, in addition to India, are dealing with the same issue. The numerous sources of pollution such as sewage discharge, industrial effluents, and agricultural runoff, as well as their potential, have been identified as the primary danger to water pollution in India [24]. Figure 8 depicts the current state of water contamination and the decline of water quality.

Water pollution may have a significant negative impact on the health of any life form that lives near a polluted water body or uses water that has been contaminated to some level. In reality, in India, contaminated water is one of the primary causes of the country's poor health, particularly in rural regions. Cholera, TB, dysentery, jaundice, diarrhea, and other illnesses can all be caused by contaminated water. Ingesting contaminated water is responsible for around 80% of gastrointestinal problems in India. At a certain point, contaminated water may harm crops and diminish soil fertility, affecting the agricultural industry as a whole and the country as a whole. When saltwater is contaminated, it has a negative influence on marine life. The most fundamental consequence of water pollution is the quality of the water, which can cause a variety of illnesses if consumed.

In India, water contamination has reached a catastrophic level. Almost every river system in India is now significantly polluted. Water quality evaluation and other elements of water pollution in major and small Indian rivers have been studied by several researchers in India. Nearly 70% of India's water is contaminated, according to experts at the National Environmental Engineering Research Institute (NEERI) in Nagpur [25].

The influence of sewage water entering the river Yamuna, bacterial pollution in Delhi, and river water quality in Agra have all been studied by researchers [26–28]. The biological characteristics of the river Yamuna are seen far inferior to those of the Ganga. A number of scientists have investigated pollution in the Ganga River. Studies were carried out at Mirzapur and Varanasi to investigate the Physico-chemical characteristics of the river Ganga [29, 30]. Both investigations come to the same conclusion: the physicochemical characteristics of Ganga water have been degrading over time and are continuing to do so. The studies point to the existence of a significant number of pathogenic and non-pathogenic microorganisms in concentrations much above their maximum permissible levels. Blue-green algae were studied in the Gomati river in India [31], and similar investigations were done in the Mahanadi river in Orissa state [32]. The Tungabhadra reservoir has also been the focus of extensive water quality investigations [33].

9 Agriculture in India

Agriculture accounts for about 46% of the gross national product in India, and it is also the people's primary occupation and the government's primary concern, with the responsibility of providing adequate food for a population that accounts for about 16% of the world's population while holding only 14% of the world's total agricultural potential [34]. India's agricultural achievements during the last four decades have been outstanding. The agricultural industry has done a good job of keeping pace with the growing food demand. Increasing land area under agricultural production has that important over time, and recent output increases have been nearly exclusively attributable to increased productivity [35]. Agriculture has made a significant contribution to overall progress in the country in recent decades. Increased production has helped feed the poor, increased agricultural revenue, and given direct and indirect job possibilities.

The success of India's agriculture may be credited to a series of events that led to the availability of farm technology, which resulted in substantial gains in production during the Green Revolution era in the 1970s and 1980s. The introduction of new varieties of crop, intensification of input usage such as fertilizers and manures in agriculture, and investments leading to extension of the irrigated area were the main sources of agricultural growth during this era [36]. Now the growth has been slowed in places where 'Green Revolution' technology had a significant influence in past. To push out the yield to new highs, use inputs more effectively, and diversify to more sustainable and higher value cropping patterns, and introduce new farming technologies. At the same time, if we are to fulfill agricultural growth and poverty reduction objectives, we must better harness the potential of rainfed agriculture and other less endowed areas with new techniques, schemes, and programs. Indian agriculture has a wide range of requirements, possibilities, and prospects due to the diverse agro-ecological settings. Future growth must be faster, more evenly dispersed, and more precisely focused. Agricultural research and scientific approach will be increasingly required to address community-specific problems; on the other hand, agricultural systems will have to position themselves in an increasingly competitive environment in order to generate and adopt cutting-edge technologies to address the problems that a large majority of poor farmers face [37].

Agriculture's long-term viability is contingent on soil management methods that provide healthy soils, and ecosystem services, as well as avoid resource degradation and ultimately enough food security. Conservation agriculture has provided a common thread for the application of five principles for sustainability—improved soil health, reduced use of agrochemical, sufficient use of water, adaptation to climate change, and doubling farmer income. The state's involvement in Indian agriculture must be based on the conservation agriculture method for food and ecological sustainability [38].

India would require 1498 BCM of water per year to meet agricultural water demand, according to the 2030 Water Resources Group Report [39]. Agriculture is expected to account for 80% of demand, with industry accounting for 13% and residential requirements accounting for 7%. India's current water supply is around 740 BCM, which is insufficient to meet this demand. As a result, unless the Ganga, Krishna, and Indus basins in India take coordinated action, most of India's river basins might face a severe shortfall of 758 BCM by 2030. Rice-wheat farming methods in Punjab and Haryana alone have resulted in a net loss of 109 BCN water yearly between 2002 and 2008 [40]. Agricultural water-productivity strategies help close the water gap by combining enhanced water application, soil moisture, and tillage management methods to produce 'more crops per drop', as well as improving crop yields through their relationship to soil health [41]. Surface flooding is the most common mode of irrigation water application in India and abroad, it is critical to enhancing the surface ridge-furrow irrigation system utilizing lay-flat gated pipes as soon as possible [42, 43]. Rolling toposequences must be pushed for pressurized sprinkler systems. Under marginal soil, water quality, and climate circumstances, a drip system works best which saves more water than gated pipes and sprinklers.

Direct dry seeding of Kharif crops before the rainy season begins can enhance production and decrease soil erosion risks. Dry rice sowing avoids puddling, saving at least 25 cm of irrigation water per hectare. Laser-assisted precision field leveling, surface seeding, dry planting, zero tillage, mulching, and other practices increase infiltration, reduce evaporation, conserve irrigation water, increase soil moisture and improve crop yield [41]. Improved soil health and reduced evaporation are still key alternatives for enhancing water productivity in areas with poor water productivity [44].

10 Irrigation Development in India

Agriculture is impossible to carry out without the presence of water. We must rely on rain to provide water for our farms, but rain does not fall throughout the year in the Indian region. India receives virtually all of its rain during the monsoon season, which runs from June to September and is not always consistent. Some areas receive modest rainfall, while others are left dry. Rainfall changes from year to year as well having huge temporal variation. In certain years, we receive a lot of rain, but in others, we don't get nearly enough. This demonstrates the need for water resource development, conservation, and optimal utilization. Indian irrigation infrastructure consists of a network of the major, medium, and minor canals originating from irrigation projects constructed on rivers, tanks, lakes, as well as well-based groundwater systems and other rainwater harvesting installations for agricultural purposes [45]. Groundwater wells can irrigate roughly 39 million hectares out of India's around 160 million hectares of cultivated land, while irrigation canals can irrigate another 22 million hectares [10]. Only approximately 35% of India's total agricultural area was reliably irrigated and the monsoon is the main source for around two-thirds of India's agricultural rainfed land. In the last 50 years, advances in irrigation infrastructure have helped India to reduce monsoon reliance, enhance food security, increase agricultural output, and generate rural job opportunities. Irrigation dams have also aided in the provision of drinking water to a rising urban as well as rural population, flood control, and the prevention of crops from drought-related damage.

10.1 History of Irrigation in India

In the early nineteenth century, according to irrigation sources, canal irrigation was 45 percent, irrigation from wells was 35 percent, tanks 15 percent, and from other sources, it was 5 percent. Famines in India in 1897–98 and 1899–1900 have prompted the British to create the first irrigation commission in 1901, specifically to report on irrigation as a way of famine prevention in India. The overall irrigated area by governmental and private works grew to 16 Mha in 1921 as a consequence of the first irrigation commission's suggestions. There was no substantial increase in tube well-irrigated areas from the beginning of the nineteenth century until 1921. Irrigation growth rates were projected to be 2.0 percent per year for government canal irrigation, 0.54 percent per year for well irrigation, and 0.98 percent per year for irrigation from all sources between 1910 and 1950.

10.2 Post-Independence Irrigation Development

India's irrigation development accelerated significantly after independence. The nation started on a large irrigation effort during the First Five Year Plan (1951–56). Hirakud, Bhakra Nangal, Chambal, Nagarjunasagar, Kosi, Kakrapar, and Tungabhadra were among the major and multipurpose irrigation projects undertaken. Simultaneously, under the Agricultural Sector, smaller irrigation initiatives, including groundwater, were given priority, with financial support from the Centre. New irrigation initiatives were established during the Second Five Year Plan (1956–61), the Third Five Year Plan (1961–66), and the three yearly plans (1966–69). During the Fourth Five Year Plan (1969–1974), the emphasis was moved to project completion, integrated surface and groundwater usage, effective management approaches, and modernization of existing systems. The Command Area Development Programme (CADP) was established during the Fifth Plan (1974–78) as a Centrally Sponsored Scheme with the goal of closing the gap between the irrigation potential developed and actual irrigation to make the best use of available resources. The program was designed as a way to bring all relevant activities under one roof in order to achieve these goals and initiated by the government with a 15 Mha of CCA of 60 major and medium projects.

Construction of new projects was continued during the year 1978–80 and the Sixth Five Year Plan (1980–85). By the conclusion of the Seventh Plan, there were 182 large and 312 medium active projects in the country. Due to huge investment and fund requirements in new projects in the period of 990–91, the Govt. of India changed its focus and started work of completing projects that were already well. This was continued up to the Eighth Plan (1992–97) and the Ninth Plan (1997–20). In 1996–1997, the Government established the Accelerated Irrigation Benefit Program (AIBP) scheme to expedite the completion of ongoing projects in the advanced stages of development. During the Eighth Plan period, irrigation potential of 2.22 Mha was developed at a rate of 0.44 Mha per year in the major and medium sectors. During the Ninth Plan period, this grew to 4.12 Mha, with AIBP accounting for 1.65 Mha (almost 40%). The importance of user engagement in large and medium irrigation systems was highlighted. Repairs and improvements to local irrigation projects were also encouraged as part of integrated micro-development. Later the sprinkler and drip irrigation systems, as well as the combined use of surface and groundwater, grew in popularity. According to the second irrigation commission's assessment, the total ultimate irrigation potential created in 1972 was 113.47 million hectares which include 58.47 million hectares from major, medium projects and

55 million hectares from minor irrigation (surface and groundwater). The total ultimate irrigation potential increased to 139.90 million hectares up to 2013–14 which include 58.47 million hectares from major, medium projects and 81.43 million hectares from minor irrigation (surface and groundwater) [46].

11 National Water Policy

The Ministry of Jal Shakti, Government of India (Formerly M/o Water Resources) has developed a National Water Policy to guide the planning and development of water resources as well as their optimal usage. In September 1987, the first National Water Policy was enacted. In 2002, it was re-evaluated and modified, and then again in 2012. It proposes connecting rivers to alleviate the country's water shortage. The major focus of the National Water Policy 2012 was to consider water as an economic good, with the goal of promoting its conservation and effective use, according to the ministry.

11.1 National Water Mission

A National Water Mission program has been developed to ensure integrated water resource management, which will aid in water conservation, waste reduction, and more equal distribution between and within states. The mission is to establish a framework to optimize water usage by increasing water use efficiency by 20% through regulatory mechanisms with differential entitlements and pricing, taking into consideration the requirements of the National Water Policy. It aims to ensure that a significant portion of urban water needs can be met by augmenting groundwater recharge through water harvesting measures and wastewater recycling. Similarly, the water needs of coastal cities with insufficient alternative water sources are met through the adoption of new and appropriate technologies, such as low-temperature desalination technologies that allow the use of ocean water. The National Water Policy is to be reviewed in coordination and cooperation with states to guarantee river basin level management plans to deal with climate change-related variations in rainfall and river flows.

The National Water Mission includes goals to improve surface and groundwater resources, storage, as well as rainwater harvesting, and fair and efficient management structures. New regulatory frameworks, as well as proper entitlements and pricing, are planned by the Mission. Its goal is to improve the effectiveness of existing irrigation systems, including the rehabilitation of deteriorated systems and, where possible, irrigation expansion, with a focus on increasing storage capacity. Water-positive, water-neutral technologies, enhancing subterranean water sources, and adoption of large-scale irrigation schemes relying on drip, sprinklers, ridge, and furrow irrigation will all be promoted through incentive structures. The

National Water Mission program has five goals: promotion of citizen and state actions for water conservation, augmentation, and preservation; focused attention on vulnerable areas, such as over-exploited areas, increasing water use efficiency by 20%; public domain for water database and an assessment of the impact of climate change on water resources [46].

12 Conclusions

After a thorough review of the number of government reports, research papers, scientific studies, popular articles, and media reports on the development and management of water resources in India, overall conclusions are drawn and summarised in this chapter to better comprehend the findings. In India, after independence water demand has grownup due to growing population, growth of agriculture, industry, and other sectors which resulted in decreased per capita water availability. Despite having abundant water resources, clean drinking water and agricultural water have always been in shortage. As a result, India must establish policies for long-term water management to bridge the widening gap between demand and supply, ensuring sufficient water for residential, agricultural, and industrial use.

A major portion of India receives rainfall from the southwest monsoon. It does, however, have a lot of temporal and spatial variation. India is blessed with a network of large, medium, and small river systems that contribute 1645 BCM of yearly water flow, with large rivers accounting for 85 percent of this total. The country has a further scope to improve its water utilization from its bestowed river systems. Groundwater is the most favored source of water in India's different consumer sectors; nevertheless, the country's growing reliance on it has led to indiscriminate extraction without consideration for aquifer recharge capacity. There is still scope at certain lares where groundwater resources might be developed further. However, urgent action is required for the augmentation of groundwater recharge in water scare and over-exploited areas in the country.

Droughts are extreme events posing threat mainly to water, agriculture, livelihood, economy, and other sectors. India has undergone a series of disastrous droughts that have affected over half of the country's geographical area, with more than half of the country classified as drought-prone. As a result, assessing droughts is critical for developing drought mitigation strategies and managing water resources. In India, floods are the most commonly occurring natural disasters. In India, the major causes of floods include a lack of capacity inside river banks to retain heavy flows, riverbank erosion, and riverbed silting. Due to fast population expansion and the growing encroachment of flood plains, about 33 million people in India are in danger of flooding, and this number is certain to rise as the country's population grows. The Indian government has adopted a variety of flood prevention and protection measures, including the creation of a National Flood-Prevention program.

A huge number of reservoirs have been built across large and medium rivers and the intensity of floods has decreased as a result of these reservoirs.

Due to its agrarian economy, India is one of the most vulnerable countries to climate change, with 70% of its population reliant on agriculture for a living. Climate change in India may affect rainfall patterns, reducing water supply and causing frequent droughts and floods. It would have a major influence on agricultural productivity, water management, and the country's general economy. In 2006, India's government announced its National Environment Policy as a critical component of the country's response to climate change, with the goal of identifying India's main vulnerabilities to climate change and laying out adaptation plans.

Water pollution is one of the severe problems in India, with organic, inorganic, biological, and toxic contaminants contaminating the majority of surface and underground water resources. The quality of water in rivers, lakes, and ponds is under rising threat as a result of population growth, urbanization, and industrialization. Untreated sewage, agricultural runoff, and uncontrolled small-scale industry are the most common sources of water pollution, with the majority of contaminated sections occurring in and near big metropolitan centers. Water quality issues must be addressed urgently to ensure clean and good quality freshwater supply for the expanding population.

Agriculture is the primary occupation of the people in the country, as well as the government's primary concern when it comes to providing enough food for a population. Over the last few decades, India has made significant strides in agriculture. India's agriculture industry has done a good job of keeping up with growing food demand. The Green Revolution period, which brought about substantial improvements in production in the 1970s and 1980s, is credited with India's agricultural prosperity.

One of the major factors contributing to India's green revolution was the rapid construction of irrigation infrastructure following independence. The government began a significant irrigation program, undertaking a number of multifunctional and big projects while also focusing on minor irrigation schemes, including groundwater, and implementing new irrigation programs. The Government of India created the Accelerated Irrigation Benefit Program (AIBP) to advance irrigation development by ensuring project completion, user involvement, minor irrigation project repairs, and improvements as part of integrated micro development, sprinkler, and drip irrigation programs, and the conjunctive use of surface and groundwater.

The Indian government has established a framework policy for the development, management, and planning of water resources through a series of policy documents, high-level committees, and missions. The Ministry of Water Resources developed a National Water Policy to guide the planning, development, and optimal use of water resources. A National Water Mission program has been developed to ensure integrated water resource management, which will aid in water conservation, waste reduction, and equal distribution between and within states.

13 Recommendations

Following are some recommendations based on a thorough study and evaluation of the state of water resource management in India, which will aid in the formulation of plans and the implementation of essential actions to meet challenges.

Despite India's enormous water resources, there has always been a scarcity of safe drinking and agricultural water. As a result, sustainable water resource development is critical in the country in order to bridge the growing gap between demand and supply, ensuring sufficient water for domestic, agricultural, and industrial use.

Groundwater is the most preferred source of water in India, and increased reliance on it has resulted in groundwater resource depletion in many parts of the country. Identification of potential areas for surface water resources development through a vast network of rivers and implementing programs for augmentation of groundwater recharge through natural and artificial means will be the key to tackle this issue.

The systematic scientific approach for assessment of water scarcity, droughts, formulate mitigation plan, assessment of extent and magnitude of floods, flood management, development of policies under climate change scenarios are key issues to tackle flood and droughts.

Agriculture is the main occupation of the people in the country and it accounts for the majority of rural livelihood. Policies, provisions, plans have to be made for assured water availability to meet irrigation demand, thereby increasing food production to meet the increasing food demand of the increasing population.

Deterioration of surface and groundwater quality is a major environmental concern in India. Immediate action is needed to implement policy options for reducing industrial water, industrial effluent treatment, and urban waste-water treatment for better handling of organic wastes.

Surface and groundwater quality degradation is a serious environmental problem in India. For improved handling of organic wastes, immediate action is required to adopt policy alternatives for decreasing industrial water, industrial effluent treatment, and urban waste-water treatment.

Acknowledgments The authors are thankful to the National Institute of Hydrology and Rabindranath Tagore University Bhopal for providing facilities, data, and information to compile this chapter.

References

1. Government of India (2006) Report of sub-group of minor irrigation, CAD and private sector and beneficiaries participation for the XI five year plan (2007–12), Ministry of Water Resources, GOI
2. Government of India (2006) Towards faster and more inclusive growth: an approach to the 11th five year plan (2007–2012), Planning Commission, GOI
3. Water Resources Development in India: Critical Issues and Strategic Options (web site:www.indiaenvironmentportal.org.in/files/Water-Assessment.pdf)

4. Website: www.nih.ernet.in; National Institute of Hydrology
5. Bobba, A. G., Singh, V. P. and Bengtsson, L. (1997) Sustainable development of water resources in India. Environmental management, 21, 3, 367–393, Springer, New York. https://doi.org/10.1007/s002679900036
6. Brown LR (2013) India's dangerous food bubble, Los Angeles times. Archived from the original on 4 December 2013
7. Pathak S, Pramanik P, Khanna M, Kumar A (2014) Climate change and water availability in Indian agriculture: impacts and adaptation, Indian J Agric Sci, 84
8. Jain SK (2012) Sustainable water management in India considering likely climate and other changes. Curr Sci 102(2)
9. National Commission for Integrated Water Resources Development (1999) Integrated water resources development – a plan for action. Report of The, Ministry of Water Resources, New Delhi
10. FAO (2013) Global map of irrigated areas: India FAO-United Nations and Bonn University, Germany
11. Rao KL (1979) India's water wealth: its assessment, uses and projections. Orient Longman, New Delhi
12. Gangwar Sneh (2013) Water resource of India: from distribution to management. Int J Inf Comput Technol, 3(8), 845–850, ISSN 0974-2239
13. Jha BM, Sinha SK (2008) Towards better management of ground water resources in India CGWB report. Central Ground Water Board, India
14. Central Groundwater Board (1995) Groundwater resources of India. CGWB Report, New Delhi
15. Raghava Rao KV (1969) An estimate of the groundwater potential of India. First Approximate Soil and Water Management Symposium, Hissar
16. Sharma KD, Pandey RP (2005) Hydrology of extremes: drought, key note paper, proc. of international Conf. On hydrological perspectives for sustainable development (HYPESD-2005), 23–25 Feb, Indian institute of technology, Roorkee, vol I, pp 223–2243
17. Riebsame WE, Changnon SA, Karl TR (1991) Drought and natural resource management in the United States: impacts and implications of the 1987–1989 drought. Westview Press, Boulder, CO, p 174
18. Webster KE, Kratz TM, Bowser CJ, Adagnuson JJ (1996) The influence of landscape position on lake chemical responses to drought in northern Wisconsin. Limnol Oceanogr 41(5):977–984
19. Mohapatra PK, Singh RD (2003) Flood management in India. Nat Hazards 28:131–143
20. IPCC: Climate change report (2007) Climate change impacts, adaptation and vulnerability. Working Group-II contribution to the Intergovernmental Panel on Climate Change Fourth Assessment Report
21. Milly PCD, Wetherald RT, Dunne KA, Delworth TL (2002) Increasing risk of great floods in a changing climate. Nature 415:514–517
22. Clark PU, Alley RB, Pollard D (1999) Northern hemisphere ice-sheet influences on global climate change. Science 286:1104–1111
23. Xu CY, Gong LB, Jiang T, Chen DL, Singh VP (2006) Analysis of spatial distribution and temporal trend of reference evapotranspiration and pan evaporation in Changjiang (Yangtze River) catchment. J Hydrol 327:81–93
24. Dwivedi AK (2017) Researches in water pollution: a review. Int Res J Nat Appl Sci 4(1)
25. Martin, P. (1998) River pollution in India: an overview. Emp News XXII (52):1–2
26. Sharma KD, Lal N, Pathak RD (1981) Water quality of sewage drains entering Yamuna at Agra. Indian J Environ Health 23(2):118–122
27. Balani MC, Sarkar HL (1965) Some observation on the pollution of Yamuna river at Okhla waterworks, intake, Delhi. Environ Health 7(2):84–86
28. Kaushik NK, Prasad D (1964) Coliform periodicity in water of river Jamuna at Wazirabad, Delhi. Environ Health 5(2):118–124
29. Shukla SC, Kant R, Tripathi BD (1989) Ecological investigation on Physico-chemical characteristics and phytoplankton productivity of river ganga at Varanasi. Geobios 16:20–27

30. Shukla SC, Tripathi BD, Nagendra P (1988) Physico-chemical and bacteriological characteristics of river Varuna at Varanasi. J Scientific Res 38:133–141
31. Prasad BN, Saxena M (1980) Ecological study of blue-green algae in river Gomati. Ind J Environ Health 22(2):151–168
32. Patra AK, Nayak L, Patnaik E (1984) Seasonal primary production of river Mahanadi at Sambalpur in Orissa. Trop Ecol 25(2):153–157
33. Rao DS, Govind BV (1964) Hydrology of Tungabhadra reservoir. Ind J Fisheries 11(1):321–344
34. Directorate of Economics and Statistics (2000) Indian agriculture in brief, Department of Agriculture and Cooperation, Ministry of Agriculture, Govt. of India
35. CMIE (1997) Directory of Indian agriculture, pp 402 Centre for Monitoring Indian Economy Pvt. Ltd., Mumbai
36. Indian Council of Agricultural Research (1998) National agricultural technology project, pp 193, ICAR, New Delhi
37. Abrol IP (2005) Agriculture in India. Centre for Advancement of sustainable agriculture
38. Gupta R, Mehra M, Sahoo R, Abrol I (2019) Indian agriculture redefining strategies and priorities. Econ Polit Weekly, LIII, pp 84–91
39. Addams L, Giulio B, Mike K, Martin S (2011) Charting our water future: economic frameworks to inform decision making, 2030 water resources group
40. Rodell M, Velicogna I, James SF (2009) Satellite-based estimates of groundwater depletion in India. Nature 460:999–1002
41. Humphreys E (2010) Halting the groundwater decline in north-West India. Adv Agron 109: 155–217
42. Rawitz E (2008) In: Chesworth W (ed) Irrigation, encyclopedia of soil science. Springer, Dordrecht, pp 369–380
43. Hassan AA, Elwan AM (2016) Gated pipes irrigation system for optimum water productivity of sugar cane in Egypt. J Am Sci 12(7):215–225
44. Molden DJ, Oweis T, Steduto P, Kijne J, Hanjra MA, Bindraban PS (2007) Pathways for increasing agricultural water productivity. Water for Food, Water for Life: A Comprehensive assessment of water management in agriculture
45. Siebert S (2010) Groundwater use for irrigation a global inventory. Hydrol Earth Syst Sci 14: 1863–1880
46. Government of India (2009) Report on National Water Mission under National Action Plan on Climate Change

Part III
Water Harvesting and Water Qualities

Drought Characterization During Monsoon Months Based on Standardized Precipitation Index (SPI) in Nuapada District, Odisha, India

Suman Kalyani Parida, Jyotiprakash Padhi, Paromita Chakraborty, and Bitanjaya Das

Abstract Frequent droughts experienced in World's different parts due to the effect of climate change. It is a challenging situation considering wide variation of conditions leadings to drought. Therefore, better scientific analysis is necessary to forecast, monitor and manage the drought. In this study, Standardized Precipitation Index (SPI) is computed because of its popularity across the World as an important indicator across space and time. Estimation of SPI 3 Aug and Sep drought value for Bodan, Khariar, Komna, Nuapada and Sinapali blocks of Nuapada District were found out for the interpretation of drought in the months of monsoon. This is achieved by using the rainfall data month wise for different blocks from 1983–2017. It was observed that, Bodan and Sinapali blocks experienced maximum number of total (moderate+severe+extreme) drought events based as per SPI 3 Aug as well as SPI 3 Sep. Highest severe drought events occurred in Nuapada and Sinapali blocks whereas maximum extreme drought events (2) detected by Khariar block as per SPI 3 Aug. Similarly, maximum events of severe (3) and extreme droughts (2) occurred in Bodna and Komna blocks respectively as per SPI 3 Sep. Also in this study, two threshold precipitation values were computed in monsoon months for the identification of agricultural drought. Agricultural drought threshold limit varied from 462.1 to 595.5 mm and 464.6 to 622.5 mm as per SPI 3 Aug and SPI 3 Sep values respectively. Maximum drought duration of 13, 18, 23, 21 and 17 months experienced by Bodan, Khariar, Komna, Nuapada and Sinapali blocks respectively. This analysis can guide different strategies on the management of water and also for planning of the crops in various blocks of Nuapada District.

Keywords Drought; standardized precipitation index; agricultural drought · Nuapada · Odisha · India

S. K. Parida · J. Padhi (✉) · P. Chakraborty · B. Das
School of Civil Engineering, KIIT Deemed to be University, Bhubaneswar, Odisha, India
e-mail: jyotiprakash.padhifce@kiit.ac.in; paromita.chakrabortyfce@kiit.ac.in; bdasfce@kiit.ac.in

S. Yadav et al. (eds.), *Water Quality, Assessment and Management in India*, Earth and Environmental Sciences Library, https://doi.org/10.1007/978-3-030-95687-5_4

1 Introduction

Droughts drew the attention of scientific experts from environment, ecology, hydrology, meteorology, geology and agriculture backgrounds because it is recognized as an environmental disaster [1]. Due to deficiency in annual or seasonal rainfall over a period of time, drought took place. Drought can take place in almost all climatic zones including the areas of high and low rainfall. Weather parameters (temperature, high wind and low relative humidity), duration, intensity, timing, onset, withdrawal and rainy days distribution during crop growth period play an important role in the drought occurrence [1]. Drought interpretation and knowledge of the historical droughts events and its impact helps in the management and planning of water resources. Therefore, knowledge of basic theories of drought will be useful for the development of models to explore the various properties of drought.

The information on the areal extent, duration and severity of drought is required for tackling the drought's adverse impact [2]. Forecasting and monitoring of drought can be performed through drought indices. Planning and management of water resources can be done by the policy makers based on the information about the drought characteristics obtained through the drought indices [3]. Therefore, various indices have been developed by the scientists for monitoring drought [4–8]. SPI is a probability-based drought index and based solely on precipitation. Better representation of irregular wetness can be performed by SPI [6]. Spatial and temporal variation in drought have been analyzed across the World using Standardized precipitation index [9–12]. Guenang and Kamga [13] studied the assessment of drought from the estimated value of SPI index by using monthly precipitation data collected over twenty four stations of Cameroon during 1951–2005. Assessment of the severity of drought performed in Australia by utilizing the rainfall threshold limit instead of SPI value [14]. In this study, SDF (severity, duration and frequency) curves were also developed using the calculated value of SPI through the technique of partial duration series. SDF (severity, duration and frequency) curves developed for various (1, 3, 6, 9 and 12) time scales using SPI index in Brazil through the assessment of meteorological drought [15]. SPI index used to assess the impact of climate change for the estimation of the dryness condition in future over 6 climatic projections in Poland [16]. Characterization of drought was assessed in the Hwanghae plain of North Korea with the help of SPI and SPEI by considering the effect of climatic change [17]. Investigation of spatio-temporal variation of drought was performed in Zambia by utilizing the computed value of SPI index during 1960 to 2016. Monthly GPCC (Global Precipitation Climatology Centre) precipitation data were used for the interpretation of drought both spatially and temporally in Zambia [18]. Spatio-temporal drought variation in Tel river was studied using multiple timescale based SPI and GIS by utilizing the precipitation data collected over the 7 stations during 1965–2008 [19]. Research works related to drought using SPI mostly focused on the interpretation of drought using various (1, 3, 6, 9 and 12) time. Authors concluded that, due to the occurrence of severe as well as extreme droughts over the time may create problems for the cultivation practices and water resources [19].

Assessment of the frequency, duration, intensity and trend of the meteorological drought due to climate change was investigated in GWB (Gangetic West Bengal) of Eastern India using SPI index geo-statistically [20]. Authors mentioned the importance of intensity, frequency as well as trend in the characterization of drought statistically that will help in the better understanding of the pattern of drought over the region. Assessment and identification of the meteorological drought using SPI, EDI and PNPI indices were investigated in Marathwada region of India during 1979–2014. Three hourly rainfall data converted into monthly rainfall value as per the requirement of the study and it was found that the characterization of meteorological drought in Marathwada region was similar based on the interpretation of all 3 drought indices [21]. Drought study will be helpful for the policy makers in making effective decisions regarding the management of drought. Few studies attempted in monsoon months for the assessment of drought in case of drought prone area in India [21, 22]. Drought Interpretation was performed through various indices of drought including SPI index but the estimation of agricultural drought threshold limit in monsoon season were not found out [22]. Therefore, this study is carried out to plan and manage the monsoonal water resources through interpretation of estimated SPI index.

2 Materials and Methods

2.1 Study Area

The drought study undertaken in five blocks (Bodan, Khariar, Komna, Nuapada and Sinapali) of Nuapada District, Odisha, India (Latitude lies between 82° 20′ E and 82° 53′ E and Longitude lies between 20° 00′ N and 21° 05′ N) having a total geographical area of 3852 km^2 (Fig. 1). This area was selected as study area because of experiencing frequent droughts. Nuapada District experiencing Western Undulating agro-climatic zone of Odisha experiencing warm as well as moist climate. District's annual normal rainfall was 1286.4 mm.

2.2 Methodology

Monthly rainfall data for the five blocks of Nuapada District were collected over 35 years (1983–2017) from the Water Resources Department, Government of Odisha, India for the calculation of Standardized Precipitation Index (SPI).

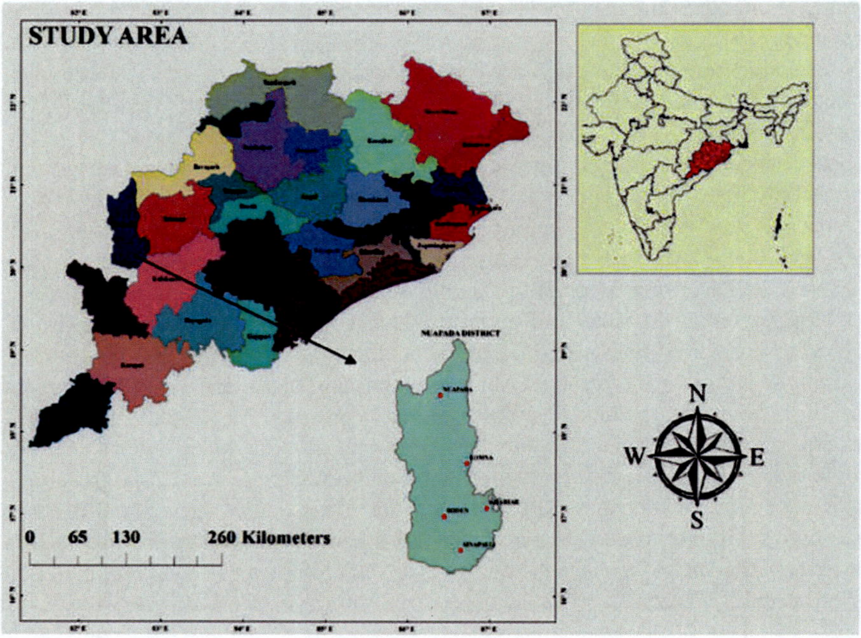

Fig. 1 The study site, Nuapada District of Odisha, India

2.3 *SPI for Drought Analysis*

McKee and his colleagues developed the Standardized Precipitation Index (SPI), at Colorado State University, which is one of the most widely used drought indices [6]. Drought event commences, when the standardized precipitation index value becomes -1.0 and terminates as soon as the value of SPI attains positive value. The major advantage of the SPI over other drought indices is that SPI allows the assessment of drought activities over various time scales and also monitoring of various categories of drought. Due to this flexibility, SPI can interpret the temporary effect on the supply of water (e.g. soil moisture) useful for agricultural production and longer-term effect on water resources such as supply of groundwater, levels of lake and reservoir and stream flow.

2.3.1 Computation of SPI

Computation of SPI is performed by fitting the PDF (probability density function) to the frequency distribution of precipitation summed over the time scale of interest. This is performed separately for each month (or any other temporal basis of the raw precipitation time series) and for each location in space. After that, individual PDF is transformed into a standardized normal distribution. The gamma distribution is defined by its probability density function is

$$g(x) = \frac{1}{\beta^{\alpha}\Gamma(\alpha)} x^{(\alpha-1)} e^{-x/\beta} \quad for \quad x > 0 \tag{1}$$

where $\alpha > 0$ is a shape factor, $\beta > 0$ is a scale factor, and $x > 0$ is the amount of precipitation. *(α)* is the gamma function which is defined as

$$\Gamma(\alpha) = \int_0^{\infty} y^{\alpha-1} e^{-y}\, dy \tag{2}$$

Fitting the distribution to the data requires the estimation of α and β. Edwards and McKee [23] suggested a method for estimating these parameters using the approximation of Thom [24] for maximum likelihood as follows:

$$\alpha = \frac{1}{4A}\left(1 + \sqrt{1 + \frac{4A}{3}}\right) \tag{3}$$

$$\beta = \frac{\overline{x}}{\alpha} \tag{4}$$

for n observations.

The resulting parameters are then used to find the cumulative probability of an observed precipitation event for the given month or any other time scale.

$$G(x) = \frac{1}{\Gamma\left(\widehat{\alpha}\right)} \int_0^{x} t^{\alpha-1}\, e^{-1} dt \tag{5}$$

Computation of SPI in August and September on a time scale of 3 months (SPI 3 Aug and SPI 3 Sep) over 35 years 1983–2017) for the interpretation of drought characteristic. Flow chart for the estimation of SPI 3 (Aug and Sep) is shown in Fig. 2.

Drought characteristics (incidence, duration and intensity) for five blocks of Nuapada District was estimated. Drought event begins, when the drought value reached to −1.0 or less and ends once the drought value return back to zero or positive value [6]. Drought incidences were computed by following the procedure mentioned above. Once the drought incidences were identified, minimum and maximum duration of drought were extracted from that result. Drought intensity was computed by dividing the cumulative value of droughts with number of months experiencing drought. Studies made by previous researchers infer that, estimation of agricultural drought threshold limit can be done as per the calculation of SPI over 2 to 3 months time scale [1, 13, 25, 26]. Therefore, threshold limit of precipitation for the agricultural drought was computed for SPI 3 Aug and SPI 3 Sep.

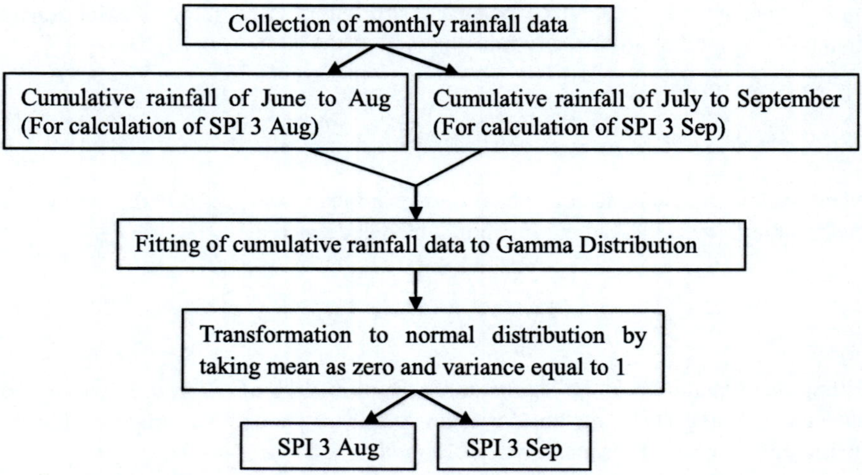

Fig. 2 Flow chart for the calculation of SPI 3 Aug and SPI 3 Sep

3 Results and Discussions

Average annual precipitation of 1155, 1292, 1364 (maximum), 1023 (minimum) and 1208 mm and monsoon average rainfall of 986, 1083, 1206, 895 and 1073 mm are observed in Bodan, Khariar, Komna, Nuapada and Sinapali blocks respectively. Amongst all blocks, the highest monsoon and annual precipitation occur in Sinapali block.

3.1 Drought Classification as per SPI 3 Aug and Sep for Five Blocks of Nuapada District

According to the classification suggested by McKee et al. [6], drought is classified into various categories (moderate (−1.00 to −1.49, severe (−1.50 to −1.99) and extreme ($\leq$ −2.00)) as per the calculated SPI 3 Aug and Sep value. Similar classification was adopted in case of five blocks of Nuapada District, and the result is presented in Table 1. Maximum (6) number of total (mod+sev+ext) droughts are experienced in Bodan and Sinapali blocks whereas minimum (4) total drought events are experienced in Khariar and Komna as per the calculated SPI 3 Aug value. Similarly, as per estimated value of SPI 3 Sep, both Bodan and Sinapali blocks experienced maximum number of droughts, and Komna block experienced lowest number of droughts.

Bodan and Khariar blocks detected with no extreme events of drought as per SPI 3 Sep, whereas these blocks experienced 1 and 2 no of extreme droughts respectively based on SPI 3 Sep. This may be due to the reduction in rainfall in June, which is

Table 1 Drought events in five blocks of Nuapada District as per SPI

	Drought (SPI 3 Aug)				Drought (SPI 3 Sep)			
Block	Moderate	Severe	Extreme	Total	Moderate	Severe	Extreme	Total
Bodan	5	0	1	6	3	3	0	6
Khariar	2	0	2	4	3	2	0	5
Komna	3	0	1	4	1	1	2	4
Nuapada	2	2	1	5	2	2	1	5
Sinapali	3	2	1	6	3	2	1	6

Table 2 Drought incidence, intensity, minimum and maximum drought duration of five blocks of Nuapada District as per SPI 3

Block	Drought Incidence	Min. Duration	Max. Duration	Drought Intensity
Bodan	18	1	13 (Sep 2002–Sep 2003)	−1.00
Khariar	19	1	18 (Oct 1992–Mar 1994)	−0.95
Komna	19	1	23 (Oct 2001–Aug 2003)	−1.04
Nuapada	19	1	21 (Nov 2001–July 2003)	−1.10
Sinapali	16	1	17 (Oct 2001–Feb 2003)	−0.92

considered in SPI 3 Aug but not in case of SPI 3 Sep. Due to the variation of rainfall spatially as well as temporally across five blocks of Nuapada District, different blocks experienced different categories of drought. Therefore, the management of water resource should be treated differently for each block.

3.2 *Drought Characterization (Maximum and Minimum Duration, Drought Intensity and Incidence) of Different Blocks of Nuapada District*

Information on the maximum and minimum duration of drought, intensity and incidence of drought extracted from SPI 3 and is presented in Table 2. Maximum drought duration of 13, 18, 23, 21 and 17 months observed for Bodan, Khariar, Komna, Nuapada and Sinapali blocks respectively during 1983–2017, whereas minimum drought duration is 1 month for all the blocks. Nuapada block detected with highest drought intensity of −1.1 over 21 months (Table 2). The highest number of drought incidences (i.e. 19) is detected in case of Khariar, Komna and Nuapada blocks whereas Sinapali block is identified with minimum number of drought incidences (i.e. 16).

3.3 Computation of Agricultural Drought Threshold Limit

Rainfall threshold limits for SPI 3 Aug as well as SPI 3 Sep were estimated for the assessment of agricultural drought in this study. First step in the calculation of the rainfall threshold limits was to compute the SPI 3 Aug and SPI 3 value. After that, linear equation (y = 267.7x + 779.3) was found out by plotting the SPI 3 Aug value on x-axis and total rainfall of June–August on y-axis in case of Bodan Block (Fig. 3). Computation of the agricultural drought threshold limit was found out by taking the x value as −1.0 in the equation developed earlier for Bodan block. Similarly, the threshold value of agricultural drought computed through the development of the linear equation (y = 250.9x + 803.7) between SPI 3 Sep (x) and total precipitation from July–September (Fig. 4) by following the same procedure as mentioned above.

Two agricultural drought threshold limits calculated as per SPI 3 Aug and Sept for five blocks of Nuapada District and presented in Table 3. The agricultural threshold limit is found to be 511.6, 595.5, 587.4, 518.0 and 461.9 mm for SPI 3 Aug, and 552.8, 572.5, 559.3, 516.4 and 464.6 mm in case of SPI 3 Sep for Bodan, Khariar, Komna, Nuapada and Sinapali blocks respectively. If the total rainfall over the months Jun to Aug and Jul to Sep occurred below the threshold limit in a particular year, that year can be declared as a drought year.

Rainfall threshold limit for the agricultural drought was different for different blocks because of spatial and temporal variation of rainfall over the years. Therefore, different strategy for water resources management should be adopted in the five blocks of Nuapada District. Agricultural drought took place on 6, 3, 1, 5 and 6 occasions based on the cumulative rainfall of June till August, whereas for 5, 3, 4, 7 and 6 occasions based on cumulative rainfall of July till September in Bodan, Khariar, Komna, Nuapada and Sinapali blocks respectively. Therefore, it can be

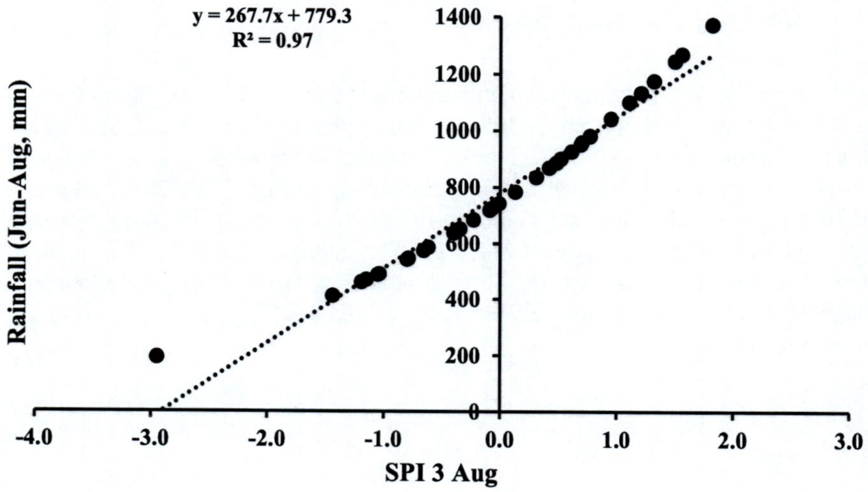

Fig. 3 Relationship between SPI 3 Aug value and total rainfall (Jun–Aug, mm) of Bodan Block

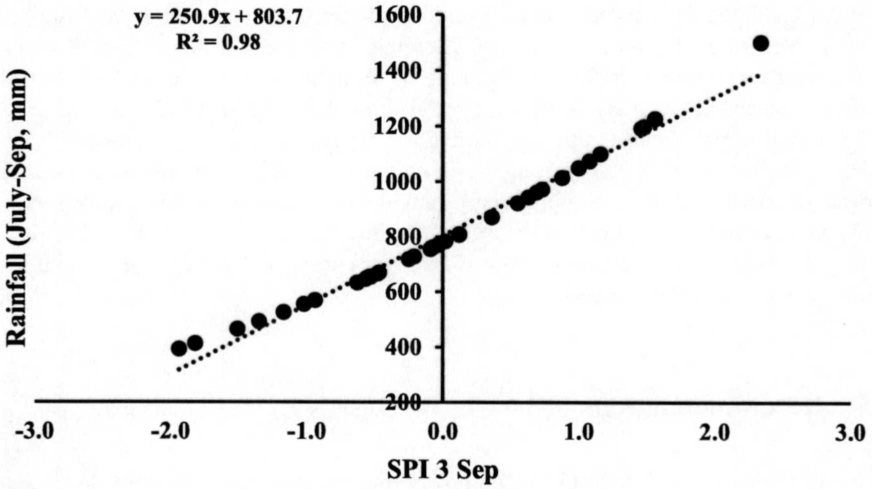

Fig. 4 Relationship between SPI 3 Sep value and total rainfall (Jul–Sep, mm) of Bodan Block

Table 3 Rainfall Threshold limit in five blocks of Nuapada District as per SPI 3 Aug and SPI 3 Sep

	SPI 3 Aug			SPI 3 Sep		
Block	Equation	R^2	Threshold value (mm)	Equation	R^2	Threshold value (mm)
Bodan	y = 267.7x + 779.3	0.97	511.6	y = 250.9x + 803.7	0.98	552.8
Khariar	y = 267.5x + 863.0	0.92	595.5	y = 269.1x + 841.6	0.93	572.5
Komna	y = 298.4x + 885.8	0.98	587.4	y = 304.8x + 864.1	0.97	559.3
Nuapada	y = 209.2x + 727.2	0.98	518.0	y = 218.7x + 734.3	0.97	516.4
Sinapali	y = 408.4x + 870.3	0.97	461.9	y = 382.6x + 847.2	0.97	464.6

concluded that the total rainfall during July to September months reduced further in comparison to total rainfall during June to August.

4 Conclusions

Thirty-five years of monthly precipitation data were utilized during monsoon months for the computation of SPI in Nuapada District's five blocks. Estimated SPI value were used for the identification of drought characteristic to suggest the strategy for water resources management. Maximum total (mod+sev+ext) drought events occurred in Bodan and Sinapali blocks where as Komna block experienced lowest cases of total drought events as per SPI 3 Aug and Sep value. Therefore, the blocks experiencing frequent droughts like Bodan and Sinapali blocks, runoff water should be harvested through water harvesting structures for utilizing the limited water resources. Drought resistant varieties of crop should also be grown in the drought areas. Maximum drought duration of 23 months observed in the case of Komna block. Different threshold limits of agricultural drought value were identified among the five blocks of Nuapada District as per the computed value of SPI 3 Aug and Sep. Bodan block experienced minimum agricultural threshold limit value in case of both SPI 3 Aug and SPI 3 Sep. Existing cropping pattern should be modified in case of drought-prone areas by growing the crops having low water requirement. Categories of drought severity was not uniform across the different blocks of Nuapada District. Therefore, the management of water resources for coping with the drought should be different for different blocks.

5 Recommendations

Drought study will be helpful for the policy makers in making effective decisions regarding the management of drought. Interpretation of the drought characterization through the drought index, SPI will be helpful for the decision makers as well as stakeholders on the effective management of limited water resources particularly in drought-prone areas experiencing frequent droughts. Major stake holder i.e. farmers should be sensitized by the Government officials for maximizing the production of crops with limited availability of water resources in the drought prone areas. Agricultural officials should give training to the farmers on water saving techniques and also on growing of the drought resistant crop varieties and low water requirement crops in the field instead of growing the same crop over the years. Therefore, water harvesting structures should be constructed to harvest the runoff water during monsoon season, which can be utilized latter for growing the crops. Irrigation projects may be taken up in the drought prone areas by the Government. Validation of drought severity in the drought years should be verified through the computation of NDVI (Normalized difference vegetation index) value from the analysis of remote sensing satellite images captured in the same drought year.

References

1. Mishra AK, Singh VP (2010) A review of drought concepts. J Hydrol 391:202–216
2. Jain VK, Pandey RP, Jain MK, Byun HR (2015) Comparison of drought indices for appraisal of drought characteristics in the Ken River Basin. Weather Clim Extremes 8:1–11
3. Dogan S, Berktay A, Singh VP (2012) Comparison of multi-monthly rainfall-based drought severity indices, with application to semi-arid Konya closed basin, Turkey. J Hydrol 470–471: 255–268
4. Byun HR, Wilhite DA (1999) Objective quantification of drought severity and duration. J Clim 12:2747–2756
5. Gibbs WJ (1987) A drought watch system. World Climate Programme WMO/TD No 193 wcp-134:1–22

6. McKee TB, Doesken NJ, Kleist J (1993) The relationship of drought frequency and duration to time scales. In: Proceedings of the 8th conference on applied climatology. AMS, Boston, MA, pp 179–184
7. Meyer SJ, Hubbard KG, Wilhite DA (1993) A crop-specific drought index for corn: I. Model development and validation. Agron J 85:388–395
8. Palmer WC (1965) Meteorological drought. Research paper no. 45, US Department of Commerce Weather Bureau, Washington, DC
9. Cheval S (2015) The standardized precipitation index – an overview. Romanian J Meteorol 12(1–2):17–64
10. Dalezios NR, Loukas A, Vasiliades L, Liakopoulos E (2009) Severity-duration-frequency analysis of droughts and wet periods in Greece. Hydrol Sci J 45(5):751–769
11. Wagan B, Zhang Z, Baopeing F, Wagan H, Si H, Ahmad I, Kabo-Bah AT (2015) Using the SPI to interpret spatial and temporal conditions of drought in China. Outlook Agr 44(3):235–241
12. Guenang GM, Komkoua MAJ, Pokam MW, Tanessong RS, Tchakoutlo SA, Vondou A, Tamoffo AT, Djiotang L, Yepdo Z, Mkankam KF (2019) Sensitivity of SPI to distribution functions and correlation between its values at different time scales in Central Africa. Earth Syst Environ 3:203–214
13. Guenang GM, Kamga FM (2014) Computation of the standardized precipitation index (SPI) and its use to assess drought occurrences in Cameroon over recent decades. J Appl Meteorol Climatol 53:2310–2324
14. Rahmat SN, Jayasuriya N, Bhuiyan M (2015) Development of drought severity-duration-frequency curves in Victoria, Australia. Aust J Water Resour 19(1):31–42
15. Juliani BHT, Okawa CMP (2017) Application of standardized precipitation index for meteorological drought analysis of the semi-arid climate influence in Minas Gerais. Brazil Hydrol 4(2):26. https://doi.org/10.3390/hydrology4020026
16. Osuch M, Romanowicz RJ, Lawrence D, Wong WK (2016) Trends in projections of standardized precipitation indices in a future climate in Poland. Hydrol Earth Syst Sci 20:1947–1969
17. Hyun Lee S, Hwan Yoo S, Jin-Yong C, Bae S (2017) Assessment of the impact of climate change on drought characteristics in the Hwanghae plain, North Korea using time series SPI and SPEI: 1981-2100. Water 9:579
18. Libanda B, Mie Z, Ngonga C (2019) Spatial and temporal patterns of drought in Zambia. J Arid Land 11:180–191
19. Mishra SS, Nagarajan R (2011) Spatio-temporal drought assessment in Tel river basin using standardized precipitation index (SPI) and GIS. Geomat Nat Hazards Risk 2(1):79–93
20. Ghosh KG (2018) Geo-statistical assessment of the intensity, duration, frequency and trend of drought over gangetic West Bengal, Eastern India, topics in hydrometerology, Theodore V Hromadka II and Prasada Rao, IntechOpen, https://doi.org/10.5772/intechopen.80037
21. Swain S, Nandi S, Patel P (2017) Application of SPI, EDI and PNPI using MSWEP precipitation data over Marathwada, India. In: 2017 IEEE International geoscience and remote sensing symposium (IGARSS), 5505–5507, IEEE
22. Sudarsan Rao A, Padhi J, Das B (2018) Assessment of drought in Balangir District of Odisha, India using drought indices. In: Singh V, Yadav S, Yadava R (eds) Climate change impacts. Water science and technology library, vol 82. Springer, Singapore
23. Edwards DC, McKee TB (1997) Characteristics of 20th century drought in the United States at multiple timescales. Colorado State University: Fort Collins, Climatology Report No. 97–2
24. Thom HCS (1958) A note on gamma distribution. Mon Weather Rev 86:117–122
25. Janapriya S, Bosu SS, Kannan B, Kokilavani K (2016) Spatial and temporal analysis of drought in Manjalar sub-basin of Vaigai in Tamil Nadu using standardized precipitation index(SPI). J Appl Nat Sci 8(2):609–615
26. Lloyd-Hughes B, Saunders MA (2002) A drought climatology for Europe. Int J Climatol 22: 1571–1592

Catchment Scale Forest-Water Interfaces for Pollution Management

Murari Lal Gaur

Abstract Forests offer an imperative role for supplying clean water for multiple utilities, and also protecting soils from vast erosion. Much of the biosphere's freshwater is delivered through forested catchments. Forests also safeguard numerous dams and even groundwater reserves by tackling siltation & contamination from various pollutants/pollution. Water and soils remain two essential drivers of the health & growth of forests. With the swelling demand for agrarian and metropolitan land (due to explosive populace & more fancy lifestyles); bulk of forests are placed in tremendous stress (further worsened by climate change). At catchment scale; the land & water policies, are frequently prejudiced to a large extent by countless apparent geo-hydological possessions like rainfall-runoff processes, soil erosion control and sediment-reduction benefits. It is truer in tropical situations like India, where region-specific understanding of such hydrological processes & functioning become indispensable for truer argument on forest-water connections at catchment scale. This chapter summarizes precise consensus on key facets of forest-water associations; by addressing relevant water quantity, quality & pollution issues therein. It accommodates broader facets like, 'forest water edges' and associated 'water quality/pollution features'. Trepidations or information gaps towards hydrological impacts of forest management along with emerging futuristic R&D issues are suitably assimilated. It also includes focused highlights towards influences of forests & their management on different streamflow factors, soil loss, stream sedimentation, water quality, landslides, and water use of diverse vegetation. The chapter is engrossed in practically vital concerns like; forest functions, forest-water interactions, relevant hydrologic determinants and their influences on water quality/quantity, and other climate change-based implications. An updated forest-water interface for India is offered along with integrated & inclusive catchment management strategies, which could be of imperative utility for relevant researchers, watershed managers, academicians/scholars, field functionaries, forest-water policy planners, and society as well.

M. L. Gaur (✉)
Department of Agril. Engg, BACA, Anand Agricultural University, Anand, Gujarat, India
e-mail: mlgaur@aau.in

S. Yadav et al. (eds.), *Water Quality, Assessment and Management in India*, Earth and Environmental Sciences Library, https://doi.org/10.1007/978-3-030-95687-5_5

Keywords Hydrology · Forest · Water quality · Forest water relations · Management · Climate

1 Introduction

Forests are essential constituents of the hydrologic water cycle; as they massively regulate channel-flow, upkeep recharging of underground water, and bestow to raincloud generation/rains via better regulated evapotranspiration. They always remain effective natural purifiers, by way of refining water and thus tumbling soil loss and sedimentation (land & waterbodies both). The foremost share (>75%) of the biosphere's reachable freshwater often arises from forested catchments; and more than 50% of the Earth's population is every time remain dependent on such zones to meet due water demands (domestic, agrarian, industrial, and ecological). Three vital sectors (Forest, Water, Energy) always convey a realistic basis on broader policy planning (global to local) for prevailing burning issues like integrated water management, water scarcity mitigation & climate transformation issues. It continually necessitates passable deliberations towards proper water interfaces in each and every forest catchment, where rain is used or reused by forest vegetation; followed by its multiple transmissions all across earth surfaces or profiles. Rising fluxes of soil moisture, unstable organic amalgams, and microbes from plant facades generate rain prompts; while the forest-driven wind force demeanour distinctive moisture toward mainland. Water fluidities chill the heat and yield clouds that rebound balancing emission from surface of earth. Likewise, the pair of 'fog' and 'cloud' intrusion by trees lures added moisture out of the air. Overall it is enhanced by further hydrologic routes like 'soil infiltration' and 'groundwater recharge'; which indeed are truly eased by trees/forests. All these changes in forest-based hydrological processes naturally diffuse water; thereby curbing water-based natural calamities (floods/droughts). This metaphysical arrangement is well portrayed by [1]. From a quantitative perspective, it is well-established fact that the forests currently cover only about 33% of Earth's surfaces [2]. In recent tiles; the tree cover deterioration (pants & coverage) is found very high. Only during 2000 to 2012; the urban growth, agricultural land adaptations, forest logging & fires resulted in the loss of almost 1.5–1.7 million km^2 of tree cover, being around 3.2% of worldwide forest cover [3]. Such deforestation and anthropogenic land-use amendments have domineering insinuations towards climate, ecosystems, water, and thus the sustainability of livelihoods and the survival of species raises long-term concerns. The UN based inferences revels that nearly 1900 million persons still live in water scarce zones. If predominant propensities continue, this number will certainly rise to about 3000 million by 2050, with up to 5700 million people existing in areas suffering water scarceness at least a single one month per year [4].

1.1 Background

Owing to the multifaceted roles of forests, the offered ecosystem services from them (mainly water-related) are often misinterpreted, underrated, and thus remains unnoticed. Forest-water interactions unswervingly pay towards clean water & sanitation and supporting/maintaining life on land. Restoring degraded land and upholding forests to achieve a fair level of hydrologic regulations towards streamflow as well as groundwater recharge; is always expected to be greatly facilitated via restoring degraded lands and upholding native forests; which in turn reduce the time required to collect water. Making forest management decisions to plunge water footprints, continually remains prominent to judge region-specific water-related ecosystem services in a better way. Forests have a very close liaison to our water resources for three explicit purposes namely; (i) regulating supply of good-quality fresh water, (ii) protection against natural perils like floods/soil erosion, and (iii) combatting desertification. A key contest is to exploit this wide range of multi sectoral forest benefits without any impairment to water resources and ecosystem function. To speak this challenge, there is an exigent need for a better understanding of the interface between forests/trees and water. Assertions that jungles offer water or else they decrease it; are not all the time candid. The actual affiliations among forest and water frequently persist on abundant relevant factors categorically; the scale (spatial & temporal), species, slope, soil, climate, forest management practices, and other alternative explicit set of conditions. Fundamentally, any forest uses water to upswing, and consequently, fast rising varieties/classes will always consume water in more quantity and also more swiftly [5]. Forest trees or likewise vegetation release ample water into the atmosphere through evapotranspiration, which frequently returns on earth as local rains [1]. Under the above sets of physical progressions, managing forests may, therefore, partake varied impacts (negative or positive) on quantity & quality of water parcels, species, vegetative distributions, and tree densities.

1.2 Forests & Water

Forests have been perceived on Globe for about 350 million years or so which touched a peak somewhere 270 to 220 million years; nearing Carboniferous period [6]. If we consider the extent of 'forests', obviously it covers about 26% of the biosphere's land surface and symbolizes a divergent biotic community, being the most illustrious vegetation community to offer plentiful resources and ecological purposes that far surpass those of added flora covers. Upholding vigorous forests continually helps for improving water and ecological pre-eminence. Better intermingling across water & soil in multiple ways; provides canopy facades that trap rains in a better way; which ultimately regulates, (i) the fractions of rainwater that ultimately reached at forest floor via through fall and (ii) the dragged water from the soil for transpiration needs.

1.2.1 Key Spectrums

Being the most imperative resources on earth; the water & forests have greater impacts on each other. They together deliver multiple deliverables like food, energy, habitat, functions (biological, chemical, physical, and socioeconomic), essential services to lives, and the optimal atmosphere [7]. For their growth/development/ survival; not only the humans but the majority of the flora & fauna too remains utterly dependent on rainwater. Water entirely sustains human societies, as environments deprived of water always create uncomfortable living conditions (hot, dry, and incongruous). Providentially, the Globe is sanctified through ample extent of water; about two third Earth's surface is concealed by water, and a stratum of water vapour (nearing 90 km thick). Though water on Earth is so plenteous in quantity and so enormous in transfer, it still imposes worries in various areas. At least 80 countries (comprises about 40% of the world's population); had experienced sporadic inadequacies of water averting them from growing adequate food to sustenance their people [8]. Water shortage not only creates glitches/embarrassments in humdrum life but also disturbs lifestyles and ethnic growth. Water-pollution complications have made the regions with such water shortages even worse. At this sensitive juncture of the problem, it is only forests that momentously support the array of complex flora & fauna by establishing an idiosyncratic microclimate to suit enhanced land uses. Presences of good forest cover always mends the optimum levels of occurrence, distribution, and transmission of water in its positive side; offering multiple direct & indirect benefits. In the majority of forest types, the vegetative physiognomies (size, canopy compactness, litter flooring, root arrangements etc.) continually persist ominously taller, bigger, denser, and deeper as compared to any other vegetation types. Such leading features make forests capable not only to deliver & conserve several natural resources but also to accomplish a variety of other favourable functions like; control of water & wind erosion, protection of headwater & reservoir, improving watershed/riparian zone, sand-dune & stream-bank stabilization, landslide & avalanche deterrence, protection of wildlife habitats & gene groups, extenuation of flood damage & wind speed, and acting as an overall big sink for atmospheric carbon dioxide [9].

1.2.2 Issues & Perspectives

Under population explosion scenarios, water and forests have arisen as the two greatest key issues of the twenty-first century, where they are deliberated not only as an indispensable entity by cultures & industries but also as the vital factor to sustain an adaptable environment. Therefore, learning of the interface between these 2 possessions (forest & hydrology) befits it as an imperative scientific field; offering basic acquaintance/fundamentals for managing rainwater and forested watersheds as well. Forests typically raise and progress well in zones having annual rains of 500 mm or higher, and they remain partly suitable for certain agrarian activities too. Worldwide about 30% of the land is covered with forests, yet this forest land spawns nearing 60% of total runoff. Undoubtedly, this is the biggest reason why utmost fraction of drinking water deliveries instigates from forested catchments. Forest destroying activities (any kind & any extent) severely influence, advantageous development/utilization of forested areas and also the forest-based water supplies. These activities may be in the shape of destroyed forest canopies, disturbed forest floors, improper rainwater spreading/movements, floods, stream degradations, deteriorated/depleted riparian zones, and inadequate land and water development actions. Such rescinding actions badly disturb water magnitude via transpiration, canopy capture losses, soil infiltration, soil water holding capability, and overland flow rapidity. It altogether results in increased soil erosion & nutrient losses, raising multiple issues about quantitative & qualitative perspectives of water, whose fundamental knowledge; adequate studies, ample information, and right management still to be attained.

1.3 Water Resources Glitches

The existence of inadequacy as well as complex uncertainties about water resources does not leftovers as a big concern only from a quantitative standpoint, but it has equal worries from the viewpoints of water quality and water distributions (in time and space). Certain catchments (or their parts) often have too much excess water (floods); while others too diminutive (droughts). Accordingly, the rainwater might not supervene at the correct time and the exact place. Also the water could not be adequately fresh enough (water pollution) for use in drinking/other human-uses across varied space-time grid in a given region. Additionally, the convolution of water usage in our contemporary civilization and the need for water for pecuniary growth make the right to detain rainwater or its apportionment an imperious issue (water rights). Not a single province on the globe is invincible on such hitches that indeed shoot from the asymmetrical dispersal of water, water agility, the discrepancy in supply & demand, and stable expansions towards prevailing populace, economic progress, poor environmental controlling, and lack of concern by the public. Besides huge gaps in between demands & supplies of water across various regions/localities; it is water quantity that ultimately decides the fate of drought or flood conditions. Forests play a decisive role in both scenarios. During drought conditions, they have sound influences on controlling procedural elements like artificial rains, translocations of stored runoff or even rains, reverting salinization, water recuperation, rainwater harvesting, household water conservation, evaporation reduction, and irrigation efficiency. Similarly, during floods forests suitably regulates essentials of floods like their forms, causes, types, nature, damages, inundation, control measures, and flood plain zoning.

2 Forest Functions at Catchment Scale

A forest happens to be a greatly prearranged natural system in which foremost features remain the plants (trees in specific), crafting a varied canopy cover to play its own specified appeal. It all together engenders several key functions (via forests, forest soils, forest water, or their interfaces) that contribute sizably towards food security and also the due healthy environment. Such forest-based functions might be capriciously clustered in 3 broad classes, (i) protective purpose, posing steadying influence for natural atmosphere (precipitation, rainwater exchange, airflow, hotness, universal/micro level climate, slowing down soil erosion), (ii) productive purpose, offering raw harvests/things (fruits, wood, herbs, mushrooms, etc.), and (iii) social purpose, for promising setting & ecological surroundings backing healthiness & restitution of society and thus augmenting better livelihoods & markets. Here in this chapter, we will remain confined mainly towards water-related functions, by encircling a few hydrological and environmental clasps. The primary elements of various popular narratives on hydrological or environmental functions may comprise many generalized influences on water quality and quantity [10, 11]. Accordingly, there remains a sizeable divide amongst scientific accord and the prevalent description of forest–water connections. The process of assembling, dispensation, and construing such evidence is often referred to as a "catchment valuation study", encompassing plentiful points of few characterized activities like,

- Magnitude, configuration, scheduling, water level, and assertion of stream inflows
- Qualitative features including physical, chemical, and biological characteristics of water
- Appearances and condition of the in-stream and riparian environment
- Physiognomies, state, and dispersal of the aquatic biota.

2.1 Hydrologic Functions & Impacts

On the earth, the most imperative ecosystems that deliver numerous hydrological services; undoubtedly remain the natural forests. These services include vital aspects like water production & protection, with which one can ensure to have qualitative uninterrupted water harvest as well as other protective services for land & soils. The quality & magnitude of all such forest-based roles are predisposed by their prevailing ecosystem structure and compositions; offering integral utility towards the local hydrological cycle. Trepidations on hydrological effects of forest handlings is an century old art & science which alike prevails current day even. Forest-based vegetation (trees, plants) consume rain-water via 2 major progressions, (i) transpiration i.e. captivating water up from soil via roots and then vanishing it from pores in leaves; and (ii) rain capture by surfaces of leaves/branches/trunks even during active rains followed by its straight evaporation. Overall, it results into a larger hydrologic possessions on countless stream flow factors (whole water yield, small flows, inundation flows), topsoil erosion, watercourse sediment deposits, water-quality, land-slippage, and improper water use by dissimilar vegetation types/species. Although there exist a firm body of methodical evidence for understanding/inferring the associations amid forests & water; still there occurs analogous as well as profoundly ingrained “popular narratives” that frequently run against the consent opinions of forest hydrologists [12].

2.1.1 Salient Hydrologic Progressions in Forest Watersheds

Under prevailing climate change scenarios, region-specific hydrological processes remain utmost complex & uncertain; and thus unavoidably necessities site-specific uses of proficient information on predominant settings in respects to climate, geology, soils, biology, pastures, livestock, human, and relevant interfaces across them under real field situations. Investigators [13] revealed prevalent folklores having lots of indecisions on hydrological working of forests. It comprises interrogations like, (i) Do forests upsurge precipitation (contrariwise, eliminated forests declines precipitation)?, (ii) Weather forest escalates runoff (contrariwise, forest amputation slashes runoff magnitude)?, (iii) Weather forests diminish inundations (in opposition, forest exclusion proliferate flood), and (iv) Do stream base-flows continually becomes amplified because of forests (on the contrary, confiscation of forests drops such base-flows)?, (v) Weather the forest cover continuously controls the stem-flows to decline extraordinary flows and upturn flows of little magnitudes (on the other hand, forest removal promotes lesser magnitudes of well-regulated watercourse flows)?, (vi) Weather forests always diminish soil erosion (equally, elimination of forests rises soil loss)?, and (vii) Weather forests continually avert/alleviate land-slides (in opposition, removal of forests rises land-slides).

2.1.2 Salient R&D Outcomes

Furthermost forest hydrology-based investigations (until the 1970s) were conceded on forests located in humid/temperate areas, and thus giving a diluent understanding of intricate hydrological processes that remains pertinent in forest catchments. A thought-provoking discussion on key hydrological impacts owing to forests [14], well reflected many regional studies on forest-water relations. Afterward, many researchers [15, 16] have adopted the concept of paired watersheds, where afterward a prolonged calibration; one watershed of the adopted pair is considered as ‘control’, while any imposed treatment (forest reaping, complete or partial clearing) is smeared to other watershed. End deliverables (hydrological results) were then categorically measured and compared upon. Observations on the above-quoted control catchment were often maintained over several years, to assess their relative hydrological performance (rainfall-runoff relationships in particular). Much of such original ground-breaking efforts were reviewed in a classic paper [17] and afterward by

[18] to assess changes in water yield from such experimental catchments with varied alterations to vegetation cover. Further, a considerable magnitude of forest hydrology research was reported at small plots/micro-catchments scales [14], which remained quite useful in gaining further insights into forest-based hydrological processes. Despite all such efforts, there remained enormous limitations to effectively extrapolate such findings at the small watershed scale. Extrapolation of foremost hydrologic progressions, such as rain-based soil infiltration, overland flows, transportation of deposited soil mass, and sub-surface moisture movements along forest hillside catchments, was found infamously problematic, where heterogeneity of forest catchments added the difficulties [19].

Looking into many research-based findings it gets well established that numerous fragments of forest vegetation use to have sizable influences on the overall hydrological budget inside a forested catchment. It gets happen via regulation & alterations of key hydrological processes via canopies, roots, biological configurations, and other physiological attributes of forest vegetation. Such vegetation remains present in multitier configurations, moreover, 3 key tiers could be grass-like stands, shrubs, and trees. Figure 1 depicts a simpler conceptual sketch in regards to vegetation-centered impacts on local hydrological elements as well as water balance constituents. Part (a) of this figure reflects the simpler architecture of dynamic fluxes in regards to water, heat, and other energy aspects of forest catchments. Moreover, from a comprehensive impacts point of view, it is not only the vegetation or vegetative elements but in a true sense, it remains the 'composite soil-vegetation-cover complex' that produces higher effects on prominent hydrological routes. As illustrated in part (b) of the same figure, the forest land floors too have equal supremacy to control rainfall excess flows (both overland and underground). Ground canopy most often has superior impacts on effective hydraulic roughness, which in turn controls the flow velocities and thus the discharge rates towards downsides (horizontally or vertically driven water fluxes). The composite portrayal of Fig. 1, recognizes the conjunctive standing of canopy and forest land floors to control the local hydrologic cycle and its diverse components.

2.2 *Environmental Functions*

Forest based ecological functions often comprise physical process based elements like; control of water and wind erosion, protection of headwaters, reservoirs, nearby watersheds under multiple lands uses and other associated critical areas like riparian zones along streams, stabilization of sand-dune/stream-banks, landslide control, safeguard of biota/gene, the assertion of inundation damage & breeze swiftness, and effectual bowls for soil or atmospheric based carbon. Lots of conventional woodlands have succeeded in attaining one or more of above ecological gatherings, while others are well-looked-after for averting a decrease or dilapidation of the biome. From water quality viewpoints there persist wide-ranging anxieties which are eventually getting prejudiced or directed by definite sets of constituents as

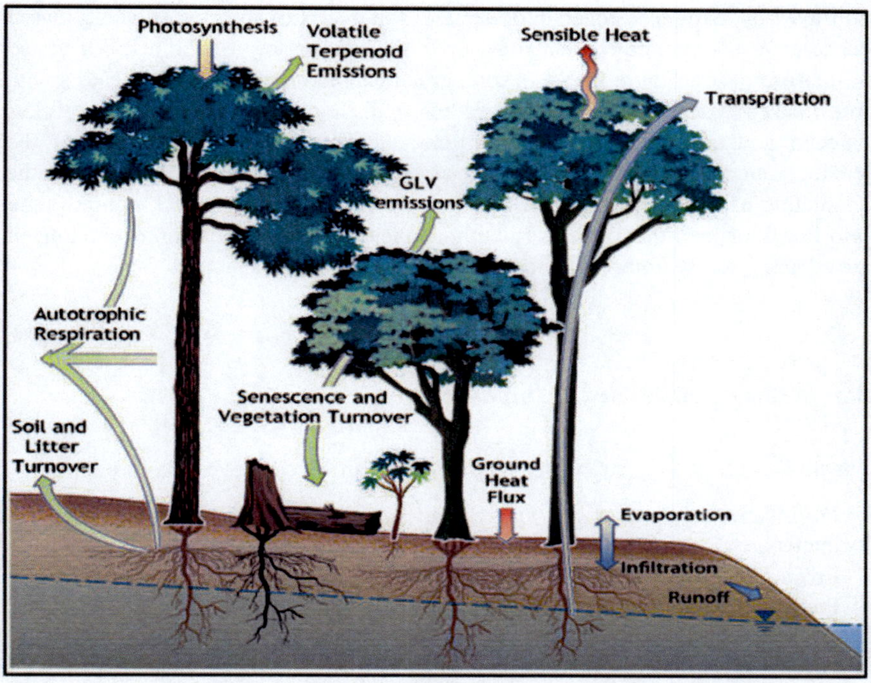

(a)

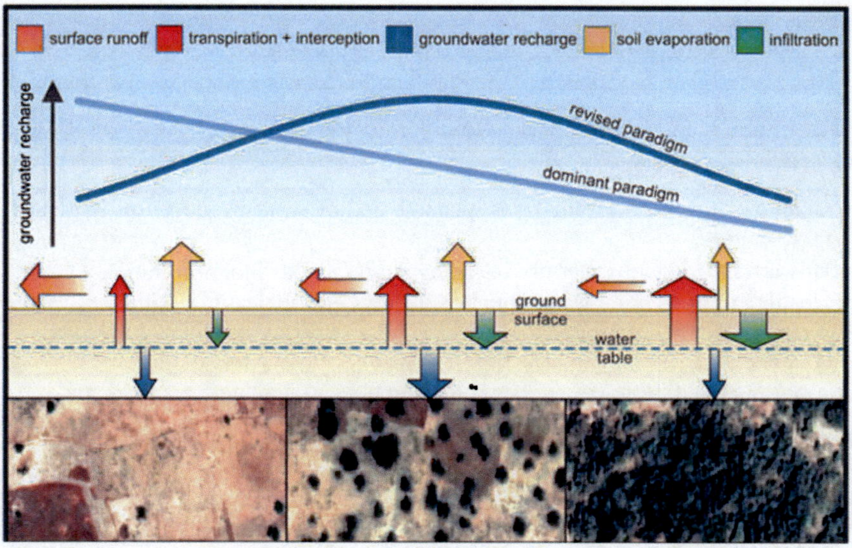

(b)

Fig. 1 Diverse effects of forest canopies on hydrologic processes [20, 21]. (**a**) Canopy based influences. (**b**) Soil-vegetation cover based influences portrayed by many researchers [20, 22, 23]. The matrix of such water quality-based concerns & answerable elements rest upon the efficacy of stakeholders for varied utilities/purposes. Figure 2, illustrates a broad panorama where key indicators with forest-based water quality are framed for 4 distinguished water use categories, namely, aquatic, recreational, aquaculture, and irrigation sectors. Some of the leading water quality concerns as well as their relevant ingredients to govern the qualitative aspects; are reviewed and put together in a categorical, concise, and prioritized manner; that remains suitably applicable for any of catchment/command area dealing forests/forest-like vegetative stands.

2.3 Other Functions & Impacts

Functionally following supplementary key roles are too achieved via forests,

- Fortification of water assets via vegetation, rough bark, and plentiful litter like matters
- Erosion control by decelerating down flow speed of runoff or wind
- Preserving land via thick system of roots
- Proposing cushioning effects to normalize bulk erosion/landslides
- Abundant impacts on native microclimate & greenhouse gas emanations
- Inclusive protection of natural habitation and a biological assortment
- Frivolous & other societal purposes in the vicinity of towns, tourism & healthiness choices
- Defensive socio-economic & ethnic scopes in the area
- Machine-driven, industrialized, market-based products for manhood & livestock.

Furthermore, if forests are not managed correctly, following could be undesirable impacts,

- Contrary influences from trees planted nearby water's edge or non-native monocultures
- Unwarranted higher water use releasing substantial evapotranspiration
- Hostile impacts on water quality (acids eutrophication, siltation, native swamping)
- Aggressive organic influences (injured spawning areas, block gills),
- Additional undesired effects (gloomy drinking water quality, killer conifers)

Reliant upon the level of forest managing, the water-based ecological influences might be either optimistic or even some time annoying too. Rewards may comprise the following,

- Flood controls/managing
- Dispersal/extenuation of contamination & pollutants
- Vindicating downstream inundating
- Declines in nutrient & insecticide harm into water

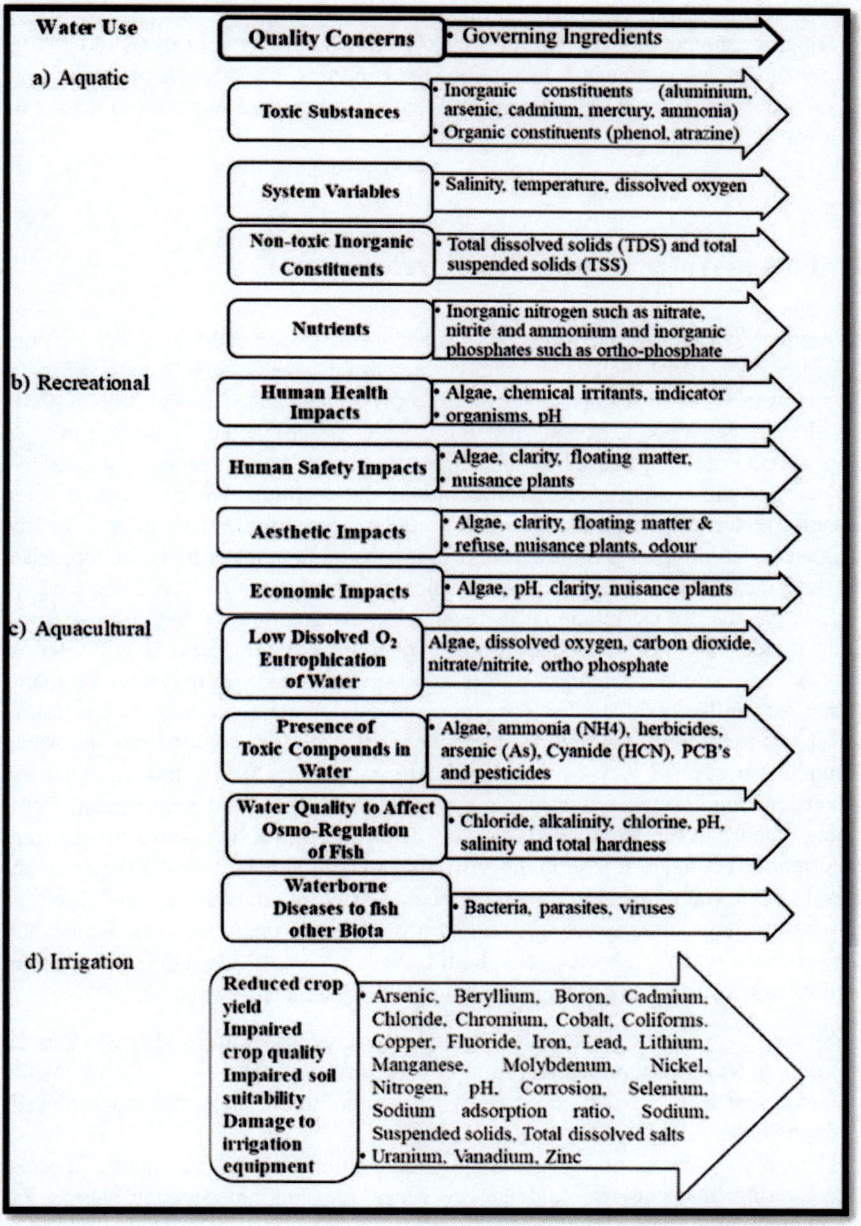

Fig. 2 Leading environmental concerns of water quality across different uses [22, 24]

- Soil defenses from consistent turbulences
- Dipping dangers of sediment transfer to water bodies (static or dynamic)
- Enhancements in fitness & habitations for hominids, livestock, aquatic life
- Environmental paybacks along with frivolous gains and other socio-economic rewards

3 Forest-Water Relations & Interactions

As evident by existing scientific consensus, the knowledge of forest-water relationships has been established as an important scientific aspect. At the catchment scale, the real water uses of forests versus other forms of vegetation; always remain pretty unalike. In general, the forests use more water as compared to littler forms of vegetation. Foremost causes to this regards happens to be deeper root systems, greater leaf area indices, and higher rainfall interception; which eventually led toward greater evapotranspiration. Due to taller flora, forests have greater surface roughness, tempting larger turbulence, leading to equally greater transpiration. Also, the tree harvesting frequently upshots increased water yield. It is interesting to realize that the utmost relational increase arises in low flow periods (dry season). Such increased water yield gets dropped over time if the forest is permitted to re-grow. At micro-scale; integrated forest management practices may have a remarkable effect in this regard. Moreover, on smaller catchments (<10 km^2), harvesting forest trees recurrently surges the peak flow rates during small to medium-sized precipitation storms. Key causes remain the rainfall features (amount, intensity, antecedent conditions), geomorphology and vegetation inside the catchment. Dominate influence of forests on all kind of low flows to promote base flows are recurrent occurrence. Yet again it is thought-provoking to comprehend that the minor catchments with trivial rains often have a restricted ability for adjusting channel flows, as compare to large catchments, big rain storms, or better managed flora. Forests are always found equally advantageous, both for water quantity & quality, which might be further boosted by espousing location-specific managerial ways like,

- Sieving & scrubbing water as leaves/root systems may trap or alter detrimental toxin, averting contaminations from inward water
- Governing silt deposits by barring water pollution, habitats, reservoir silt depositions.
- Defending habitations by shielding propagation grounds for aquatic species, delivering nutrients & coldness to water for sinking chemical needs for aquaculture
- Swelling flora density, to destroy the energy of falling raindrops for preventing splash erosion & high speed movements of overland flows.
- Escalating rain by boosted evaporated water vapours & prolonged raincloud covers.
- Efficaciously gripping rainwater for averting erosion and inundating.

3.1 Fundamental Roles of Water

Rainwater has greatest significance to life and the contiguous atmosphere, via its numerous vital functions (biological, chemical, physical, socioeconomic). Biological perspectives, articulates that water remains not only a vital necessity of life, but it happens to be a wider habitat of life in terms of wetlands, estuaries, ponds & lakes, streams & rivers, oceans; and even sometimes even as a means of therapy for illness (hydrotherapy). From chemical viewpoints, it is a universal solvent of substance at one side, while a versatile medium during multiple chemical reactions on the other side. Similarly, from physical standpoints, water always serves as a broker of climate, as a proxy of obliteration, a latent shape of energy, a methodical customary for possessions, and sometimes as a medium of transportation too. Socio-economically, it includes vital efficacies in terms of; a key source of comfort, a stimulus of creativity, a perilous issue of world peace & regional solidity/clashes, robust medium for agrarian & industrial production (hydroponics, fish/aquatic culture, industrial operations).

3.1.1 Properties, Dynamics & Hydrologic Framework of Water

Being the most vigorous & omnipresent constituent on Earth, water is often designated as a wonder of Mother Nature, the echo of knowledge, and the lifeline of development. Numerous water possessions are used as mentions and principles on sectoral possessions (physical, hydraulic, chemical, and biological) of other substances [23]. Physical properties encompass its states (3 no's), latent heat, saturated vapor pressure, vapor dispersal, heat capability, thermal conductivity. Similarly, the hydraulic properties of water may embrace its density, pressure, buoyancy, surface tension, capillary rise, viscosity, Reynolds number, shear stress, stream power, and many other entities. From chemical standpoints, the water properties like its molecules, formation, and chemical reactions remain the prime ones; while biological properties encompass temperature, dissolved oxygen, pH, conductivity, sediment in water.

The true dynamics of water is regulated by the hydrologic cycle, which includes distinguished physical processes (interception, depression storage, infiltration, percolation, evaporation, transpiration, overland flows, channel flows, subsurface runoff, etc.), to governs the real budgets of water & energy at catchment scales. Forests play a dominant role to govern all these features in a variety of ways. Historically, on a global scale, the quantity of existing total water was supposed to be around $1.384 \times 109\ km^3$, which may be simply reflected as if Earth's surface is virtually filled with water to a depth of about 2.7 km [25]. About 97.4% of this was well-thought-out to be saline water which mostly exists in oceans, leaving merely less than 3% as freshwater out of which more than 77% is locked in icecaps/glaciers. As an incessant process with no beginning or end, the hydrologic cycle plays a critical role to regulate such inclusive dynamics of water (Fig. 1a). From a water balancing point of view, rainfall, runoff, and evapotranspiration happen to be the key players across the whole domain of the hydrologic cycle (Fig. 1b). Any deforestation activity frequently extinguishes these important hydrological key players; which in turn imposes diverse risks or costs on people.

3.2 Forest Hydrology

Forest hydrology educates water resource planners and managers regarding the prevailing distribution, storage, movement, and quality of water and associated hydrological processes in forest-dominated ecosystems (Fig. 3). Such knowledge is always regarded as the foundation of modern integrated watershed management conceptions. It may greatly facilitate the planning and execution of various developmental activities by adhering to truer paths and storages of water bodies in any forest area. A correct scientific understanding in this regard is compulsory for every well-versed difference of opinion on forest-water interactions. From the time rain touches the soil's surface, and till it returns back to the atmosphere via transpiration and evaporation; it often gets prejudiced to some degree by human actions. In the majority of forest-dominated watersheds, the active components like evaporation, transpiration, moisture movements, rainfall excess flows, saturation excess flows, partial area saturation-based interflows, soil infiltration in multiple ways, and groundwater recharges; have their own specified and pre-designated sequence of happening. Figure 4 is synthesized by visualizing & accommodating various kinds of activity charts; that is most often remains present during rains falling on forest catchments. Part (a) of this conceptual figure throws some light on sequenced hydrologic processes, while part (b) illustrates a simplistic connectedness among few prime elements of the water cycle. The physics of handling rainfall via vegetative stands of forests used to be a very complex, uncertain, and less understood phenomenon. Part (c) of the above figure portrayed some of the leading influences of canopy & forest land floors during different active phases of various hydrologic functions. Dynamics, as well as routes of water (on the ground, below ground & above ground), are very much visible in this figure. All these functions & processes are of prime importance, whenever any effort is made to understand forest-water interactions and an effectual water productive forest management.

The mechanisms of the water cycle remain most vulnerable to alteration by land-use configurations; which greatly regulates the use of water by vegetation (evapotranspiration) or even the differential flow paths for water moving over & inside the soil surface. Certain diverse but well-established principles [20] by prevailing hydrological reactions are itemized below; which eventually produce region or location specific changes in forests.

- Fractional or entire exclusion of canopy declines intercepted rain and rises net magnitudes of rainwater that arrives at land surface.
- Incomplete or pervasive elimination of canopy decreases transpiration.

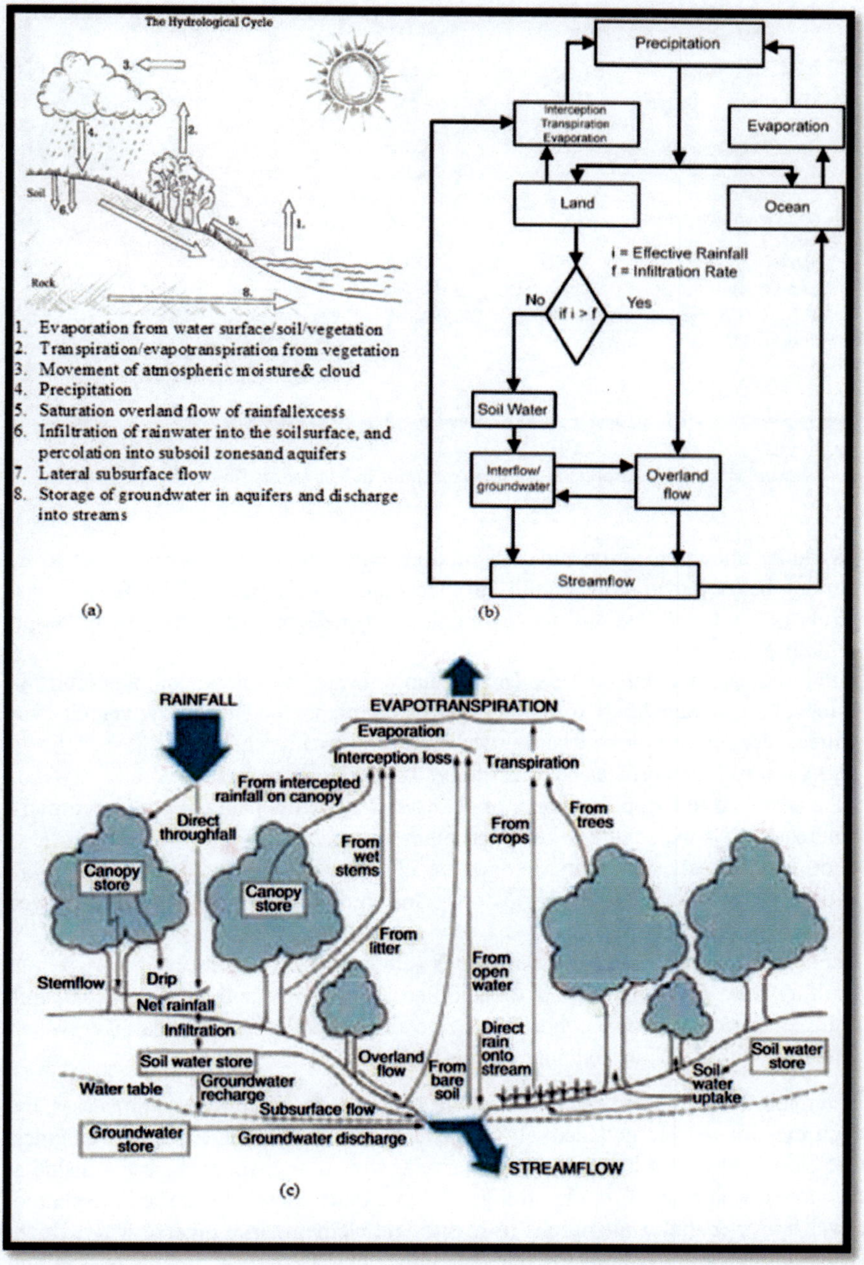

Fig. 3 Conceptual connectedness of hydrologic processes & elements in a forested catchment [26, 27]

Class	Area Km²	% of TGA
• Very Dense Forests	98158	2.99
• Moderately Dense Forests	308318	9.38
• Open Forests	301797	9.18
• Total Forest Cover (*It includes 4921 Km² under Mangrove Cover*)	708273	21.54
• Scrub	45979	1.40
• Non-Forests	2533217	77.06
• Total Geopgraphic Area (TGA)	3287469	100.00

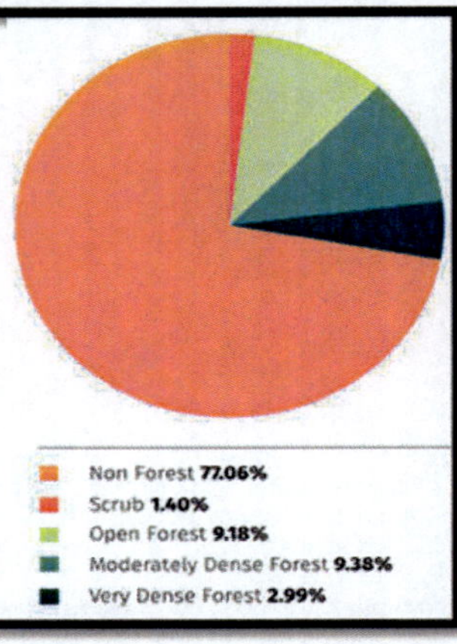

Fig. 4 Recent scenario in respects to the areal extent of Indian forests [28]

- Reduced rain interception and transpiration could offer amplified wetness of soils, giving better moisture availability to plants, and groundwater yields [26]
- Enlarged soil wetness and root strength deteriorations might lessen land slope stability
- Impervious woodland surfaces (roads) and reformed hill slope contours (cut/fill-slopes) are always likely to modify water flow paths, increased flow velocity via direct fast delivery of overland flows to nearby streams/channels.
- Impervious land surfaces or forest floors surely increase soil erosion.
- Transformed hill slope contours or changed water flow paths along forest roads increase mass worsening of land and other resources.
- Forest chemicals (fertilizers, herbicides, insecticides, fire retardants) may unfavorably disturb aquatic ecologies once smeared straight into water bodies/wet soils.
- Enduring uses of chemical compounds (atmospheric deposition of nitrogen & sulfur) may certainly be led to acidified forest soils and thus depleted soil nutrients; which produce hostile effects on prevailing forest health & water quality, with multiple toxicities on aquatic creatures.

In the majority of forested catchments there remain numerous physical processes, by which the water cycle gets actively influenced both in time and space. It includes basic causes or actors like fire, forest harvest, silviculture, roads, or trails inside a given forest watershed. From hydrological standpoints, these adverse activities often deliver major & minor alterations in hydrologic elements in a diverse way, which invites ample understanding of relevant indicators. Degree & period of straight hydrologic effects of forests on catchment outputs encompasses various pointers as catchment's outputs like; harvested rainwater, uttermost flows, short flows, erosion, landslides, sedimentation, water temperature and water chemistry. There exist wide research gaps to realize predominant sets of forestland evolutions that characteristically and often alter the hydrology of such forested catchments. Among leading sets, the entities like forest-fire, silviculture & forest based harvest, and forest based paths & roads plays governing roles; which needs to be understood at regional scale.

3.3 *Determinants of Forest-Water Connection*

Connections across forests and water always remain the biggest research gap, where the majority of factual information remains poorly or even non-spoken facet for most of the field functionaries or even policy planners. There remain many invisible determinants of change (gravity, soil pedology, climate change) at varied scales (temporal & spatial). Some of them have extensive influence, while others have a recurrent/endless kind of influence on woodland hydrology. Past studies [1, 29] have very well established the facts that, during tree evapotranspiration most of the precipitation that they receive, are very small overtures on their leaf surface. As a regulator of water use in forests, tree's leaf area can serve as a substitution for evaluating forest water productivities/yields [30]. Therefore, leaf area invariably plays a vital role towards quality of forest water owing to causes like soil erosion and stream turbidity. Also, the forest canopy captivates abundant rain energy during falling raindrops. Owing to this the net erosive power of the drops is amply reduced. It could be considered a valid positive erosive scope of the rainwater owing to forests. Canopies of forest tees further protect quality of water by dipping temperature of streams and upholding greater dissolved oxygen during hot months. Other factors (past land utilization, land-slope, parent material of soils etc.) also regulates quality of streams or water therein. Owing to above reasons, the leaf area index (LAI) is mostly considered as a solitary single determinant of changes whatever occurs in relation to forest water interactions. Nowadays under an era of smart satellite data and equally smart sensors and sciences (IT/ICT), the remote sensing-based normalized difference vegetation index (NDVI) and equivalent clear-cut area (ECA) percentage are evolving as a substitute indicator to enumerate & appraise forest vegetation changes from the lookout of water and hydrology. In current days these are being accepted as promising cohesive indicators that facilitates in quantifying the nature & extent of forest disturbances (spatial & temporal) by accommodating recovery aspects via vegetation & hydrological elements [31].

4 Forest Water Quality and Climate Change

Under prevailing climate change scenarios, the water quality-related paybacks from forests are continually at the forefront. Forests or forested catchments (irrespective of their level of management) are customarily considered advantageous for protecting water quality therein. It happens due to their inherent specific competencies for sizable ascendency to deal stabilization of steep slopes, landslide damage, quality drinking water supplies, and many other location-specific ecological amenities. From a water quality standpoint key affirmative forest features that effectually control an array of land and water issues. More preciously it includes entities like siltation, turbidity, steadiness of streambanks, chemicals or pesticides, streamflow, eutrophication, acids, water colour, and many dissolved elements. Perversely some few specific tree shelters arrest air contaminants, which many a time lead to accelerate polluted water parcels in different parts of catchment. Insinuations of global & local climate change with its connections like; sea-level rise, coastal imbalances, land degradations, soil erosion/landslides offer another threat to water resources inside any forested catchment.

4.1 Forests, Erosion, and Water Quality

Forested catchment often experiences, curtailed soil loss and thus sediment entering into streams owing to certain foremost physics on real ground. It includes refining of soil arrangement & steadiness, amassed soil infiltration; dipping of quick runoff over land surface; and offering an active shelter to face high-speed wind. There exist vast prevalent tales on functional linking across soil & nutrient losses, forest chopping, imports & exports of pollutants ('to & from' water bodies) in many areas. A sample widespread tale is that "trees in forests reduce erosion and conversely, the removal of trees upsurges erosion of water and soil". With this firm fact, a good forest catchment always reduces hillslope erosion and thus produces high-quality water being free of sediment & other pollutants. Certainly, the prevalent riparian vegetation (complex edifice of grasses, shrubs, trees) plays a leading role to cope with foremost water quality parameters. Numerous constructive influences of consistent strength of the roots of forest trees are recognized by investigators [24, 29] displaying high weight to forest-water relationships.

4.2 Vital Indicators of Water Quality

Articulating water quality is undeniably a comprehensive topic, moreover, by restricting it only up to catchment's runoff standpoint, we can enlist few elementary indicators (given below) to reflect the prevalent dynamics of forest water across several fragments of forested catchments [22, 26].

(a) Water Temperature: It is exaggerated by entities like stormwater runoff, air temperature, ground-water influxes, turbidity, and disclosure to sunlight.
(b) pH: Being an effective extent of a solution's acidity via an integer of hydrogen ions, it decides the apparent water quality in forests. Basically from a chemistry point of view, in any parcel of water, small numbers of water molecules (H_2O)

will break apart into hydrogen ions (H+) and hydroxide ions (OH-). Supplementary compounds entering these water parcels may often react with these ions, ensuing in a lively disproportion in the numbers of these ions. When added hydrogen ions react, more hydroxide ions are left in solution and the water becomes rather basic; while on the other hand, when more hydroxide ions react, added hydrogen ions are left and water happens to be frequently acidic. From an applied standpoint fresh-water aquatic creatures prefer a pH ranging from 6.5 to 8.0.

(c) Turbidity: It remains one of the strong measure for quantifying that in what way soil-particles are suspended in water, and thus affect clarity of water to illustrate the level of suspended sediment and erosion. Turbidity increases abruptly during and after rains, causing sediments to be massively carried into the watercourses. Raised turbidity often results in: temperature rise in water, lowering of dissolved oxygen, and abridged light reaching towards aquatic floras. Such higher magnitudes of turbidity decrease the ability of aquatic plants to properly photosynthesize, releasing sound harmful effects on prevailing fish gills and eggs.

(d) Conductivity: It is continually accepted as the strongest indicator to measure the ability of a solution (stream water here) by passing an electric current, for demonstrating the concentration of dissolved electrolyte ions in water. Noteworthy upsurges or reductions in such conductivity are often taken as a truer indicator for confirming that contaminating discharges have entered the stream water, resulting in visibility on ions of nitrate, phosphate, and sodium. It is measured in micromhos per centimeter (μmhos/cm) or micro siemens per centimeter (μS/cm), whose safer value (150 to 500 μS/cm) in freshwater streams is expected to care aquatic life [22, 26].

(e) Dissolved Oxygen: It is usually considered as a quantum of oxygen molecules (O_2) existing in any parcel of water. Since plants and animals cannot sprightly use the oxygen (part of water molecule H_2O); hence they persist reliant on liquefied O_2 for respiration. Oxygen repeatedly enters streams either from the surrounding air or as a creation of photosynthesis via aquatic plants. It is commonly measured in mg/L, and it reaches its peak during the day. It needs to be always more than 2 in any case, however, the best values lie in between 5–10 mg/L to support aquatic life.

(f) Nitrate: Though Nitrogen (N_2) is plentiful on earth (about 80% in air), the majority of plants cannot use it in its original form. Legumes and blue-green algae have decent ability to transfigure such N_2 gas into nitrate (${NO_3}^{-1}$), easily grasped by plants. These Nitrates some time enter channels from their natural sources like decomposing plants, animal waste, and human sources (sewage or fertilizer). From a quality point of view, their normal levels remain less than 1 mg/L, moreover, concentrations over 10 mg/L always hurt freshwater (drinking or aquatic).

(g) Phosphate: It is characterized by Phosphorus, which is continually indispensable to attain proper growth by plants, along with facilitation of key growth and metabolic reactions in animals and plants. Major sources of phosphate embrace fertilizers, animal wastes, detergents, sewage, distressed land, and salts used on snow-covered roads in winter. Bigger streams are typically considered delicate to respond with phosphate at levels approaching about 0.1 mg/L; while smaller streams may react even at the lowest levels of 0.01 mg/L or less. Values more than 0.1 mg/L surely harm any river system, flowing across forests.

4.3 *Water Quality Monitoring*

Issue of water quality includes a comprehensive area for integrated catchment management for fulfilling its intrinsic objectives. While small forest areas may be readily monitored for water quality or other stream characteristics, it is more apposite that a similar mechanism of the bigger catchment is preciously addressed equivalently. Water sampling from streams and the elucidation of chemical analyses raises some issues, which always remain highly correlated with contamination happening via non-point sources of sediments or chemicals at catchment scales. It remains the solitary reason, why the specific focus & value of water quality assessment is emerging as an inevitable area of research for catchment managers and forest policy planners. A strong need is always advocated to develop the aspects of water quality monitoring in forests at catchment scales. Long terms studies in countries other than India have sizably predisposed these approaches [6, 11, 26, 32] by adopting varied calibration phases (8–10 years) with varied post-treatment analysis over a comparable broad spectrum of time. Moreover, still, the basic question that remained unanswered is: How to address the water quality from the catchment's perspective?. In the majority of conventional approaches, the water quantity is truthfully assessed using a calibrated weir allowing sediment or nutrient loadings to be calculated. Several such systems have been further established by researchers [24, 33], which allowed better development of models for real understanding of processes involved therein.

4.3.1 Categories of Quality Data

A critical factor in the design and implementation of water quality monitoring is to collect below given categorized data by pre-determined objectives,

- Stream Loading: Quantity of material moving past the sample point per unit time (e.g. kg, or kg/ha for the catchment).
- Concentration Change: Changes in components of water quality over time after a management treatment or disturbance.
- Grab sampling: It involves point sampling on some basis for analysis, management treatment, or general comparisons for specific purposes.
- Instantaneous Spatial Assessment: It involves intensive manual sampling at a huge number of points in a short period to deliver Spatio-temporal variation in a tiny discrete time frame, where concentrations can be effortlessly transformed into relative units.

- Observational Assessment: It assesses streams for problem areas and reasons (roads, stream bank collapse) and could include surveys for bed-load movement and cross-sectional surveys.
- Specific Surveys: It indeed is an assessment of stream health in addition to water quality sampling, which may include periodic measurements of stream bank, bed-load movement, riparian zone vegetation health, invertebrate analysis, etc.

4.4 Catchment Water Quality Management

Catchment water quality managing is always considered a tedious and multifaceted task, because the water quality, water quantity, and aquatic ecosystem components inside a natural catchment; remain interdependent and critically linked via a complex set of biological, physical, and chemical interfaces. Water quality gets continually changes as and when effluent is added to the river/stream, which gets further amended as the river/stream flows downstream. There could be some basic mechanisms/elements (given below) which persist dynamic shares for any water quality controlling plan.

- Setting an applied description of essential water quality, say it's all period suitability for use as irrigation water for overall vegetation/crops without yield loss.
- Explanation of the point or area where the above relations remain functional.
- Setting an extreme permissible concentration/range of concentrations or statistical terms.
- Portrayal on catchment-based water quality status and goal for next 5 years or so.
- Pure narrative portrayal of point-sources or non-point sources needed to dispose wastes.

4.5 Inference of Climatic & Environmental Variations

Under the prevailing era of climate change, its variability along with connected wider issues like rising of sea-level affects water resources in variety of ways. Often these segments highly interact with land-use changes and thus aggravate larger influences on the quantity & quality of water. Site-specific realizations of affiliations across land resources, water, forest, pollution hazards, and other relations with climate change is perilous to craft activities that curtail damaging effects. From a wider viewpoint, any kind of change in climate or even micro-climate can easily lead to larger variability in rainfall and greater rates of sediment/soil erosion. These fluctuations can further threaten water quality by swelling storm water runoff, supplementary erosion/sedimentation, intimidating source waters, and thus growing occurrences of injurious algal blooms. Deviations in the water quality of rivers, lakes, or streams greatly regulate the water quality of available drinking source waters. Changes in ocean acidification could lead to threaten or degrade estuaries. Hence, the decisive aim shall have to be to reduce and decelerate the polluted surface runoff with key water quality indicators of forested catchments. We need to get adopted via active interfaces like,

- Predominant climate change may offer negative impacts towards water service actions, water quality, and biome security exertions
- Prevailing climate change also confound water efficacy processes via added recurring and strong drought events/storms, speedier sea-level rises, more salt-water imposition, etc.
- Frequent/extreme droughts and flooding storms, portend to sizeably affect water quality in rivers, lakes, streams; where water quality parameters need to be well-thought-out.

5 Updated Forest-Water Facade: Indian Scenario

Archeologically, at the global level, the natural plants always remained there (for more than 370 million years), out of which about 0.08 million species still exist well, inhabiting about 3.5-billion-hectare area worldwide, which include nearing 250 million ha of commercial plantations too [34]. Woodlands always offer enormous ecological, social, and pecuniary paybacks to peoples; but if they are not managed optimally they vastly interrupt the hydrologic progression in contradictory ways (favorable or vice versa). This functional mirror image of forest hydrology rests more valid for temperate and semi-temperate nations like India, where a sharp rise in grown for water demands; even when the prevailing native rainfall forms deviate massively with parallel attenuation in forest covers.

5.1 Situation of Forests in India

Taking into account the proportional acreage under forests (24.4% of its land), India ranks tenth in the world. The country accounts for only 2.4% of the world surface area, but still effectually endures the needs of around 17% humans and 18% livestock population of the world. While accomplishing this giant target, the major role is played by water that remains occurred, stored, circulated, and variably utilized across various tiers of forest-land interfaces. Some other conservative estimates reveal the total forest cover of India somewhere around 708,273 km^2 (21.54% geographical area of the country) [35]. This comprises a range of fractions/types of forests as exemplified in Fig. 4, being self-understandable to portray that the extent of dense forests barely 3% of total geographic extent. Amongst them, some of the definite forests are having massive high values towards conservation aspects of water and land. One such specimen is bamboo-based forests/plantations. The country happens to witness richest bamboo resources in the World, second only to China; witnessing nearing 15.69 million hectares bamboo plantations and coarsely projected 2868 million bamboo culms offering an equivalent weight of about 17.412 million tonnes [35]. However, such bamboo grown forest zones remain extremely dispersed through various states in India, with uppermost acreage in north-eastern regions. Bamboo forests have established their supremacy; as they always recognized as a persistent, multipurpose plantation resource, that decidedly manages (quantitative as well as qualitative) the runoff water generated from such forest lands. Bamboo based interventions are found highly effective to deal land degradation, erosion, sediment movements, stream bank equilibria, and many other water and land preservation aspects (point, plot and catchment scales) [36, 37].

5.1.1 Suitability of Bamboo to Regulate Hydrological Processes

The classic green leaves, dense canopy, and numeral culms are some of the peculiar characteristics of bamboo; that positively regulates and facilitates good crown interception during all kinds of rains. Sinking drops of rain on bamboo trees effortlessly amend their route with reduced velocity, and thus declines harmful effect of raindrops towards direct soil erosion. It gets happens because of numerous rain-interceptions by multiple shoot layers and the higher number of bamboo-culms. As vast magnitudes of leaves naturally dropped on the ground surface below trees, thus creating underground porous net structure; which ultimately improves permeability, water holding capacity, and many other soils fixing capacity [37] have well revealed, that bamboo forest soil shows stronger anti-scours and anti-erosion, establishing the prominence of bamboo-forests in terms of an effectual vegetative measure for conserving rain and upholding detached soil particles in variety of watersheds [38] have described hydrological physiognomies of bamboo forest in South China, displaying that annual runoff ratio in their research watershed) with bamboo, the plantation was about 54.8% of rainfall, but in contrast the proportion for rapid runoff (direct runoff + surface runoff), remained to the tune of 0.8% only. Even for reclamation of problem soils (gullies, ravines, wastelands) there exist a plethora of land and water conservation technologies with options of various permutations and combinations of mechanical & vegetative measures; comprising bamboo and other forest-based interventions at the catchment level [36].

5.2 State of R&D on Forest Water

The last 150 years could be well termed as an effectual plantation era, where the necessity for woods or forest-based raw ingredients steered to the creation of forest planting; even on grasslands and degraded regular jungles too. It overall transformed or at least predisposed hydrological belongings in forest areas or forest-dominated natural catchments. The most frequent adverse effects encompass qualitative as well as quantitative deteriorations in water parcels or whole runoff water volumes that are generated in various streams or at outlets of catchments/micro-catchments. Such sound effects have established the fact that the net magnitude of above-quoted water-based deteriorations, remained highly dependent upon the integrated management and the stages of forest growth and falling rains as well.

5.2.1 India's Forest Hydrology Experimentations

For Indian or alike situations, there subsist massive works on hydrological understanding of forests via historical R&D efforts; initiating from 'forest-hydrological experiments' to further key hydrologic facilities, experimentation on paired-catchments, and research based information from eco-hydrological observations on diverse catchments under forests; located in different parts of India. A brief-expression on foremost efforts are as follows,

(a) *Paired Catchment Studies*: The prominence in early periods was fundamentally kept on the role of afforestation in dwindling peak runoff rates from catchments. One major exemption was a peculiar study on soil moisture at different depths under Acacia, Eucalyptus, grassland, and fallow land (up to 90 cm) in a semi-arid forest area [16], which exposed that transformation of grasslands into blue gum forest plantations reduced annual water yield by 16% during first rotation and 25.4% during second rotation. The first-generation coppiced blue gum plantations [32] emphasized the command of sub-surface water pathways and their implications on low flows in streams. Undoubtedly these were the first methodical studies in India to exhibit the possessions of changeable transpiration demands of trees as a function of size, class, and age of trees; that too at catchment scale. The study concluded that Eucalyptus was drawing moisture from much deeper layers in the soil, but its inferences were not unified into the design of further catchment-based studies. [39] shared valuable outcomes on determinations towards hydrological appraisal of a pair of small forested catchments (Sal forest clear-felled in 1955 and allowed to regenerate and described as "brush") near Dehradun after steering 8 years long calibration (1961–1968). One forested watershed was then clear-cut and re-afforested with Eucalyptus trees. The flow-recording gauge did not function in the first year after cutting but, subsequently, they assessed that reforestation resulted in a 28% reduction of annual streamflow and a 73% reduction in peak flow rates. Scientific reviews from few hydrological responses such as soil and water loss through runoff, interception, infiltration [40, 41], and soil moisture under different forest influences in experimental basins of India; recognized that the overall moisture regime of a forested catchment can be corrected with reforestation and controlled grazing; to improve land and water productivity and also mitigating effects of droughts/floods. Here in all such limited hydrological studies, the noticeable highlighting remained on peak-flows reductions and rarely was there any weight on low flows or dry-season flows. Furthermore, the size of the catchments studied was often too small concerning the prevailing climate to sustain perennial flows.

(b) *The Himalayan Experience*: In India Himalayas remained the region with dominant forest land uses. One of the pioneer study group of ecosystem ecologists was located in the Himalayas ("Kumaon school") which indeed was a historical R&D platform in India. It undertook, a comprehensive functional & applied view of hydrologic processes, sediment production, and nutrient cycling, all the complex vegetation and litter characteristics in Himalayan area [42]. This group of experts was the first to report that rainfall represents an important source of nutrient cations to ecosystems. The average overland flow was frequently found below 1% of the total incident precipitation, indicating that these forest catchments behaved more or less like "subsurface flow systems", ascribing sediments to large landslide zones, rather than overland flow zone. [15] studied some facets of rainfall apportioning into an interception, stem-flow, and throughfall across prevailing species in the central Himalayas; concluding that rain-interception from conifers was as high as 30%, compared to 20% or lower in broad-leaved species. Such interception values were established to of significant extent. Overland flows were always detected lower for all forests; making forest hydrology subjugated by subsurface water flow systems. Moreover, in other words, after deforestation, the pathways could easily shift more towards overland flow. Field-based soil infiltration rates as reported in these Himalayan studies were generally high, and thus occurrences of overland flow were low and infrequent. Further linkages in between forest, hydrology, and ecology were too established by such limited R&D efforts, enlightening an upsurge in overland flow while reductions in base flow. In other studies, in the upland hills of northeast India, it is reported that afforestation increased base flow by 20–40% over pre-afforestation levels. Such studies remain extremely limited for regions other than hilly or mountainous forest trees [40].

(c) *Eco-Hydrology Based Lessons*: There had been vast exertions towards ecological understanding & its perfections via most recent interventions like 'Green India Mission' and many others; pushing superior prominence on 'forest-water interface' from qualitative and pollution points of view. A sequence of studies using arduous measurements & modeling in the last 2 decades (dry & humid Western Ghats forests); have added plentiful information & knowledge on region-specific hydrologic processes and linked hydrology from many evergreens as well as deciduous forests. These include the first observations on the significance of large macro-pores or soil pipes [43], being a discrete feature of humid forests, in determining runoff responses (subsurface quick). Results provided a substantial misperception in regards to overland flow during superficial scrutiny of runoff hydrographs. Such process-based conclusions fetched into question on the relevance of pretentious "average" infiltration characteristics of any land cover. Contrarily, [8] suggested that afforestation activities could sizably harm soil pipes, resulting in extensive overland flows, an energetic mechanism for engendering surface runoff. Hydrologic measurements in the dry forests east of the Western Ghats were introduced (~2003) by two projects; (i) Indo-UK and Indo-French combined projects steered by the National Institute of Hydrology-Belgaum center and (ii) Indian Institute of Science Bangalore project for accommodating even the role of protected streams with check dams, to create water-holes supposedly to trap sediment [44]. Hydrologic responses/services and ground-water recharge-based research efforts under 3 ecosystems were too undertook and scrutinized in general, by using instrumented catchments (7 to 23 ha), revealing higher frequency and longer duration of low-flows under natural forests, as compare to disturbed land covers.

5.2.2 Ecologic Effects of Forest Conversions

Transfiguring forests into agrarian land use (crops) greatly influence topsoil, lessen evapo-transpiration and soil-infiltration, and upsurge over-land flows. Irrespective of post reaping treatment, typical hydrologic deviations owing to forest removal; often led to bigger stream flows with equally higher peak rates of flows. Furthermore, there always remain other massive consequences of such forest conversions (like sediment deliveries, channel congestions, floods). From a real physics point of view, forests continually alleviate soils; otherwise, top-soil is readily battered on subsequent creeping vegetation removal on land surface. Transportation or dislocation of such eroded soil mass; could happen either on flood plains (i.e. overland planes beside stream banks) or even inside streams and their adjacent riparian landmasses. The ultimate possessions of historic land use transformations to agrarian land use (row crops), loss of soil, and succeeding shifting or accumulation of detached sediments were continually considered of deep importance by previous investigators. Under prevailing climate change era, the net special effects of such conversions on water quality or water chemistry; are of great impact. For majority of cases, the uninterrupted forest-watersheds were usually related with minute concentrations of most ions inside the stream-water. The net spread of macro-nutrients or nutrients in hefty quantities (N, P, K) from incessant forest catchments; is often considered negative, viewing whole of forest biomass. Table 1 delivers a few plausible influences of forests in regards to the ecological status of forested catchments.

5.3 *Efforts & Scope Toward R&D Essentials*

Discrete beliefs across 'scientific consensus v/s popular narrative' of forest–water interactions; offers a surfeit of R&D spheres; where due efforts are still to be commenced to make sure the harnessing of fabulous unexploited potential of forests & water in conjunction. Some of the scanty efforts in this direction are enumerated below to provide food for thought for bridging prevailing research gaps and facilitating decision-makers for sprouting effective forest-water implantation plans.

Table 1 Offerings from forests to uphold water-based ecological needs

Water-based environmental prerequisites	Plausible contributions of forests
1. Fine oxygenated water having no impurities	✓ Properly planned and succeeded woodlands may act as a trap or bowl for pollutants and also soil defending ✓ Vegetative cushion in riparian zones (along streams) have a vital role in seizing sediments, nutrients and insecticides
2. Factual light grasp in water bodies to care aquatic life	✓ A flexible compactness of forest cover is always a crucial factor offering correct steadiness of sunlit and shadow
3. Sort of regular features/habitats (puddles, flicks, blocks, swamps, ponds, back-water streams, dry river terraces, river linked plains, banks)	✓ The binding act of tree roots aids towards firming and steadying stream banks, dipping soil loss, collapsing of stream banks, land mass slides/failures
4. Regionally apt specific vegetation (type/extent)	✓ Innate riparian suggests an epitome cover for caring river morphology
5. Typical range of acidity & alkalinity	✓ Forest canopies, facilitates arrest of pollutants (acids or alike entities) in atmosphere, thus a drop in pH of stream water
6. Pertinent contributions of organic matter/nutrients	✓ Variety & season wise variability in mass of leaf-litter, root based bacterial progressions, maintains the optimal energy & nutrient flows for active biological working of aquatic biomes. ✓ Branches, foliage, earthly invertebrates that drops into water; from canopies of trees, sizeably assists as diet for water organisms
7. Natural range in volumes, depths and velocities of runoff water	✓ Abridged water movements may obstruct fish entrance declining existing habitation aimed at fresh-water lifecycle ✓ Woodlands may lessen water currents, nonetheless such result could be enhanced by upright forest plan & managing.

5.3.1 Indian Missions

There remained massive kinds of technology missions, that are recently launched by the Government of India in the sectors of the forest, agriculture, water, energy, and rural development. In all such missions; forests are given a justifiable home treating them as a sensitive link for integrated development & management of land and water resources together with agrarian & non-agrarian productivities. Description of all such missions (on watersheds, irrigation water, forest conservation, bamboo, land & water management, rain-fed areas, floods & drought-prone areas, special erosion-prone areas, streams corridors, riparian & rangeland improvements, etc.) is out of scope here; however, out of 8 most apposite missions, one namely 'National Green India Mission (GIM) rests much affiliated with subject matters of this chapter. It used to be a prime element under National Action Plan on Climate Change (NAPCC), where country is scheduling to afforest and reinstate nearing ten million hectares of forests (prevailing "degraded" forest and un afforested areas) to impound carbon and normalize atmospheric CO2 concentrations. Even right now, we distinguish very petite about the interactions and also the trade-offs between carbon & water cycles. An important issue that remains is the location-specific answering of the question "How it is prejudiced by phenological reactions that are being driven by changes in climate (historical or futuristic)?". As per recent estimates [35], the entire carbon stock is estimated to be 7083 million tonnes, reflecting an upsurge of 38 million tons of carbon stock as compared to the 2015 assessment. India's Intended Nationally Determined Contribution (INDC), demonstrates the concept of added carbon-sink (CO_2 equivalent) of nearing 2.5 to 3 billion tons via supplementary woodland/tree cover by 2030. Considering the sensitivity on this vital issue, the need of the hour is to deliver more emphasis on appropriate R&D efforts and experimentations effort to facilitate policy and management portfolios of forest carbon & forest water sectors; altogether in conjunction at the micro-scale of time and space [45].

5.3.2 The Way Forward

India and its scientific establishments are missing unresolved openings to convert into global leaders in the area of investigational hydrology as well as eco-hydrology, particularly considering vast extensive past of services from Indian forest officials & premier institutes engaged therein. All such expert organizations have a multidisciplinary mode of scientific working; which could be made sizably streamlined to care forest-water interactions and their region-specific management plans. Even when some sites were well set up by enthusiastic investigators in past, they were hardly upheld beyond 5–10 years. Hence, in the current era, we need to raise an ethos of partnership and joint functional experimental sites to reinstate forest hydrologic research; having regional as well intercontinental worth. The anti-climax to preserve longstanding observing spots in divergent ecological/climatic areas could be a solitary biggest intentions on nation's feebleness on prognosis imminent hydrology and water cycle based answers towards land-use-cover change effects, microclimate alteration effects, and many other inclusive features/special effects on natural resources. The whole linkage of rainfall-runoff measuring stations upheld by leading agencies like 'Central Water Commission: CWC', R&D agencies at central and state levels, academic institutions, and even NGOs; needs to be exposed & shared with other advanced academic and R&D establishments. These organizations needs to get connected and thus harness their best potential to craft big data and thus big reliable information on forest-water interactions at macro & micro scales.

6 Catchment Managing Approaches

Integrate management of water resources (quantity & quality) acts as the nucleus of success sphere of any 'ideal catchment supervisory plan'; where quality & quantity of water always remains interlinked/interdependent to govern both the land phases (soil-vegetation cover complexes & aquatic ecosystem). Noticeable pointers like turbidity/siltation, stream-bank steadiness, insecticides, acids, coloured water, disbanded-oxygen, biological carbon; demonstrates pivotal starring role for determining sensitivity level of specific region with varied level of water-based or forest-based sectors. It altogether makes the evaluating/checking/gauging/handling of catchment scale water quality; an exceedingly tiresome task. Below given decision-making objectives remains highly valid for accomplishing region-specific scheduling & implementation of ground actions on relevant conservation aspects,

(a) Plummeting over-land flows via better canopy-based interception or transpiration
(b) Developing porous soils throughout the organic limit and tree roots
(c) Decelerating velocities of flows overland via better litter coverage
(d) Decreasing terminal speed of rain-drops via canopy-interception
(e) Augmenting topsoil masses for soil binding by reinforcement via roots.

From intentional supervisory deliberations, one needs to always correctly recognize and apprehend many governing contrivances within watershed; that truly direct outlines the qualitative standpoints on forest-water. It includes numerous knobs like rain-interceptions (canopy & litter), rainwater via though-fall & stem-flows, evaporations from exterior parts of trees, evapotranspiration, hotness from tree-canopy & roots, soil-infiltration & other deeper water travels inside soils, dynamics of overland planes & streams, etc. If one looks into an array of straightforward practices that often remain the causes for leading contaminations, utmost ones may be termed as (i) cuttings of trees, (ii) roads/tracks inside forests, and (iii) land-use amendments & fire incidences inside forests. It altogether advocates that, any management strategies of forests or forested catchments all the time need to be aligned in a way to ensure ample scope for conserving/developing land and water resources therein. Certainly it is going to offer better and judicious planning towards in-situ rain-water conservation/harvesting along with effectual recycling of harvested rainwater across diverse parts of forested catchments. Such options may comprise, swelling prospects for infiltration, stretching period of overflow attentions, lessening speeds of flowing water, generating a superior & larger quantities of water storing features, and decreasing vaporization in water bodies (surface or underground). A widespread band of few possible possessions is echoed in Table 2; offering projected end effects by leading hydrologic processes and related quantifiable changes in real ground situations.

Table 2 Explicit possessions of discrete hydrological progressions in forest watershed

Hydrologic progressions	Form of variations	Explicit hydrological paraphernalia
1. Tree based rain-interception	• If lessened	• Smaller soil moisture level • Larger runoff in minor storms • Amplified runoff water harvests
2. Leaf litter storing of rains	• If lowered litter • If not changed • If amplified litter	• A smaller amount of water stores • No alteration • Water storing upsurges
3. Vegetative transpiration	• If provisionally eliminated	• Base-flow gets increases • Soil moisture gets increases
4. Soil-infiltration	• If reduced • If increased	• Overland flow & stream flow surges • Base-flows upsurges
5. Streamflow	• If changed in variety of ways	• Escalation in majority of eco-systems • Reduction in snowfall • Diminution in mist drip structures
6. Runoff base flow	• If altered in variety of ways	• Gets reduced with high infiltration • Gets upsurge with lesser soil-infiltration • Uncertain summertime flows (+ve/−ve)
7. Runoff storm flow	• If amplified	• Greater volumes • Larger peak flow rates • Shorter time to peak flows

6.1 Surface Water Acidification or Sea Salts

Forest managing practices; routinely shock superficial water acidification on forest land floors in numeral ways. Among key reasons, one of the well-established schools of thought remains the intrinsic greater capability of forest-tree canopies for seizing more Sulphur/Nitrogen pollutants from the atmosphere as compared to any other vegetation types. Hence, the forest streams persist more impacted in this regard; needing appropriate regulatory actions. In the majority of situations, the new tree planting always stimulus acidification by base cation uptake; rummaging of acid-deposits, sifting & absorption of sea based salts, dehydrating of soil masses, and sometimes the creation of acidic-litter deposit on land surface. There may be other distinctive effects via crop husbandry, land-drainage, paths, manure usage, tree-chopping or reaping. Another mode in which forests might aggravate acid depositions remain via acceptance of base-cations like calcium, magnesium, sodium, and potassium; all from layers of soil. Forest vegetation frequently necessitate few base-cations to facilitate their growing. If removal of such substances surpasses the rate of their replacement from varied source points (soil-mineral enduring, leaf-litter, dead-wood, atmospheric deposition); it certainly reduces overall cushioning capability of land surface. Here forest canopies might be considered efficacious for augmenting sea-salt, aerosols from the sky, or whatever leftovers utmost from coastal storms.

Soil dehydrating & litter-horizon plays perilous roles at this juncture; because parching of organic-soils may lead to amplified oxidation for soil organic matter, dispensing organic acids or stored sulphate. It altogether lessens the pH of surface drainage water-parcels, nevertheless, it seldom clues on any adversative possessions on their reception into water channels. Tree type & species, cultivation, drainage & roads, and intensive forest felling too contribute here.

6.2 *Detecting Sensitive Areas & Framing Suitable Work Plans*

Each surface water body (lake, stream, river) has an exclusive catchment zone. A few of them are more susceptible than others to make changes in water quality. It happens due to prevent the diverse state of soil-vegetation-cover complexes, underlying geology, land use, or other functional causes. Such vulnerable zones or sub-catchments might be well sensed by beholding below given thinkable fundamentals,

- Areas are sensitive to acidification for which criterion could be aquatic zone, geology, pH readings, water hardness, water alkalinity, etc.
- Areas are sensitive to erosion based upon norms like soil types, steepness of slopes, erosion-prone subsoils, geology, etc.
- Sensitive buffer zones based upon the prevailing state of stream banks & vegetation conditions, leaf litter, steepness of stream banks, etc.
- Sensitive pockets from drainage, soil-vegetation-cover complex, sedimentation, etc.
- Pockets having specific pollutant deliveries/loads, extensive compactness (roads), etc.

Cautious preparation and supervision on the above areas will surely offer extenuation against potential negative effects while maximizing the positive aspects (aquatic biodiversity enhancement, creation of appropriate riparian ecosystems, etc). Some useful key principles could be as follows,

- Avoiding ground preparation within the buffer zone
- Caring out ground preparation only after ensuring little risk related to heavy rainfall.
- Avoiding disturbing existing drains/channels.
- Creating suitable drains and sediment traps along with collector drains at an acute angle to the contour with convenient gradients (0.3–3%) to minimize flow velocities
- Ensuring tapering-out of lively drainage channels afore their entering on buffer zone

6.2.1 Influences of Forest Roads on Quality of Forest Water

Uncapped path-ways/roads in forest areas have long been recognized as a momentous source of water contamination as well as soil erosion. Contributions of such exposed points/locations/areas pose a great challenge for catchment managers/functionaries; as it has a very high impact on catchment in-stream exports of sediment/nutrients. This particular aspect has been very uninterestingly taken care of by water-based researchers at catchment scales. [46] publicized that entire sediment & nutrients load delivered by any forest dominated watershed having natural configurations of paths, roads and other approach ways; is always expected to dispense broad pair of vigorous causes; either the 'roads-related sources' or the 'non-roads related source's. Furthermore, the comparative impact of forest pathways on such exports largely rest on contextual loads from other alike source-points too. Sediment yields from above described definite paths/roads is well connected to prevailing stream network configurations (via overland flow paths). Astonishingly, even in the case where such catchment scale impacts are minimal, pollutant impacts may still be substantially higher locally in distinct stream reaches. Now, it becomes paramount vital to accommodate all such related issues, which have either direct or indirect connectivity towards a giant matrix of functional indicators to reflect qualitative & quantitative characterization of forest water. These indicators are well enumerated in preceding segments of this chapter.

6.3 Eutrophication

Forests, in their good management state (land, water & vegetation) are always considered beneficial for defending pollution and more specifically the quality of water existing therein. However, under current climate change conditions, sometimes even the well-managed normal jungles too could pretence impending regionalized intimidations via connected interfaces across elements namely the forest canopy, water, and atmosphere. Such evidence well exists in scientific literature, experimentally establishing the fact that forests might help or in some cases threaten qualitative ingredients of forest water [47] via plentiful altercation of atmosphere based ammonia under prevailing flora facades. Here the process of eutrophication plays a vivacious role to regulate vibrant relations across forest trees & water.

7 Knowledge Gaps & Research Needs

Water does not come from the tap, rather it is forests that are doing a lot for manhood and the environment. Under prevailing scenarios, there is nothing better than a tree providing myriad environmental services. Accordingly, the domain of forest-water interactions remains extremely large with countless potential/opportunities for overcoming environmental threats. The focus of this chapter was to project a comprehensive portrait with significant dots, which could be considered as knowledge gaps in forest-water research. So far the majority of past studies on forest-water relations are conducted on a limited number of forested watersheds, being inadequate to address dominant issues of drastically altered land use and climate change. In this regard, there remain a plethora of factors to regulate the real water productivities of forests. Under present era of smart solutions & smarter management, below given location-specific R&D efforts on forest water hydrology; are expected to offer some of the credible nodes for bridging addressable knowledge gaps in the forest-water domain,

- Considering 'Big-data' based efforts on forest-water interferences and relevant intrusions
- Smarter/progressive models/modelling endeavours on forest catchments (pure/mixed)
- Connecting verdicts of key operational sectors like; water-supply reservoir storages, inter basin transmissions of water, land cover alterations, river flows, and conceptual trade-offs across carbon impounding and water resources sectors.
- Fetching verified R&D results on better linkages of forest water flows with its physics

Also, the forest-based policy and management frameworks need to consider the above-quoted larger-scale relationships in a comprehensive way by assimilating the water footprints of respective forests and forested catchments. There is a strong need to enable the development of resilient and robust forest-water policies with suitable management strategies that could have the capacity to learn & adapt effectual solutions in context to climate-based ever-increasing pressures on forest & water resources. In such a multifaceted situation, it remains extremely important to opt for an interdisciplinary approach (being inclusive & integrated) to ensure need-based relevant messages, actions, and policies as well. In a truer sense, while framing forest-water dialogues or working strategies, instead of mere 'forest & water', a nexus of 'forests, water, food, and energy needs to be taken in holistic manner [28].

7.1 *State of Knowledge on Forests & Water*

Environmentalists have sound awareness on jeopardies when proper land & water management on forest catchments remain deprived of a sound & region-specific scientific understanding. Ignoring even the key structures or functioning of forest water progressions; always results in natural disasters not only in forest catchments but more in downstream locations or catchments. The functional understanding of forest-water interactions and the salient hydrological processes involved; are of uttermost importance because of two big reasons. Firstly, the majority of hydrologic elements & processes remain exceedingly varied, peculiar, uncertain, and non-uniform; when they are matched with catchments having non-forested land uses. Secondly, the truer forest-water connections regulate the overall natural forest systems in a region, by governing the transfer of energy, matter, or information across vast species and their biotic exchanges with water. Even as on date, there exist large gaps in our relevant understanding of such forest-water systems and their key operational functioning like on,

- Correct information/knowledge on biotic & structural multiplicity of forests to arrest intrinsic physical/serviceable connections between water, land-cover, and biological diversity.
- Appropriate know-how about utilities, flexibility, and variation of forest ecologies, keeping in view the queries like; How weather alteration will upset forest ecologies and their structure/output and potential in regard to active carbon-sequestration, and further; How it will intermingle with biotic (insect, pathogen, pests) and abiotic (drought, floods, fires) features?
- Adequately assessing the sustainability of forest management by considering features like forest productions, natural resource management, socio-economic conditions, and comprehensive protection approaches with need-based tailor-made viable management rules & resolutions.
- Accurate judging of multi-functionality of forests and their Spatio-temporal scaling aspects on effects of landscape structure on water balance, eco-system functioning, and overall resilience.

7.1.1 Environmental Understanding

Important eco-friendly amenities delivered by the forests are utmost recognized in present era; when facets like conservation & protection of water/land resources, carbon-sequestration, soil health/quality, biodiversity, and other favourable settings for aquatic & human life; are being considered noteworthy at diverse scales. All such ecological amenities are indeed adequately inflated by numerous kinds of management plans/executions on forests, relevant R&D knowledge/feedback, and manners in which small or even degraded forests might be scientifically treated and managed at catchment scales [45]. A need is always arising to ensure improved knowledge based understanding & reckoning of critical collective effects on forestation or deforestation; keeping central focus towards region specific bio-diversity, rainfall harvesting/conservation and, proper managing of carbon, quality of water & soil, and many other conservational ecologic amenities. Here some of the additional vibrant aspects might be enlisted as, (i) getting regionally conversant with actual expected effects of whole-tree reaping v/s stem-only reaping, (ii) location-specific data/information on evapotranspiration needs of woodlands, (iii) dispersed hydrologic modelling for watersheds under forests at finer spatio-temporal scales, (iv) conclusive impacts of land-use deviations on water productivity inside given forested catchments, (v) influences of key limitations (hydrology & oxygen based) on forest growth and water quality, (vi) understanding CO_2 efflux versus water fluxes, and (vii) inclusive sustainability standpoints in general forest actions/managing to fetch enhanced water productivity.

7.2 *Knowledge Gaps and Scope Towards Futuristic R&D*

In the current era of smart tools, technologies, and modelling; an improved thoughtful understanding, reliable and big-data, comprehensive information and applied knowledge is still to be gained via the amalgamation of targeted field and relevant modelling educations, to correctly outline a few imperious operational ground issues like,

- Enumerating influence of up-land forest dominated micro-catchments on quantitative & qualitative aspects of water parcels observed therein
- Ground analysis of relevant models for auxiliary quantified impacts that floodplain of any forested catchment might offer to mitigate large flood events.
- Measuring truer possessions of beset forest plants on catchment based dispersed contamination, soil infiltration sinks, vegetative buffering along riparian zones, predominant path-ways for point sourced or non-point sourced pollutant.
- Evolving superlative management practices for flood-plains in forest catchments.
- Counting actual water-use and evaporation guestimates for a broader variety of forests.
- Enumerating special effects of floods & diffused pollution adjusting drainage systems.
- Counting fiscal prices & paybacks of forest impacts on water services, productivity, and conservation values.
- Raising better location specific climate change based water-use impact models
- Location specific observing on long-standing forest effects on the freshwater milieu.

Further, from quantitative considerations there exists a vast scope for steering location-specific R&D interventions on issues like (i) forest hydrology for the interrelation of flow regimes and water quality, (ii) forest eco-hydrological research, (iii) catchment scale hydrological modelling approach for holistic water quantity management, (iv) managing forest streams & rivers, (v) integrated watershed management by accommodating point source and non-point source pollutants including silt, (vi) dealing sensitive areas/zones of forest catchments, (vii) carbon dynamics in forested catchments. In light of these challenging slots, there remain several hydrological nodes in a hydrologic cycle which invites specified attentions to comprehend and solve problems that are anticipated to get emerged owing to changes in some of the hydrologic processes after various kinds of disturbances (forest felling, land-use change, roads, soil conditions, climatic factors, etc.) in given forest.

8 Summary and Conclusion

For offering a wider and deeper perceptive on forest-water interfaces, the fundamental hydrological functioning via momentous processes & elements is discussed in a preeminent way. Prospective forest and water managing related stratagems are retained as strategic end focus of this specific chapter; by including pertinent challenges (climate, human & cattle population growth, etc.) for sprouting truer knowledge spheres of water-forest acquaintances. This reflects management approaches for water & forests; appealing water as utmost ranked entity; in comparison to other forest-oriented objectives (biomass accumulation, sequestration of carbon, local climate effects). Such water-based prioritizing in forest catchments could offer better & sound sustainable tactics for the attainment of longstanding forest healthiness on one side while inclusive human wellbeing on the other side.

Unpolluted & copious water is not only an unexpected biome service obtainable from forests; rather it obliges as a base input to governing climatic and overall socio-economic disquiets at a regional scale. The location-specific answer to a common question: 'How and in what silhouette the water enters the forest floors? ', happens to be an important indicator to sizeably governs the soil-moisture profile and its definite obtainability on forest lands. Once water enters the forest floor, numerous physical progressions initiate to play their collective role, vide which the soil moisture build-up gets proficient for optimum use by forest vegetation. Some of the prime processes could be high concentrations of organic matter, filling of land depressions (micro or macro), the extent of water spreading, etc. Excess water (beyond soil storage capacities) often discharges an affirmative net outcome, as it gets slowly drained through the land masses located at lower elevation and thus congregate to form minor gullies & streams, rivers. It in turn hypothetically supplement the nearby underground water bearing geologic formations. Hydrologically based studies [29] have established that water is very rarely well-thought-out, 'foremost researchable entity' while attempting integrated management of forest catchments. Nevertheless, as worldwide weather, atmospheric temperatures and microclimate erraticism endure to expand the links across forests and water yields kept exceedingly altering. Numerous hydrological compassions have recognized the fact that arriving rain is very first cast-off by vegetation and further excess rain magnitude to saturate the soil column. Only afterward (when the above two situations are satisfied) then only the runoff-water begins to drain as streamflow, from forest catchment or ecosystem. Further variations in regards to air temperature & potential evaporation define and regulates magnitudes & trends of net groundwater & stream flows from forested catchments. Besides, if shifting climatic outlines lessen rainfall, such stream-flows are likely to be further abridged as compared to historic conditions at the same location. These reductions could be amply understood and moderated if forest mortality reduces forest plant water demands. It altogether established the vital prominence of the dual-managerial approach namely the 'water-regulated forests' and the 'forest-regulated water'.

The chapter has deeply scrutinized & deliberated a range of water-forest connected concerns, disputes, themes, applied conceptual frameworks (physical/hypothetical) along with a plethora of strategies that might effectually respond to some of these challenges, with below provided leading inferences,

- An inclusive approach towards water-sensitive forests/forest catchments and their land & water management needs to be invariably recognized; which indeed establishes the prominence of critical water zones, water source areas, riparian/wetland areas, and surrounding buffer zones therein. It altogether might have the greatest impact on prevailing socio-hydrologic settings, for hydrological recognition and delineation/prioritization of such strategically important areas/zones with relevant assessments of threats and possible remedies; ensuring overall positive offerings towards the forest-water system.
- More frequently the information and records on comprehensive realization of various tied socio-hydrological systems are found insufficient [29], henceforth a need is always arising for improved observing towards espousing smarter tools & techniques (modelling, big data, sensors, ICT) together with prevailing local explanations & alternate traditional knowledge systems.
- Carbon sequestration in upright woodlands and absence of ample practical knowledge of land-scale effects are two swelling concerns to govern the forest water resources and risk of policy failures in this regard.
- Other imperative needs may comprise issues like; to enlarge the way water and forest field functionaries or managers are skilled; bringing them organized in a new united approach ensuring that futuristic forests might be explicitly bring about both for water as well as other key natural resources; to harness enhanced gains.
- Upkeep of upright & better environmental standing of water-bodies or water parcels in forest catchments; might be ensured stress-free by smearing judicious forest development plans for conserving high-quality drainage waters keeping the depressed level of nutrient/pesticide/sediments therein.
- Judging decreases in water use and upsurge/diminution in runoff water yield; when any younger forest matures, might be considered additional primacy aspect to yield a deeper understanding of forest water associations.
- Conservation of water yield (specifically base flows) across larger parts of forest catchments offers great potential for attaining improved water productivity of forests.
- Superimposing clay & sandy soils and their hydrological/ecological impacts; all need to be deliberated captivating them as a sensitive zone both for forests & water resources.
- Evaluating lessening in forest runoff water volumes, base-flows, and inconsistency of minor & higher floods; across years may offer high-value elucidations and feedbacks for scheduling water equilibriums and their inclusive utilizations at micro/macro levels.

9 Recommendations

Though the subject and theme of the present chapter is massive and wide, but still at the end; the following could be considered as a comprehensive set of spirited recommendations,

- Managing forests could be an easier & effective task once a comprehensive ‘forest & water agenda’ be thought of keeping sustainable water productivity and effectual conservation aspects as two big targets or goals.
- Preserving and augmenting the watershed-based studies in forest areas and forested catchments must be given further importance, priority, and R&D-based budgetary provisions.
- Evolving technologies with smarter elements like GIS & remote sensing, sensors, IT & ICT tools, artificial intelligence, and big data need to be appropriately unified while conceiving either the forest management plans or the water resources plans for any forest region.
- Assimilated advanced models & modeling approaches must be given high priorities for forests and forested catchments by accommodating their prevailing hydrological, climatic, meteorological, geophysical, socio-economic, and environmental deliberations.
- Creating ample R&D infrastructure and conductance of operational research on smaller watersheds (both short term & long term) will certainly pave a path for gaining better water productivities from natural or man-made forests.
- Forests and forest-based catchments or landscapes happen to be historical creatures with a variety of abandoned bits of traditional knowledge in regards to effectual natural resource management. It will be certainly a prolific exertion to plan or incorporate such elements in forest policy or forest & water-based development planning.
- Popular beliefs (wrong or right) on forest-water-climate-human domain needs to be understood at real ground level at regional scales under prevailing climate & micro-climate settings; to define, designate and decide on disagreements over the active roles of forest during the prevailing hydrologic cycle, natural hazards, natural resource depletion, socio-economic levels, and many other unforeseen fronts.
- More idealistic goals need to be framed & executed to safeguard active integration of forest with location-specific land & water development and managerial plans; ensuring a safe & sustained water supply along with due forest-related products & environmental services.
- A mission mode and network-based approach might be well-thought-out as one of the best option to gain benefits of forest-water based natural systems, as it may incorporate experts, end clients, policymakers, ground functionaries, and researchers for a giant common goal where forests can deliver their dual active role; both from production as well as protection aspects.
- Strategic R&D plans in the above regards could encompass vital elements like (i) site-specific policy, collaboration & institutional growth (ii) physical process

understanding & research, (iii) perception raising, capacity augmentation, and smart communications, (iv) tallying larger dimensions (local to global), (v) looking for evidencing role of forest-water on climate-based extenuation & alternative strategies, a (vi) effectual mechanism to ensure socio-economic incentives to end clients, and (vii) establishing synergies with relevant stakeholders & working groups for promoting forest-water agenda at grass-root level.

- Last but not least, the all-inclusive forest-water strategy must be focused to improve upon set goals in regards to science, practices, economics, policies, and capacity building; for enhanced forest-water productivity.

Acknowledgments Besides the author's views & work, this chapter incorporates views & findings of many leading experts and agencies who had worked on forest-water based issues; which is gratefully acknowledged.

References

1. Ellison D et al (2017) Trees, forests and water: cool insights for a hot world. Glob Environ Chang 43:51–61
2. FAO (2016) State of the World's Forests 2016. Available at http://www.fao.org/publications/sofo/2016/en/ Accessed on 23.12.2018
3. Riitters K, Wickham JD, Wickham Jennifer K, Costanza JK, Vogt P (2016) A global evaluation of forest interior area dynamics using tree cover data from 2000 to 2012. Landsc Ecol 31(1): 137–148
4. WRI (2018) Tackling global challenges. Accessed at https://www.wri.org/strategic-plan/tackling-global-challenges. Accessed on 26-11-2018 at 14:00
5. Filoso S, Bezerra MO, Weiss KCB, Palmer MA (2017) Impacts of forest restoration on water yield: a systematic review. PLoS One 12(8):e0183210
6. Burch GJ, Bath RK, Moore ID, O'Loughlin EM (1987) Comparative hydrological behaviour of forested and cleared catchments in southeastern Australia. J Hydrol 90(1–2):19–42
7. Andreassian V (2004) Waters and forests: from historical controversy to scientific debate. J Hydrol 291:1–27
8. Miller GT Jr (1999) Environmental science: working with the earth. Wadsworth, Belmont, CA
9. Chang M (2002) Forest hydrology: an introduction to water and forests. CRC Press, ISBN:0849313635, 9780849313639, 392 p
10. Calder IR (1998) Water use by forests, limits and controls. Tree Physiol 18(8–9):625–631
11. FAO (2008) Forests and water. FAO forestry paper 155, Rome: Food and Agriculture Organization of the United Nations
12. Wagener T, Sivapalan M, Troch PA, McGlynn BL, Harman CJ, Gupta HV, Kumar P, Rao SC, Basu NB, Wilson JS (2010) The future of hydrology: an evolving science for a changing world. Water Resour Res 46:1–10
13. Hamilton LS (1985) Overcoming myths about soil and water impacts of tropical forest land uses. In: El-Swaify SA, Moldenhauer WC, Lo A (eds) Soil erosion and conservation. Soil Conservation Society of America, Ankeny, IA, pp 680–690
14. Nuzhat Q, Qazi L, Adrian B, Rai SP, Chandra PG (2017) Impact of forest degradation on stream flow regime and runoff response to rainfall in the Garhwal Himalaya, Northwest India. Hydrol Sci J 62(7):1050–1066
15. Negi GCS (2002) Hydrological research in the Indian Himalayan Mountains: soil and water conservation. Curr Sci 83:974–980

16. Samraj P, Sharda VN, Chinnamani S, Lakehmouan V, Haldorai B (1988) Hydrological behaviour of the Nilgiri sub-watersheds as affected by bluegum plantations, part I. The annual water balance. J Hydrol 1003:335–345
17. Bosch JM, Hewlett JD (1982) A review of catchment experiments to determine the effect of vegetation changes on water yield and evapotranspiration. J Hydrol 55:3–23
18. Brown AE, Zhang L, McMahon TA, Western AW, Vertessy RA (2005) A review of paired catchment studies for determining changes in water yield resulting from alterations in vegetation. J Hydrol 310:28–61
19. Mathur HN, Ram B, Joshie P, Singh B (1976) Effect of clear-felling and reforestation on runoff and peak rates in small watersheds. Indian Forester 102(4):219–226
20. "Hydrologic Effects of a Changing Forest Landscape". Accessed on 04.09.2019 [online]. Available at http://dels.nas.edu/resources/static-assets/materials-based-on-reports/reports-in-brief/forest_hydrology_final.pdf
21. Ilstedt U, Bargués Tobella A, Bazié HR, Bayala J, Verbeeten E, Nyberg G, Sanou J, Benegas L, Murdiyarso D, Laudon H, Sheil D, Malmer A (2016) Intermediate tree cover can maximize groundwater recharge in the seasonally dry tropics. Sci Rep 6, Article number: 21930
22. Behar S (1997) Testing the waters: chemical and physical vital signs of a river by. Montpelier, VT: River Watch Network. ISBN 0787234923
23. Gaur S (2018) An updated review on quantitative and qualitative analysis of water pollution in west flowing Tapi River of Gujarat, India. In: Singh V, Yadav S, Yadava R (eds) Environmental pollution. Water science and technology library, vol 77. Springer, Singapore
24. Spooner J, Line DE (1993) Effective monitoring strategies for demonstrating water quality changes from nonpoint source controls on a watershed scale. Water Sci Technol 28:143–148
25. UNESCO (1978) World water balance and water resources of the earth, UNESCO Press, Paris. Accessed at http://www.fao.org/library/library-home/en on 10-12-2018
26. Grant GE, Tague CL, Allen CD (2013) Watering the forest for the trees: an emerging priority for managing water in forest landscapes. Front Ecol Environ 11:314–321
27. Water Cycle Summary [Online]. Available: https://upload.wikimedia.org/wikipedia/commons/1/19/Watercyclesummary.jpg
28. FAO (2015) The International Forests & Water Dialogue Final Report (Ed: Elaine Springgay). Available at http://www.fao.org/forestry/44018-070bfc15ad79c1d91523033f7ecdaf2ac.pdf. Accessed on 06.09.2019
29. USDA (2018) Management option for dealing with changing forest-water relations. Available at https://www.srs.fs.usda.gov/pubs/chap/chap_6_2018_mcnulty.pdf. Accessed on 07.09.2019
30. Caldwell PV, Miniat CF, Elliott KJ, Swank WT, Brantley ST, Laseter SH (2016) Declining water yield from forested mountain watersheds in response to climate change and forest mesophication. Glob Change Biol 22:2997–3012
31. Wei X, Zhang M (2010) Quantifying streamflow change caused by forest disturbance at a large spatial scale: a single watershed study. Water Resour Res 46(W12525):1–15
32. Sikka AK, Samra JS, Sharda VN, Samraj P, Lakshmanan V (2003) Low flow and high flow responses to converting natural grassland into blue gum (*eucalyptus globules*) in Nilgiri watersheds of South India. J Hydrol 270:12–26
33. Stoneman GL (1993) Hydrological response to thinning a small jarrah (Eucalyptus marginata) forest catchment. J Hydrol 150:393–407
34. UNESCO (2017) Forest management and the impact on water resources: a review of 13 countries. UNESCO international hydrological Programme: international sediment initiative. HP – VIII/*technical document* N° 37 Latin America and the Caribbean (Eds: Pablo et al.). Accessed at http://www.unwater.org/unesco-publication-forest-management-impact-water-resources/ on 08-12-2018
35. ISFR (2017) Indian State of Forest Report 2017. Available at http://fsi.nic.in/isfr2017/isfr-forest-cover-2017.pdf. Accessed on 09.12.2018 at 22:05 hrs

36. Rao BK, Gaur ML, Kumar G, Kurothe RS, Tiwari SP (2013) Morphological characterization and alterations in cross section of different order streams of Mahi ravines. Ind J Soil Conserv 41(1):20–24
37. Singh AK, Kala S, Dubey SK, Rao BK, Gaur ML, Mohapatra KP, Prasad B (2014) Evaluation of bamboo-based conservation measures for rehabilitation of degraded Yamuna ravines. Ind J Soil Conserv 42(1):80–84
38. Wang Y, Liu Y (1995) Hydrological characteristics of a moso-bamboo (Phyllostachys pubescens) forest in South China. Hydrol Process 9(7):797–808
39. Gupta VK, Waymire E, Wang CT (1980) A representation of an instantaneous unit hydrograph from geomorphology. Water Resour Res 16:855–862
40. Gaur ML, Kumar S (2018) Preliminary investigations on localized rainfall interception losses under real field observations. In: Singh V, Yadav S, Yadava R (eds) Hydrologic modeling. Water science and technology library, vol 81. Springer, Singapore, pp 21–36
41. Gaur ML (2003) Characterizing surface infiltration variability in a small forest watershed. Indian Forester 129(3):341–348
42. Pathak PC, Pandey AN, Singh JS (1985) Apportionment of rainfall in central Himalayan forests (India). J Hydrol 76(3–4):319–332
43. Putty MRY, Prasad R (2000) Understanding of runoff processes using a watershed model: a case study in Western Ghats in South India. J Hydrol 228:215–227
44. Krishnaswamy J, Bunyan M, Mehta VK, Jain N, Karanth UK (2006) Impact of iron ore mining on suspended sediment response in a tropical catchment in Kudremukh, Western Ghats, India. For Ecol Manag 224:187–198
45. Gaur S, Gaur ML (2017) Carbon capture through Forests & Vegetation – an environmental inkling. Int J Adv Res Rev 2(1):39–52
46. Lane PNJ, Sheridan GJ (2002) Impact of an unsealed forest road stream crossing: water quality and sediment sources. Hydrol Process 16:2599–2612
47. Shah AA, Rieley JO (1999) Influence of tree canopies on the quantity of water and amount of chemical elements reaching the peat surface of basin mire in the midlands of England. J Ecol 77: 357–370

Assessment of Groundwater Quality in and Around Nemawar, Madhya Pradesh, India

Sunil Kumar Sharma

Abstract For the water quality studies of Nemawar and adjacent areas of Madhya Pradesh, 15 groundwater samples were collected. The collected groundwater samples were analysed. Water samples were collected from Narmada River and from various agricultural fields between Karond and Nemawar. The collected water samples were analyzed for different physico chemical parameters during premonsoon and post monsoon period. The study area is a part of Harda district of Madhya Pradesh. There are different types of physic-chemical parameters to be analysed. These parameters are pH, total dissolved solids, Hardness, Chloride, sodium, calcium, Electric Conductivity, Nitrate, Phosphate are determined. The suitability of groundwater sources in the study area for irrigational purposes was analysed by comparing the values of different water quality parameters with the guidelines provided by Ayers and Westcot.

Keywords Hydrochemistry · Groundwater · Kelley's ratio

1 Introduction

Subsurface water is an essential for the life and it is a significant resource of water supply in both the municipal and rural areas. It is precious for us and global ecosystems. Sub Surface water is very important sources for human being and social and economical growth [1]. The existence of groundwater is valuable for us [2]. The sources of freshwater are limited, but demand of this resource is very high. Some of the region where surface water is not available, subsurface water is suitable for all the needs of peoples. Sub surface water is a crucial component of all animal and vegetable matters. The most serious worldwide ecological problem is the deterioration of water quality. The leading nonpoint source of water contamination is Agriculture [3]. The important substance which is responsible for groundwater

S. K. Sharma (✉)
Department of Civil Engineering, Rabindranath Tagore University, Bhopal, Madhya Pradesh, India

S. Yadav et al. (eds.), *Water Quality, Assessment and Management in India*, Earth and Environmental Sciences Library, https://doi.org/10.1007/978-3-030-95687-5_6

pollution is nitrate [4]. The quality of groundwater in an area is governed by various usual processes like the rate of rainfall, weathering and erosion of rocks and soil. Various activities of human beings are also affecting the quality of water. The nitrate contamination of sub surface water is a universal problem and it is a extensive pollutant of all type of water [5]. Groundwater consumption has great significance. Clean and clear water is of basic significance to the continued existence, safety and progress of human life. Availability of groundwater for drinking purposes depends upon the quality of that water. It has been recognized as one of the important threats to the resources of subsurface water around the globe [6]. The movement of groundwater is governed by the various properties of contaminants, the underground water and the ecological system by which the polluted subsurface water is flowing. Groundwater is globally used for various types of demand like domestic, irrigational and industrial. Household and industrial wastes, return flow from irrigation and leachate from landfill sites are some factors which change the quality of groundwater. Different type of land use practices, geology, patterns of rainfall, and some climatic factors influence the use of groundwater [7]. Occurrence of several ionic contents further than some restrictions may make it incompatible for irrigation, municipal or manufacturing use. Hydrogeochemistry of underground water is significant for the growth and efficient management of the groundwater resource. Now a days our surroundings have become polluted, impure, undesirable, and therefore unsafe for the human health. Many workers have worked on the quality of various types of water souirces and a number of valuable reports and research papers have been published by the scientists and researchers [8–18] from time to time. Depletion of ground water table due to over exploitation of ground water, unauthorized sewer lines, industrial waste water lines connected to the unlined irrigation canals [19]. The underground water conservation region is developed on the basis of hydrochemistry value (eC) to reach progress in underground water management. Critical, secure damage and vulnerable are four conservation zones [20]. The model can include potential of pollutant transport from sampling locations where contaminants have been reported [21]. For the socio-economic development of India, 90% of extracted subsurface water is used for the purpose of agriculture [22]. In present investigation 15 groundwater samples collected from the study area are subjected various laboratory analysis.

2 Study Area

The present area of investigation is the basin of river Narmada which passes through district Dewas, Madhya Pradesh. The study area is located between longitudes 77°10″ to 77°00″ E and latitude 22°30″ to 22°35″ N. Most of the population of this area belongs to agriculture (Fig. 1).

This area is well connected to district headquarters as well as Bhopal, the state capital of Madhya Pradesh, by bus with National Highway No. 12. The study area can be approached by all weather and fair-weather roads. The all weather roads are

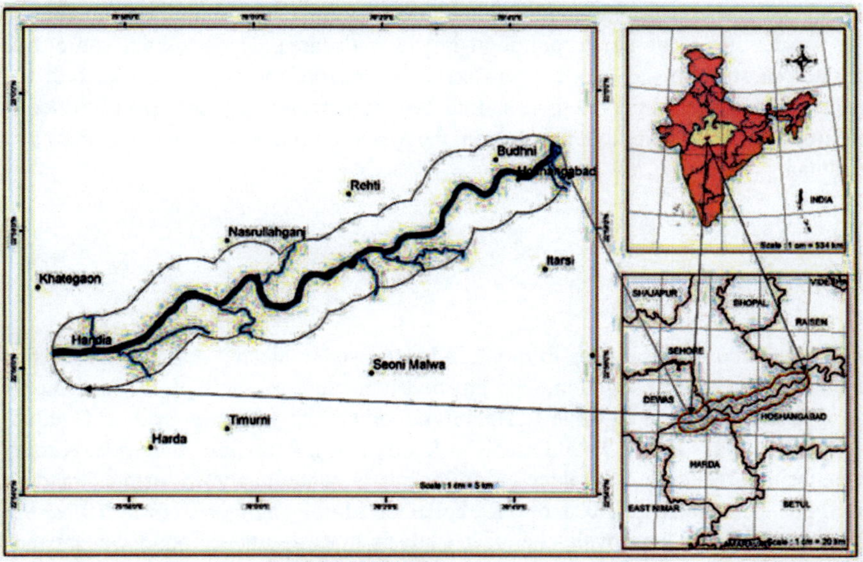

Fig. 1 Location Map of the Study Area

limited in their extent. However, the fair weathered roads provide a motorable approach to maximum localities. The climate of the area under investigation is sub-tropical monsoonic type characterized by long hot summer and mild short winter. During summer, days are hot and nights are cooler. The area falls in the belt of high temperature. The mean maximum temperature varies from 24.1 °C to 46.8 °C, in the month of May; which is the hottest month of the year; while the minimum temperature recorded in winter season ranges from 7.3 °C. to 32.2 °C in December. Most of the rainfall occurs from the southern and western monsoon. The beginning of the monsoon normally takes place in mid of the June. In the month of July and August, maximum rainfall occurs. About 1345 mm rainfall occurs annually. The weather of the area is quite warm and the cold climate lasts only for about two month's i. e. December and January. River Narmada is the major source of water in the area of study. The water of River Narmada is used for agricultural and drinking purposes. Most of the study area is covered by alluvial soil. Calcareous sand, silt, clay and conglomerate are the common rock types. The drainage pattern is generally dendritic to sub-dentritic in nature in dissected flat terrain and mountainous area, while, it is parallel to sub parallel in the plain. Streams in the basaltic terrain are mostly controlled by fractures and joints present in the rock formations. The quality of subsurface water of the area is mainly determined by various natural processes like velocity of underground water, lithology, quality of recharge waters and interface among other types of aquifers [23]. The application of NPK fertilizer contributed to raise the toxic levels in the surface water bodies. Further, it is suggested that the streams with more than 1 mg/L of Nitrate may be investigated to determine the actual cause of contamination [24]. The farming practises and agricultural waste are the major sources of nitrate pollution in the study area. The potential sources of nitrate pollution in the area are fertilizers. For restore the aquatic ecology of the polluted area the proper method should be implemented [5]. Industrial effluents, anthropogenic activities and runoff from the agricultural field are the major source of pollutants [25].

3 Methodology

Fifteen groundwater samples from groundwater were collected during both seasons from the area of investigation. The physicochemical characters of collected water samples are presented in Table 1. The physico-chemical properties such as Conductivity (EC), pH, sodium, Total Dissolved Solid (TDS), Potassium, Chloride, Nitrate, Magnesium, Calcium, Sulphate and Phosphate were analyzed using Standard Methods. The water quality index for collected samples ranges from 37.10–93.93. It is reveals from the physic-chemical analysis that the underground water needs several treatments before utilization [26].

From the above mentioned characters some parameters are recorded at the time of sampling. pH meter is used for the analysis of pH of water samples. Electrical Conductivity is calculated by conductivity meter. The bicarbonate was analysed by titrating analysis. For this methyl orange indicator was used. Calcium and Magnesium content from the water samples was calculated from Ca hardness and mg hardness respectively. Other parameters like Nitrate, Sulphate, Chlorides, Sodium and Potassium are analysing with the help of various methods [27]. Values of various cations of subsurface water samples which are collected from the field are shown in Figs. 2 and 3. While values of various anions are presented in Figs. 4 and 5.

4 Results and Discussion

4.1 Water Quality Parameters

pH The pH of water samples collected from the area ranged between 7.4 to 7.9 in the summer season and 7.0 to 7.7 in the winter season. As per the classification of Ayers and Westcott the pH of water samples is slightly alkaline.

Electrical Conductivity It is a significant decisive factor in the suitability of water for the purpose of agriculture. The conductivity of the study area was 380 to 910 Micromho/cm^2 in summer season while whereas it was 340 to 880 Micromho/cm^2 in winter season. Electrical conductivity lies in the range of medium salinity zone (250–750 micromho/cm^2).

Table 1 Hydrochemistry of the collected water samples of the area

Pre Monsoon																Post Monsoon													
					Concentrations in mg/l														Concentrations in mg/l										
					Cations					Anions									Cations					Anions					
Well no.	Name of sampling station	pH	$ECX10^6$ at 25 °C	TDS	Na^+	K^+	Ca^{++}	Mg^{++}	T.H. as $CaCO_3$	HCO_3	Co_3^{--}	Cl^{--}	NO_3	SO_4	PO_4	pH	$ECX10^6$ at 25 °C	TDS	Na^+	K^+	Ca^{++}	Mg^{++}	T.H. as $CaCO_3$	HCO_3	Co_3^{--}	Cl^{--}	NO_3	SO_4	PO_4
1	3	4	5	6	7	8	9	10	11	12	13	14	15	16	17	18	19	20	21	22	23	24	25	26	27	28	29	30	31
1	Karond	7.7	650	416	26	3.5	38	16	160	212	Abs	46	18	24	0.27	7.2	630	403	21	2.4	33	15	147	160	Abs	41	12	20	0.25
2	Pipliya	7.9	710	454	18	4.1	40	14	157	110	Abs	71	59	18	0.31	7.6	680	435	13	3.4	36	13	143	96	Abs	66	52	13	0.28
3	Akawliya	7.5	590	378	15	5.2	36	23	186	162	Abs	76	40	12	0.28	7.2	570	365	10	4.3	32	22	172	130	Abs	72	36	7	0.26
4	Murjal	7.8	630	403	33	3.9	32	18	154	122	Abs	80	62	20	0.2	7.1	600	384	27	3.4	27	17	140	110	Abs	76	57	15	0.18
5	Jamner	7.8	690	442	45	7.3	47	18	160	180	Abs	50	34	19	0.28	7.3	650	416	40	6.7	30	17	145	169	Abs	44	28	14	0.26
6	Dait	7.5	910	582	36	1.8	70	21	206	165	Abs	89	29	25	0.28	7.2	880	563	31	1.3	43	20	190	144	Abs	83	22	21	0.26
7	Rehtai	7.4	850	544	27	1	40	26	207	136	Abs	46	39	27	0.21	7	820	525	22	0.6	36	24	189	121	Abs	41	32	24	0.19
8	Mandhi	7.6	430	275	14	2	36	29	211	116	Abs	85	57	23	0.17	7.4	410	262	9	1.8	32	28	197	100	Abs	87	50	20	0.14
9	Tigali	7.6	620	397	15	2.9	34	25	190	105	Abs	94	40	27	0.23	7.5	590	378	10	2	30	24	195	92	Abs	80	34	23	0.21
10	Satdev	7.6	730	467	12	6.2	28	26	176	108	Abs	65	47	43	0.28	7.3	700	448	8	5.6	23	25	161	98	Abs	60	42	38	0.25
11	Chichli	7.6	640	410	17	4.7	44	21	196	136	Abs	61	22	46	0.41	7.2	610	390	12	3.9	40	20	182	120	Abs	57	16	41	0.39
12	Guraria	7.6	730	467	37	5.6	58	21	233	140	Abs	90	41	52	0.22	7.3	690	442	31	4.8	54	20	218	125	Abs	83	37	46	0.18
13	Kundgaon	7.5	700	448	20	3.6	40	19	178	106	Abs	96	31	25	0.32	7.1	670	429	14	2.7	36	18	163	95	Abs	90	26	20	0.29
14	Sawasari	7.7	480	307	18	5.1	30	22	165	110	Abs	84	21	36	0.27	7.5	440	282	11	4.3	27	20	150	97	Abs	79	18	31	0.24
15	Nemawar	7.9	380	243	14	2.2	34	11	132	175	Abs	25	22	20	0.52	7.7	340	218	10	1.6	30	10	115	160	Abs	21	15	13	0.55

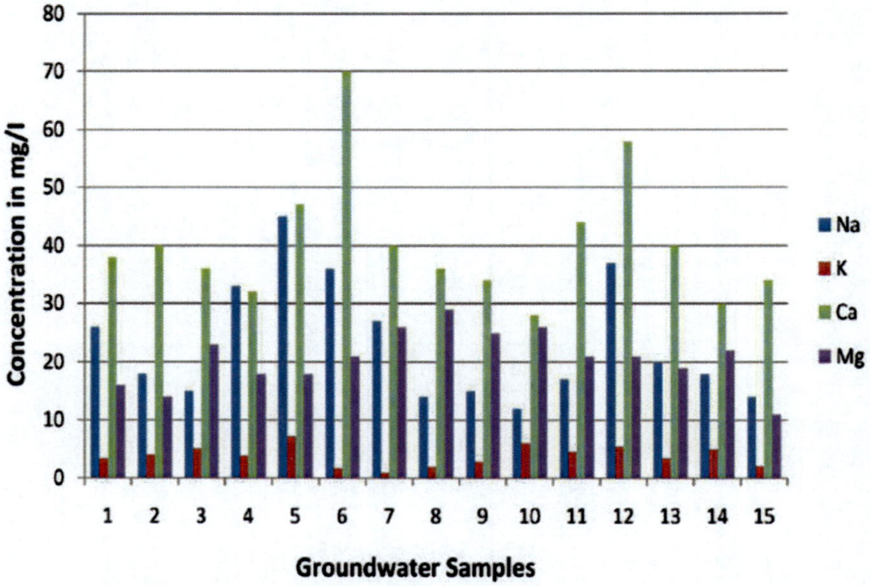

Fig. 2 Values of the cations of samples collected during pre monsoon

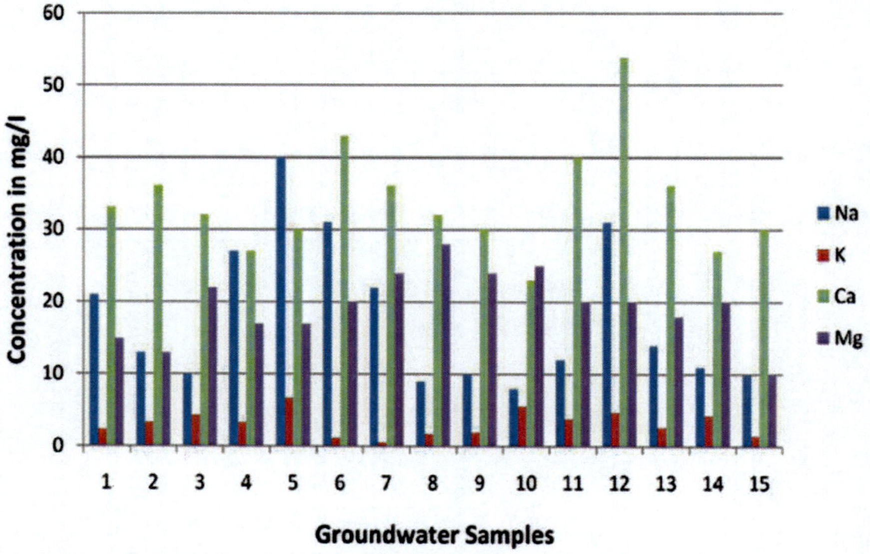

Fig. 3 Values of the cations of samples collected during post monsoon

Chloride Chloride content more than the permissible limit in underground water sample is because of the contamination from chloride rich waste water and community waste. The chloride content in the water samples from the study area was

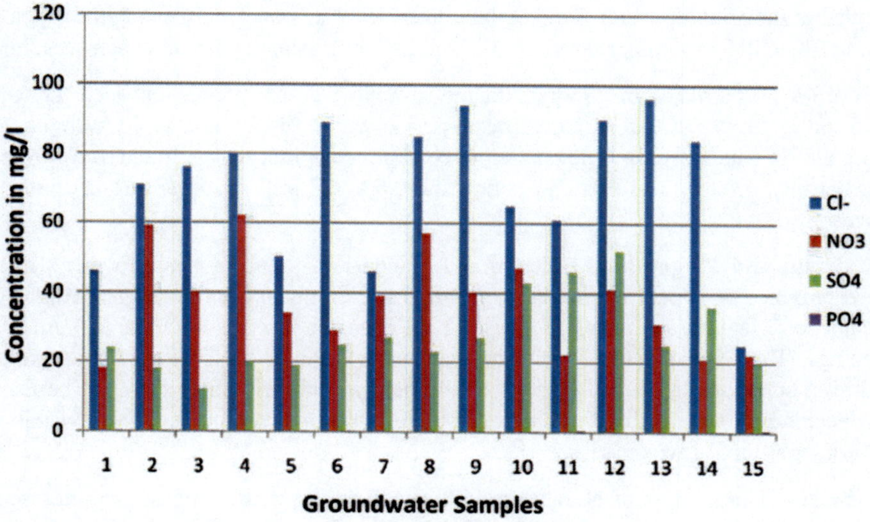

Fig. 4 Values of the anions samples collected during pre monsoon season

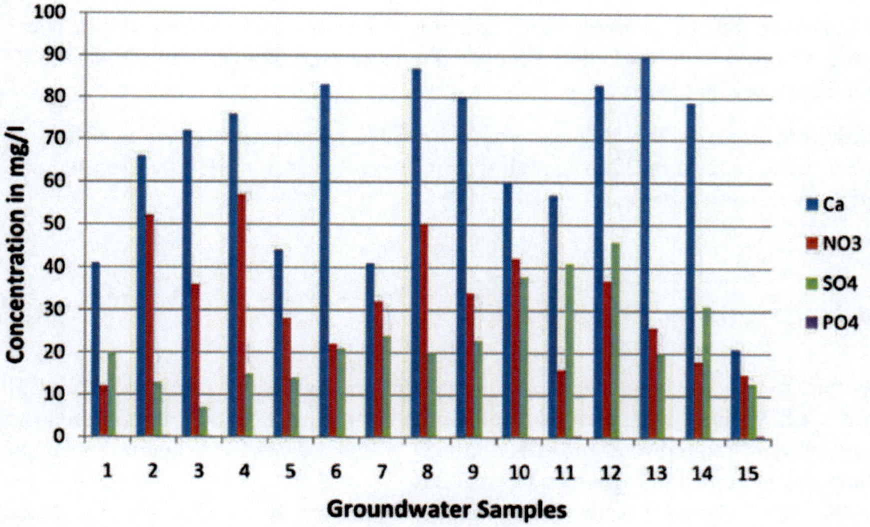

Fig. 5 Values of the anions of samples collected during post monsoon

between 25 to 96 and 21 to 90 milligram per litre respectively in summer season and in winter season.

Total Hardness In water this parameter is due to calcium, magnesium and Eutrophication. The underground water which has much amount of hardness is not suitable for drinking water. The total hardness level in water samples of study area was 132 to 233 in summer season and 140 to 218 milligram per litre in winter season.

Sodium and Potassium Sodium in the groundwater samples is between 12 to 45 milligram per litre in summer season period while it is found 8 to 40 milligram per litre in winter season. The potassium content in summer season in the study area is between 1.0 to 7.3 milligram per litre and 0.6 to 6.7 milligram per litre in winter season.

Calcium and Magnesium Calcium and Magnesium found in ground water with the presence of variety of rocks, waste water and effluents from industries. Higher value of Magnesium makes the water unpleasant and act as laxative to human beings. The Mg concentration in the water samples was 11 to 29 milligram per litre in summer season and 10 to 28 milligram per litre in winter season. The Ca concentration was 28 to 70 milligram per litre in summer season and 27 to 54 milligram per litre in winter season.

Nitrates The sources of Nitrate are fertilizer factories, matters of animals; refuse vegetables, discharge from societies and industries. The values of nitrate ranges from 18 to 62 milligram per litre in summer season while 12 to 57 milligrams per litre in winter season.

Phosphate Most important plant nutrient is Phosphorus. The values of phosphate ranges from 0.17 to 0.41 milligram per litre in summer season while 0.14 to 0.55 milligram per litre in winter season.

Sulphate Most of the sulphate ions are soluble in natural water. The values of phosphate ranges from 12 to 52 milligram in summer season while 7 to 46 milligrams per litre in winter season.

4.2 Irrigational Water Quality

In this investigation the specifications as suggested by various researchers [28–32] are used. Various parameters like sodicity, chlorinity and salinity are assessed the suitability of groundwater for irrigation [33]. Values of various irrigation specifications are presented in Figs. 6 and 7 as follows:

Sodium Adsorption Ratio (SAR), Magnesium hazard, Kelly's Ratio, and RSC are the well-known specifications. With the help of these specifications the suitability of underground water is determined (Table 2). The quality of water with above specifications is presented in Table 3.

As per the classification based on SAR, The SAR of the analysed underground water samples presented in Table 3; it undoubtedly shows that the underground water of the area belongs to Low Sodium waters in both the season.

The value of KR, below one is fit for the use of agriculture and more than one is not fit for agricultural practices [34]. It is clear From the Table 2; the Kelly's ratio

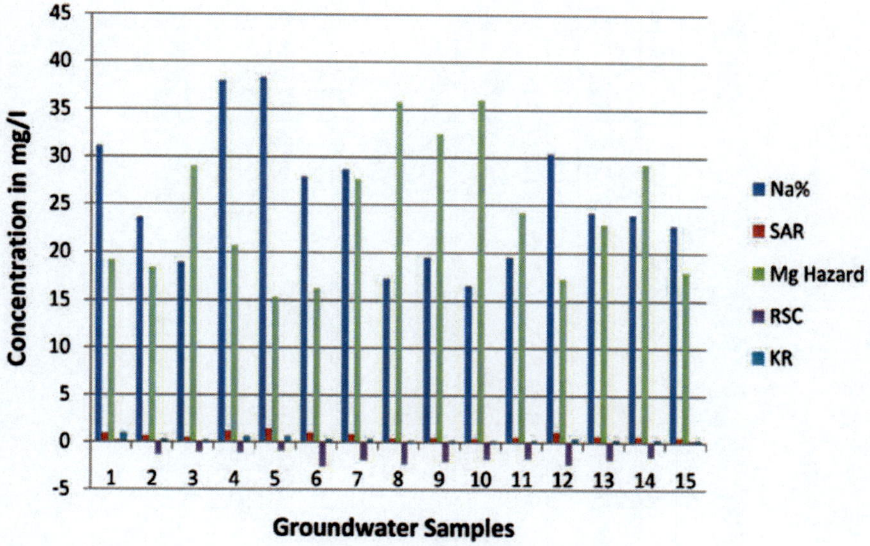

Fig. 6 Values of the various irrigational specification of groundwater (pre monsoon)

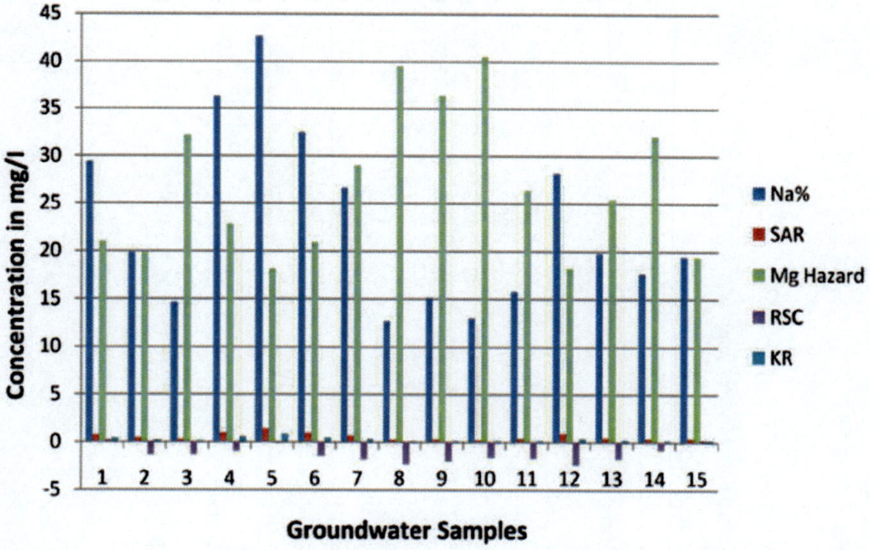

Fig. 7 Values of the various irrigational specification of groundwater (post monsoon)

below 1 in both the season. Table 3 reveals that all samples belong to suitable class. Thus, the groundwater is fit for agricultural practices. From the Table 1 we found that the natural water varies from 380 to 910 µmohs/cm and 340 to 880 µmohs/cm in summer season and winter season respectively and it belong to Medium to High

Table 2 Irrigational specifications of water samples

Sr. No.	Name of sampling station	Pre Monsoon					Post Monsoon				
		Na %	SAR	Mg Hazard	RSC	Kelly Ratio	Na %	SAR	Mg Hazard	RSC	Kelly Ratio
1	2	3	4	5	6	7	8	9	10	11	12
1	Karond	31.13	0.89	19.16	0.25	0.98	29.41	0.75	21	0.26	0.44
2	Pipliya	23.65	0.62	18.39	−1.35	0.33	19.87	0.47	19.87	−1.3	0.26
3	Akawliya	18.94	0.48	29.04	−1.04	0.25	14.64	0.33	32.21	−1.28	0.18
4	Murjal	37.97	1.15	20.71	−1.08	0.66	36.29	1	22.84	−0.95	0.61
5	Jamner	38.36	1.42	15.34	−0.88	0.69	42.68	1.45	18.14	−0.13	0.85
6	Dait	27.95	0.97	16.3	−2.52	0.39	32.52	0.98	20.98	−1.43	0.49
7	Rehtai	28.72	0.81	27.65	−1.91	0.41	26.63	0.7	29.05	−1.79	0.36
8	Mandhi	17.28	0.42	35.8	−2.28	0.21	12.71	0.28	39.54	−2.26	0.15
9	Tigali	19.5	0.47	32.5	−2.04	0.25	15.15	0.32	36.36	−1.96	0.18
10	Satdev	16.62	0.39	36.01	−1.77	0.22	12.98	0.27	40.58	−1.6	0.17
11	Chichli	19.6	0.52	24.22	−1.7	0.26	15.81	0.38	26.35	−1.67	0.2
12	Guraria	30.42	1.06	17.26	−2.33	0.47	28.23	0.92	18.21	−2.28	0.42
13	Kundgaon	24.21	0.65	23	−1.82	0.34	19.8	0.47	25.45	−1.72	0.26
14	Sawasari	23.96	0.6	29.29	−1.51	0.35	17.65	0.39	32.1	−0.76	0.23
15	Nemawar	22.87	0.53	17.97	0.27	0.31	19.37	0.4	19.37	0.3	0.25

Table 3 Classification of irrigational specifications of water samples

Irrigational specifications	Range	Class	Type of water			
			Pre Monsoon		Post Monsoon	
			No. of samples	%	No. of samples	%
EC	<250	Low	0	Nil	0	Nil
	250–750	Medium	13	87%	13	87%
	750–2250	High	02	13%	02	13%
	>2250	Very high	0	Nil	0	Nil
	Total		15	100%	15	100%
Sodium adsorption ratio	<10	Low	15	100%	15	100%
	10–18	Medium	0	Nil	0	Nil
	18–26	High	0	Nil	0	Nil
	>26	Very high	0	Nil	0	Nil
	Total		15	100%	15	100%
Kelly's ratio	< 1	Suitable	15	100%	15	100%
	1–2	Marginal	Nil	Nil	Nil	Nil
	> 2	Unsuitable	Nil	Nil	Nil	Nil
	Total		15	100%	15	100%
Magnesium ratio	< 50	Suitable	15	100%	15	100%
	>50	Unsuitable	Nil	Nil	Nil	Nil
	Total		15	100%	15	100%
RSC	<1.25	Safe	15	100%	15	100%
	1.25–2.50	Marginal	Nil	Nil	Nil	Nil
	>2.50	Unsuitable	Nil	Nil	Nil	Nil
	Total		15	100%	15	100%
Soluble sodium percentage (SSP)	<20	Excellent	9	60%	9	60%
	20–40	Good	5	33%	6	40%
	40–60	Permissible	1	7%	NIl	Nil
	60–80	Doubtful	Nil	Nil	Nil	Nil
	>80	Unsuitable	Nil	Nil	Nil	Nil
	Total		15	100%	15	100%

Salinity class. Table 3, further shows that 87% of natural water in summer season and winter season belongs to Medium salinity class. Sodium ions replace the exchangeable calcium and magnesium ions [35].

The sodium percentage in the natural waters varies from 12 to 45 milligram per litre in summer season and 8 to 40 milligram per litre in winter season. From the Table 3 it is clear that 60% of water samples in summer season and winter season belong to excellent class and 33% of water in summer season and 40% in winter season belong to good class. It means the maximum number of water sample is suitable for purpose of agriculture.

The magnesium ratio of natural water varies from 15.34 to 36.01 in summer season and 18.14 to 40.58 in winter season. From Table 3, it reveals that the entire water sample is in suitable class. So the groundwater has no magnesium hazard. More amount of magnesium in subsurface water affects the quality of soil. High Magnesium content converts the soil into alkaline nature. The alkaline nature of the soil decreases the yield of crop [36, 37].

RSC of natural water varies from −2.52 to 0.25 in the period of summer and −2.28 to 0.26 in the period of winter. From the Tables 2 and 3, it seems that all waters belong to Safe class and it is clear that it is suitable for irrigational practices. Values of various irrigation specifications of the subsurface water samples of the study area are presented in Figs. 6 and 7. The underground water is influencing by various natural processes such as exchange of cations, rainfall and dissolution of minerals [38, 39]. The quality of accessible underground water was degraded extremely by various activities [40]. The quality of subsurface water degrades because of the geological and chemical reactions beneath the Earth and when it is supplied during irrigation [41].

5 Conclusions

On the basis of various irrigational specifications, it is concluded that the agricultural water quality of the collected water samples was analysed on the basis of Bicarbonate hazard, Salinity hazard and Sodium hazard. The samples collected from the area of study are analysed and compared with various specifications like Kelly's Ratio, RSC and Magnesium ratio. The majority of water samples of the study area are found suitable and fit for the purpose of agriculture.

6 Recommendations

On the basis of the hydro chemical and agricultural studies, the following recommendations are made for the planners and agriculturists:

- It is recommended that the groundwater in which the nitrate concentration exceeded more than 50 mg/l should not be used for infant feeding and by the habitants. The continuous use of such waters may result to Blue baby disease and cancers in human beings.
- The potassium concentrations in few of the natural water of the study area have exceeded 10 mg/l, such water should not be used by the patients of hypertension and an alarming note to disaffect may be circulated to habitants.
- The phosphate concentrations in few of the natural water of the study area exceeded 03 mg/l, such water should not be used by the patients of hypertension and an alarming note to disaffect may be circulated to habitants

- It is recommended that, the subsurface water having high concentration of nitrate, potash and phosphate, should be used for irrigation as a substitute of the fertilizers. This recycling may lead to improvement in the quality of subsurface water and further pollution may be checked.

References

1. Umamageswari TSR, Sarala Thambavani D, Liviu M (2019) Hydrogeochemical processes in the groundwater environment of Batlagundu block, Dindigul district, Tamil Nadu, conventional graphical and multivariate statistical apporoach, applied water. Science 9(1):1–15
2. Balamurugan P, Kumar PS, Shankar K, Nagavinothini R, Sajil Kumar PJ (2020) Impact of climate and anthropogenic activities on groundwater quality for domestic and irrigation purposes in Attur region, Tamilnadu, India. Desalin Water Treat 208:172–195
3. Khan MN, Mobin M, Abbas ZK, Alamri SA (2018) Fertilizers and their contaminants in soils, surface and groundwater. In: DellaSala DA, Goldstein MI (eds) The encyclopedia of the Anthropocene, vol 5. Elsevier, Oxford, pp 225–240
4. Zhang Q, Xu P, Qian H (2019) Assessment of groundwater quality and human health risk (HHR) evaluation of nitrate in the Central-Western Guanzhong Basin, China. Int J Environ Res Public Health 16(4246):1–16
5. Jhariya DC, Shandilya AK, Dewangan R (2012) Nitrate pollution in the groundwater around Sagar town, Madhya Pradesh, India. International Conference on Chemical, Ecology and Environmental Sciences
6. Lalitha Vijaya B, Sai TK (2017) A study on assessment of groundwater quality and its suitability for drinking in Vuyyuru, Krishna (dist.), Andhra Pradesh. Int J Eng Devel Res 5(2):1662–1668
7. Dhiman SD (2014) Groundwater quality assessment for irrigation use in Rajkot district. J Indian Water Resour Soc 34(1):34–39
8. Annapoorna H, Janardhana MR (2015) Assessment of groundwater quality for drinking purpose in rural areas surrounding a defunct copper mine. Aquat Proc (Elsevier) 4:685–692
9. Bansal J, Dwivedi AK (2018) Assessment of groundwater quality by using water quality index and Physico chemical parameter: Review Paper. Int J Eng Sci Res Technol 170–174
10. Jain RK (1993) Study of the effects of Excessive use of fertilizers on the quality of Groundwater in Barna Command Areas Distt. Raisen (M.P.), unpublished Ph. D. Thesis, Barkatullah University Bhopal
11. Rubia K, Jhariya DC (2017) Groundwater quality assessment for drinking purpose in Raipur City, Chhattisgarh Using Water Quality Index and Geographic Information System. J Geol Soc India 90:69–76
12. Chegbeleh LP, Akurugu BA, Yidana SM (2020) Assessment of groundwater quality in the Talensi District, Northern Ghana. Sci World J 2020:1–24
13. Bhat MA, Wani SA, Singh VK, Sahoo J, Tomar D, Sanswal R (2018) An overview of the assessment of groundwater quality for irrigation. J Agric Sci Food Res 9(1):1–9
14. Panpan X, Feng W, Qian H, Zhang Q (2019) Hydrogeochemical characterization and irrigation quality assessment of shallow groundwater in the Central-Western Guanzhong Basin, China. Int J Environ Res Public Health 16:1492
15. Parashar VK (2001) Excessive use of fertilizers and its impact on the quality of surface and subsurface waters around Hoshangabad area. Project report submitted to University Grants Commission, Central Region Bhopal
16. Parashar VK (1994) Study of quality of surface and subsurface waters and their agricultural utility in the lower alluvial plains of Narmada Valley between Khalghat and Haribphal Distt. West Nimar M.P., Ph.D. Thesis Barkatullah University Bhopal

17. Ravikumar P, Somashekhar RK, Angami M (2011) Hydrochemistry and evaluation of groundwater suitability for irrigation and drinking purposes in the Markandeya River basin, Belgaum District, Karnataka State, India. Environ Monit Assess 173(1–4):459–487. https://doi.org/10.1007/S12517-013-0940-3
18. Ravi R, Aravindan S, Shankar K, Balamurugan P (2020) Suitability of groundwater quality for irrigation in and around the main Gadilam river basin on the east coast of southern India. Arch Agric Environ Sci 5(4):554–562
19. Hanumantharao C, Koteswararao M, Kalyan T (2018) Groundwater quality assessment for drinking purpose in Vijayawada region, Andhra Pradesh, India. Int J Eng Adv Technol (IJEAT) 8(5):2147–2152
20. Putranto TT, Widiarso DA, Susanto N (2017) Assessment of groundwater quality to achieve sustainable development in Semarang coastal areas. IOP Conference Series: Earth and Environmental Science, https://doi.org/10.1088/1755-1315/79/1/012001
21. Shekhar S, Sarkar A (2013) Hydrogeological characterization and assessment of groundwater quality in shallow aquifers in vicinity of Najafgarh drain of NCT Delhi. J Earth Syst Sci 122(1): 43–54
22. Ray RK, Syed TH, Saha D, Sarkar BC, Patre AK (2017) Assessment of village-wise groundwater draft for irrigation: a field-based study in hard-rock aquifers of Central India. Hydrogeol J 25:2513–2525
23. Sridhar SGD, Balasubramanian M, Jenefer S, Shanmugapriya P, Naveenkumar R, Ragunath P (2017) Assessment of groundwater quality in different parts of Thiruvallur District of Tamil Nadu, southern India. J Acad Indust Res 5(11):161–164
24. Maghanga JK, Kituyi JL, Kisinyo PO, Ng'etich WK (2013) Impact of nitrogen fertilizer applications on surface water nitrate levels within a Kenyan tea plantation. J Chem 1–4
25. Shraddha S, Rakesh V, Dixit S, Praveen J (2011) Evaluation of water quality of Narmada River with reference to physicochemical parameters at Hoshangabad city, M.P., India. Res J Chem Sci 1 (3):40–48
26. Perween S, Fatima U (2015) Study of groundwater quality by the assessment of physicochemical parameters and water quality index in Aligarh, Uttar Pradesh. J Chem Pharm Res 7(5): 761–771
27. Asadi E, Isazadeh M, Samadianfard S, Ramli MF, Mosavi A, Nabipour N, Shamshirband S, Hajnal E, Chau K-W (2020) Groundwater quality assessment for sustainable drinking and irrigation. Sustainability 12:177. https://doi.org/10.3390/su12010177
28. Ayers RS, Westcot DW (1994) Water quality for agriculture, FAO irrigation and drainage, paper 29, Rev 1
29. Eaton FM (1950) Significance of carbonate in irrigation water. Soil Sci 69(2):123–133
30. Kelley WP, Brown SM, Liebig GF, jr. (1940) Chemical effects of saline irrigation waters on soils. Soil Sci 49:95–107
31. Paliwal KV (1972) Irrigation with saline water I. A. R. I. monograph no. 2, (new series), New Delhi, 198
32. U. S. Salinity Laboratory Staff (1954) Diagnosis sand improvement of saline and alkali soils. USDA Handbook No 60, 160
33. Saraswat C, Kumar P, Dasgupta R, Avtar R, Bhalani P (2019) Sustainability assessment of the groundwater quality in the Western India to achieve urban water security. Appl Water Sci 9:73
34. Panneerselvam B, Paramasivam SK, Karuppannan S (2020) A GIS-based evaluation of hydrochemical characterisation of groundwater in hard rock region, South Tamil Nadu, India. Arab J Geosci 13(17):1–22
35. Esmeray E, Gökcekli C (2020) Assessment of groundwater quality for drinking and irrigation purposes in Karabuk province Turkey. Environ Earth Sci 79(13):1–17
36. Gautam SK, Maharana C, Sharma D, Singh AK, Tripathi JK, Singh SK (2015) Evaluation of groundwater quality in the Chotanagpur plateau region of the Subarnarekha River basin, Jharkhand state, India. Sustain Water Qual Ecol 6:57–74. https://doi.org/10.1016/j.swaqe.2015.06.001

37. Singh SK, Srivastava PK, Pandey AC, Gautam SK (2013) Integrated assessment of groundwater infuenced by a confuence river system: concurrence with remote sensing and geochemical modelling. Water Resour Manag 27(12):4291–4313. https://doi.org/10.1007/s11269-013-0408
38. Nagaraju A, Thejaswi A, Sreedhar Y (2016) Assessment of groundwater quality of Udayagiri area, Nellore District, Andhra Pradesh, South India using multivariate statistical techniques. Earth Sci Res J 20(4):E1–E7
39. Molekoa MD, Avtar R, Kumar P, Minh HVT, Kurniawan TA (2019) Hydrogeochemical assessment of groundwater quality of Mokopane area, Limpopo, South Africa. Using Statistical Approach Water 11:1891. https://doi.org/10.3390/w11091891
40. Shabya C, Shobhana R, Keshaw PR, Kumar SP, Suryakant C, Patel KS, Matini L (2016) Assessment of groundwater quality in Central India. J Water Resour Protect 8:12–19
41. Rawat Kishan S, Kumar SS, Kumar GS (2018) Assessment of groundwater quality for irrigation use: a peninsular case study. Appl Water Sci 8(233):1–24

Hydrological Response of Factors Affecting Rainfall Water Discharge and Water Balance: A Case Study of Tons Watershed

Pankaj Chauhan and Rizwan Ahmad

Abstract Water plays a vital role in natural resources & and our ecological systems. Primarily it is the medium that transports dissolved and suspended solids, and governs the rate at which these solids are eliminated from the our ecosystem and often defined as the output flux. A thorough awareness of the hydrological characteristics is required if one has to obtain a proper knowledge of the biological, chemical and physical phenomena that work within an ecosystem. Many natural and anthropogenic factors are there which affect the hydrological responses of the study area. The Himalayan region has exercised a great influence on the environmental conditions. The human interference in the Himalayan natural environmental conditions often gives these dynamics processes catastrophic proportions, thereby resulting in natural calamitie apart from irreversible loss which affects our ecosystem's natural equilibrium. It is necessary to acquire a better understanding and appreciation of the natural dynamics processes of the mountain watershed and better integration of man and his activates into his environment so that the natural balance of the ecosystem of an area could be sustained. It is also necessary to understand natural processes within the regional ecological systems, the human impacts on these processes andecological responses. There is an urgent need to discover the consequences for future planning towards sustainable development. The natural landscape around the study area falls in the southern slope of Mussoorie syncline lesser Himalaya physiographic belt. The topography is rugged. The relative relief and slopes are controlled by neotectonic activities, nature of lithology and exogenetic processes. The general pattern of the factors is already given in the previous chapter. The steeper slope and high relief are associated with the Main Boundary Thrust (MBT). The prominent parallel and dendritic and semi dendritic pattern, implying structural control, is exhibited by deformation stages in the Himalaya region. The averagerate of erosion is 1.02 mm/y. Out of the total sediment budget, 32.62% bed

P. Chauhan
Wadia Institute of Himalayan Geology, Dehradun, India

R. Ahmad (✉)
Interdisciplinary Department of Remote Sensing & GIS Applications, Aligarh Muslim University, Aligarh, India

S. Yadav et al. (eds.), *Water Quality, Assessment and Management in India*, Earth and Environmental Sciences Library, https://doi.org/10.1007/978-3-030-95687-5_7

loadwas measured, which was the most contribution in denudation for load type and 39.31% suspended load and only 28.07% was dissolved load generates thedenudation rate in Tons watershed. Quantified water balance values we found in positive trend in the catchment. Average three years value of water balance value was quantified 0.50, as minimum values was 0.39 and maximum 0.56 found in the Tons watershed. The generation and flow of sediment load in the Himalaya region may be the result of rainfall variability, land use, slopes, natural hazards, neotectonics activities.

Keywords Water discharge · Water balance · Hydrological response · Water budget

1 Introduction

The disappearance of river valley civilization has been attributed to the changing course of rivers. The scope and magnitude of such changes have been investigated and extensively studied [1–3] and [4]. Hydrology is one of the working disciplines within the larger framework of hydro geological sciences, itself a subcategory of earth sciences study. Of particular importance to hydrology are the hydrologic cycle and its components in which water is circulated through various pathways flowing freely on the surface and sub-surface areas. However the hydrological cycle is just one of the examples of the role water plays in the process of earth systems along with its physical influence, applied through such occurrences like currents and floods. A recent review reveals that most result indicates that deforestation causes increased water and sediment yield [1, 5]. Whereas, the results obtained from small catchments studies showed clearly how hydrology responds to land-use changes and human activities, uncertainty remains about waterdischarge and sediment load for large mountain river basins [6, 7]. The overflow of the stream in rivers has great harm to the people and their property living around. The most serious effect of flooding is the washing away of the topsoil cover [8]. Since water on the surface of the earth may be found in a variety of circumstances. Mountains and high lands are often called the world's natural 'water towers' as they provide essential freshwater for the population. In Himalaya, seasonal rivers water and their sediment data analysis indicate significant changes.

Many attempts have been made to analyze streamflow and sediment transfer data from the Himalayan region [9–11]. stated that flooding and sedimentation problems in India and Bangladesh were caused by geomorphic characteristics of the rivers. The average annual yield of sediment by the river is estimated to be between 14 and 64 thousand million tons [12, 13]. Classified the worldriver into 7 categories based on sediment yield in tonnes/km^2/year. The highest annual sediment yield exceeded 200 tones/km^2 in the South-East area with the lowest yield in the Arctic regions. Vegetation and steepness of the slope considerably influence the sediments load. Other aspects that contribute to the sediment load are the intensity of cultivation and lithology of the regions. The Hwang Ho of the north China probably has the maximum suspended load comprising of 40% sand, silt and clay by weight during high discharge [14, 15]. The river Ganga carries a maximum sediment load of 2.4 billion metric tones per year much greater than that carried by any other river of the world [8]. The climate of an area mainly determines the rainfall and consequently the vegetation cover which is directly related to sediment production.

The Himalayan Rivers as per the study of [16] has maximum sediment yields. The world's oceans receive 8000 million tonnes of sediments per year, out of which Indian rivers i.e. Ganga and Brahamputra together contribute nearly one-fourth [17]. Erosional surfaces marked the long period of quiescence such as erosion surface buried in geological past from the regionally extensive unconformity surface. The presence of these low relief features at a higher elevation indicates tectonic uplift in the areas. Landslide causes large mass movements of sediment due to an immediate event and therefore provides details of the exact timing of the event. Many debris and landslides contain organic matter suitable for river transport. Gully erosion due to river channel trenching is a problem that threatens vast tracts of the world's agricultural land. The damage is greatest in valley plains of the Himalayan agricultural zones and most seriously it threatens precarious subsistence-oriented agricultural systems in the developing world. Humid climatic regions and mountains generate the most runoff, upwards of 100 cm in places and havethe highest river discharges [18], have estimated 0.9 mm/y accelerated erosion rates for the Himalaya. The average rate of erosion in the Himalayan Ganga basin was estimated to be 0.9 mm/y. Recently [19, 20], have estimated erosion rates of 1.7 and 0.10 mm/y in the Central Himalaya, respectively. Similarly, in the Nana Kosiwatershed in the same region, the erosion rate estimated about 5.66 mm/y forthe different land-use conditions [19, 21]. These accelerated erosion rates are five or six times higher than the average erosion rate of 1 mm/y for the Himalaya. River morphology is the direct concern of the hydrology engineers, whoprovide detailed information pertaining to the amount of sediments denuded bythe river and the volume and velocity of water that moves through the channel. Hence, the history of water flow and sediment transport as river basin has quantitatively analyzed the behavior of the river, can be explained more or lessprecisely [22].

Floods generally occur in the rainy season between July to September [23], suggests that urban and degraded watershed producesimpermeable surfaces of tarsi, tiles and concrete, there is a tendency for the floodrunoff to increase in comparison with the natural forest environment. The previousstudies suggest that the removal of vegetation generates five or six-fold increase in the floods. In the mountainous region climatic factors play asignificant role in generating a hydrograph. The rainy season starts from July toSeptember and 90% of the annual rainfall occurs mainly in this season.

2 Water Discharge

For our study area, water discharge is monitored at Guchhupani near Robbers caves. The water discharge data were collected at natural gauge with water level was constructed by the irrigation Department of Uttarakhand. The velocity was measured

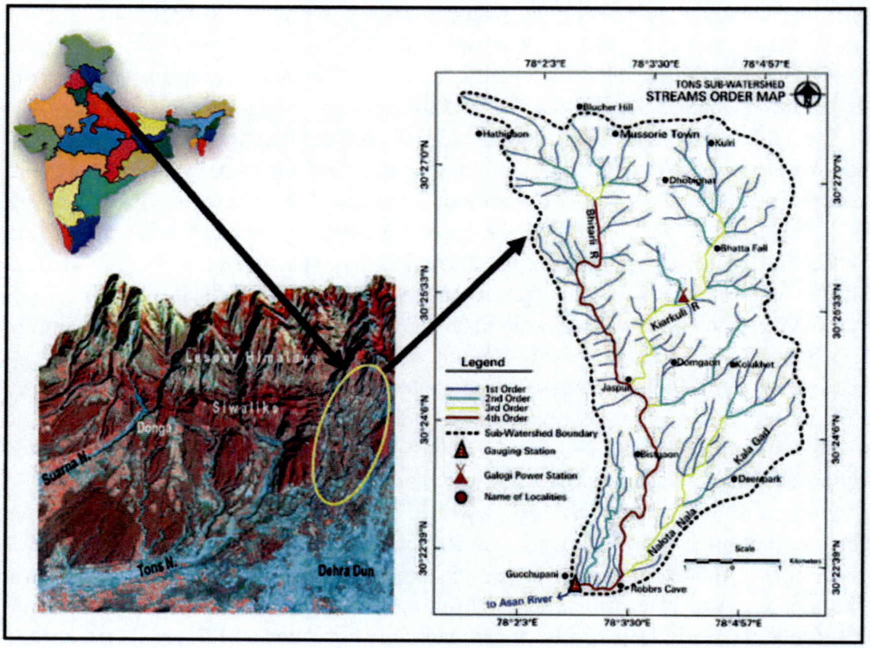

Fig. 1 Location map (after Chauhan et al. 2017) of Tons river basin, showing topography and the river systems in the Doonvalley (*left panel*). Streams order and discharge site in the basin (*right panel*)

by the float method. The water samples were collected weekly or monthly basis in the normal months and daily or hourly during the rainy season. The peak runoff data was monitored onan hourly basis and more than three water samples were taken for the analysis. The peak flood of the Tons watershed was also recorded separately and explained with the respective year. The monthly, seasonally and annually (2008–2011) water discharge data were computed and were analyzed for the finding of conclusions. The results were depicted in tables and figures. The frequency of floods during the monsoon period have been recorded separately and represented by the figures and tables. The monthly, seasonally and annually data (July 2008–June 2011) is computed and were analyzed (Fig. 1).

2.1 First-Year Water Discharge (Jul 2008–Jun 2009)

The preliminary results reveal that water generating capacity of watersheds varied with months as well as with the seasons. The monthly average of water discharge between July 2008 to June 2009 can be seen in (Table 1 and Fig. 2). The water discharge and runoff capacity is directly related to monsoonal rainfall and its

Table 1 Monthly rainfall, water discharge, sediment budget and bed load of Tons watershed (2008–2009)

			Water Discharge				Sediment load in (Tons)					
Month	Rainfall in (mm)	% of rainfall	m^3	%	m^3/s	(l /s)	Suspended	%	Dissolve	%	Total	%
Jul-08	545.80	35.71	8,899,968	25.24	3.32	3322.87	10998.07	30.91	7862.97	31.87	18861.04	31.30
Aug-08	482.00	31.54	7,801,311	22.13	2.91	2912.68	9968.83	28.02	6955.54	28.19	16924.37	28.09
Sep-08	147.20	9.63	4,655,902	13.20	1.80	1796.26	7507.24	21.10	5093.90	20.65	12601.14	20.91
Oct-08	59.50	3.89	2,812,172	7.98	1.05	1049.94	1825.97	5.13	811.17	3.29	2637.14	4.38
Nov-08	15.80	1.03	1,289,470	3.66	0.50	497.48	889.60	2.50	621.65	2.52	1511.25	2.51
Dec-08	0.60	0.04	1,022,155	2.90	0.38	381.63	263.61	0.74	250.97	1.02	514.58	0.85
Jan-09	3.20	0.21	1,236,595	3.51	0.46	461.69	236.65	0.67	258.20	1.05	494.85	0.82
Feb-09	36.00	2.36	838,954	2.38	0.35	346.79	117.72	0.33	154.94	0.63	272.66	0.45
Mar-09	5.50	0.36	1,312,738	3.72	0.49	490.12	156.78	0.44	152.93	0.62	309.71	0.51
Apr-09	19.50	1.28	1,295,959	3.68	0.50	499.98	143.85	0.40	131.05	0.53	274.90	0.46
May-09	58.40	3.82	2,000,012	5.67	0.75	746.72	257.64	0.72	368.00	1.49	625.64	1.04
Jun-09	154.80	10.13	2,094,472	5.94	0.81	808.05	3211.07	9.03	2012.45	8.16	5223.52	8.67
Total	1528.30	100%	35,259,708	100%	13.31	13314.21	35577.03	100%	24673.77	100%	60250.80	100%
Average	127.38	–	2,938,309	–	–	1109.52	2964.75	–	2056.15	–	5020.90	–

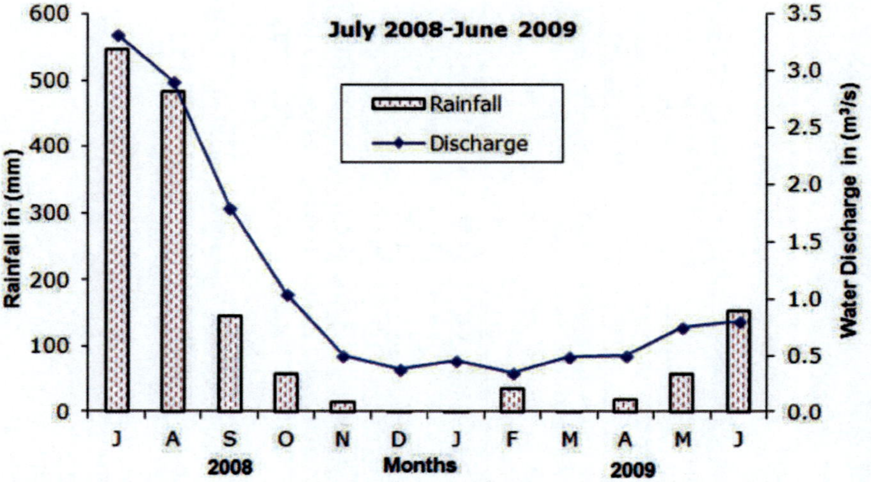

Fig. 2 Monthly average of rainfall and water discharge (Jul 2008–Jun 2009)

intensity within the whole year. The first-year results showed that the average annual flow of water discharge was 1118.08 l/s (12 m^3/s). The highest discharge of water was recorded in the month of July 2008 (3.32 m^3/s) followed by August 2008 (2.91 m^3/s). Thelowest water discharge was recorded in the month of February 2009 (0.35 m^3/s) followed by December 2008 was recorded (0.38 m^3/s) water discharge (Table 1).

The water discharge has been recorded a low amount in the months of December 2008& February 2009, is due to irrigation water used for in the cultivated land of the study area (Table 2) also shows seasonal variation of water discharge data. It reveals that about 68.55% of water discharge is found in the rainy season out of the total water discharge of the first year. A very low percentage of water discharge was found in winter and summer seasons i.e. (12.44%) and winter season (19.01%) respectively (Fig. 3).

The first year hydrograph is closely related to rainfall bars (Fig. 3). The highest rainfall was found in July 2008 which received 35.71% of annual rainfall. The second highest rainfall was recorded in August 2008 (31.54%). It showed that the rainfall was higher in July as compared to August 2008. Whereas water discharge was higher in the month of July 2008 than water discharge was also higher. The average water discharge rate is 27.16 l/s/km^2/year in the first year for the Tons watershed.

The rate of the first monsoon in first year's peak discharge is represented in (Fig. 4) the figure shows that discharges curve rise proportionally as the rainfall increases on 19 July, 31 July, 12 August and 17 August in 2008, the growth of water discharge was also recorded in peak days, i.e. (0.45 m^3/s), (0.51 m^3/s), (0.29 m^3/s) and (0.35 m^3/s) respectively as peak rainfall days.

Table 2 Seasonal rainfall, water discharge, sediment budget and bedload of Tons watershed (Jul 2008–Jun 2009)

			Water Discharge				Sediment load in (Tons)						Bed load	Total
Season	Rainfall (mm)	% of rain fall	m^3	%	m^3/s	(Liter/s)	Suspended	%	Dissolve	%	Total (D + S)	%	Total	D + S + B
Rainy	1234.5	80.78	24,169,353	68.55	2.27	2274.29	30300.11	85.17	20723.58	83.99	51023.69	84.69	27589.65	78613.34
Winter	55.60	3.64	4,387,174	12.44	0.42	423.15	1507.58	4.24	1285.76	5.21	2793.34	4.64	–	2793.34
Summer	238.20	15.58	6,703,181	19.01	0.64	635.93	3769.34	10.59	2664.43	10.80	6433.77	10.68	–	6433.77
Total	1528.30	100%	35,259,708	100%	3.33	3333.36	35577.03	100.00	24673.77	100%	60250.80	100%	27589.65	87840.45

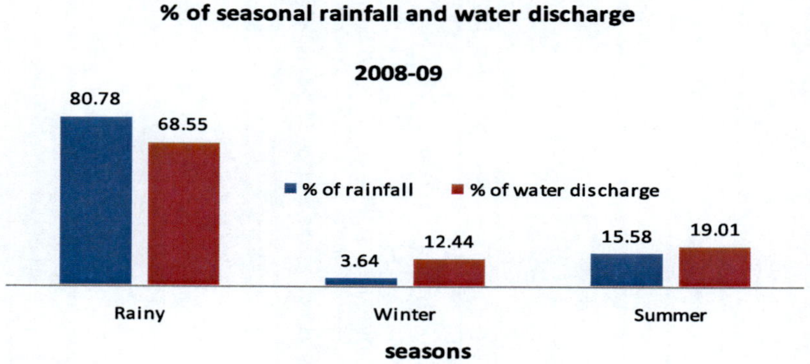

Fig. 3 Percentage of seasonal rainfall and water discharge (Jul 2008–Jun 2009)

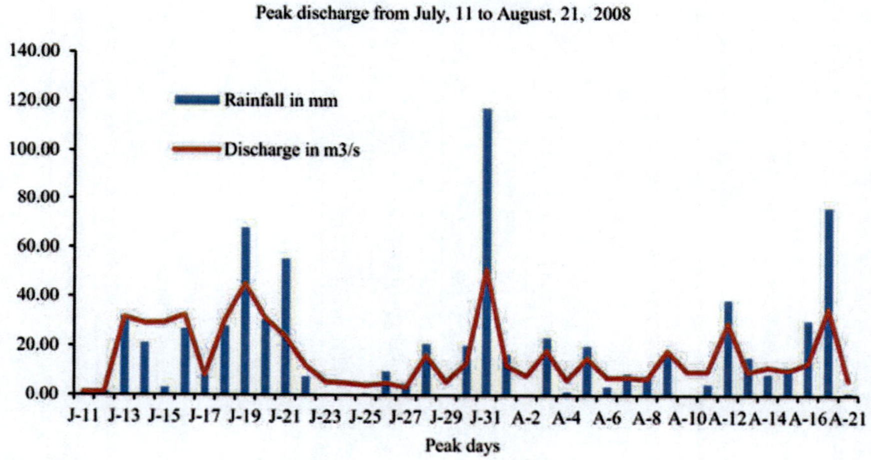

Fig. 4 Peak rainfall and water discharge (11July, 2008 to 31 August,2008)

2.1.1 Second-Year Water Discharge (July 2009–June 2010)

The mean monthly rainfall and water discharge data of 2009–2010 are presented in (Table 3). The annual average water discharge is recorded to be 1013.16 l/s (1.01 m^3/s). It is observed that the mean monthly water discharge including decrease as compared to 2008–2009. The decrease average rainfall was recorded low, (105.43 mm/month) in 2009–2010 as compared to the 2008–2009 (127.36 mm/month). The rainfall was constant during the winter season in the month of December 2009 (0.00 mm), but the minimum discharge was recorded in the month of November (0.30 m^3/s). The maximum discharge was recorded in the month of August 2009 (2.84 m^3/s) and the second highest discharge was recorded in July 2009 (2.61 m^3/s) Fig. 5.

Table 3 Monthly rainfall, water discharge, sediment budget and bedload of Tons watershed (Jul 2009- Jun 2010)

			Water Discharge				Sediment load in (Tons)					
Month	Rainfall in (mm)	% of rain fall	m^3	%	m^3/s	Discharge (l/s)	Suspended	%	Dissolve	%	Total	%
Jul-09	362.50	28.65	6,979,968	21.85	2.61	2606.02	7959.84	27.25	5987.21	27.26	13947.05	56.02
Aug-09	375.00	29.64	7,601,411	23.79	2.84	2838.04	8993.05	30.78	6862.21	31.25	15855.26	63.68
Sep-09	212.40	16.79	5,055,903	15.82	1.95	1950.58	6511.23	22.29	4987.26	22.71	11498.49	46.18
Oct-09	126.20	9.97	2,612,171	8.18	0.98	975.27	2416.23	8.27	901.21	4.10	3317.44	13.32
Nov-09	0.50	0.04	789,471	2.47	0.30	304.58	442.21	1.51	462.31	2.11	904.52	3.63
Dec-09	0.00	0.00	817,595	2.56	0.31	305.26	212.35	0.73	241.29	1.10	453.64	1.82
Jan-10	10.30	0.81	921,155	2.88	0.34	343.92	221.23	0.76	255.63	1.16	476.86	1.92
Feb-10	60.50	4.78	838,964	2.63	0.35	346.79	223.32	0.76	201.24	0.92	424.56	1.71
Mar-10	1.70	0.13	1,312,736	4.11	0.49	490.12	145.23	0.50	150.14	0.68	295.37	1.19
Apr-10	0.50	0.04	1,295,949	4.06	0.50	499.98	112.36	0.38	124.21	0.57	236.57	0.95
May-10	25.00	1.98	1,632,092	5.11	0.61	609.35	265.32	0.91	375.45	1.71	640.77	2.57
Jun-10	90.60	7.16	2,093,472	6.55	0.81	807.67	1711.07	5.86	1412.45	6.43	3123.52	12.55
Total	1265.20	100%	31,950,887	100%	12.08	12077.58	29213.44	100%	21960.61	100%	51174.05	100%
Average	105.43	–	2662573.92	–	–	1006.47	2434.45	–	1830.05	–	4264.50	–

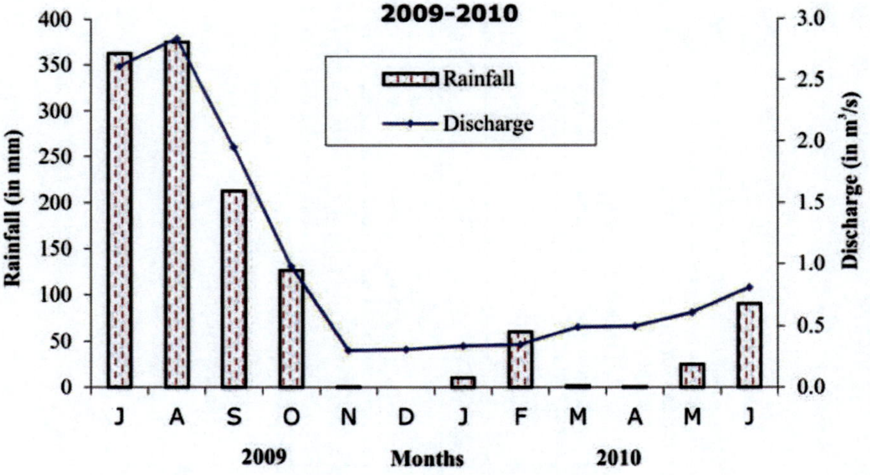

Fig. 5 Average monthly rainfall and water discharge (July 2009–June 10)

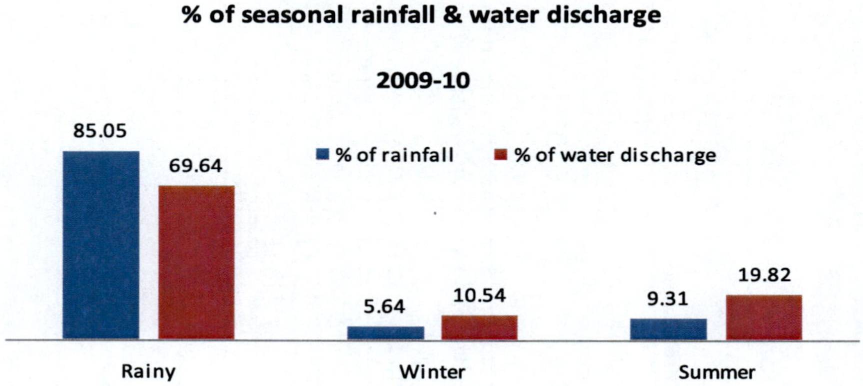

Fig. 6 Average seasonal rainfall and water discharge (July 2009–June2010)

The second-lowest month of water discharge obtained in the month of December 2009 (0.31 m^3/s), and January 2010 (0.34 m^3/s). The main reason for this low discharge in Tons watershed is the use of water for irrigation by the farmers. The discharge bars are also related directly to rainfall (Fig. 5). The seasonal water discharge and rainfall were recorded in 2009–10 (Fig. 6) in rainy season total rainfall was recorded (1076.10 mm), winter season (71.30) and the season of summer was (117.80 mm) in the year 2009–2010.

Table 4 shows that the seasonal water discharge was recorded (69.64%) in the rainy season, (10.54%) in the winter season and the summer season was recorded (19.82%) discharge out of the whole year in 2009–10. But due to water use, farmers

Table 4 Seasonal rainfall, water discharge, sediment budget and bed load of Tons watershed (2009–2010)

			Water Discharge				Sediment load in (Tons)						Bedload	Total
Season	Rainfall In (mm)	% of rainfall	m^3	%	m^3/s	Discharge (L/S)	Suspended	%	Dissolve	%	Total (D + S)	%	Total	D + S + B
Rainy	1076.10	85.05	22,249,453	69.64	2.09	2093.63	25880.35	88.59	18737.89	85.32	44618.24	87.19	24514.23	69132.47
Winter	71.30	5.64	3,367,185	10.54	0.32	324.77	1099.11	3.76	1160.47	5.28	2259.58	4.42	–	2259.58
Summer	117.80	9.31	6,334,249	19.82	0.60	600.93	2233.98	7.65	2062.25	9.39	4296.23	8.40	–	4296.23
Total	1265.20	100%	31,950,887	100%	3.02	3019.33	29213.44	100%	21960.61	100%	51174.05	100%	24514.23	75688.28

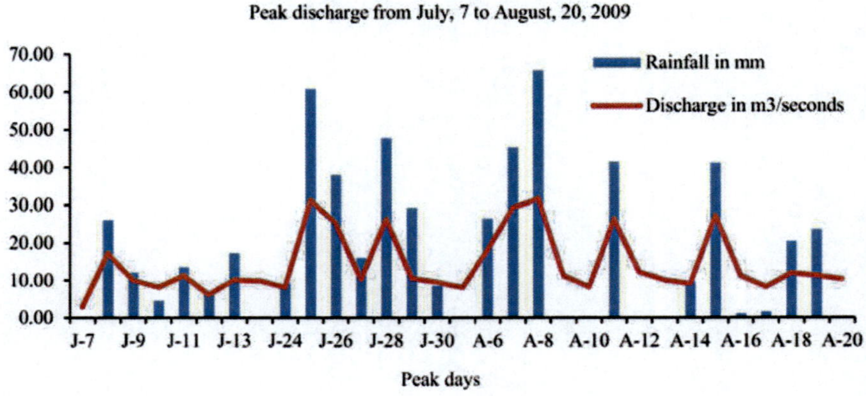

Fig. 7 Peak rainfall and water discharge (7st July, to 20thAugust, 2009)

and villagers in summer & winter seasons, rainfall percentage was recorded higher than water discharge in 2009–10 (Fig. 6).

Figure 7 shows peak discharge (0.31 m^3/s), (0.32 m^3/s) and (0.26 m^3/s) respectively corresponding to three peak floods days. And rainfall recorded on 25th July, eighth August and 11th August 2009 is, 60.80 mm, 65.70 mm and 41.40 mm respectively.

2.1.2 Third-Year Water Discharge (2010–2011)

The third-year hydrological data is tabulated in (Tables 5 and 6). It is noted the mean monthly discharge of third-year varied from the previous two years, and recorded rainfall and discharge approximately two times higher than 2008–09 and 2009–10, due to heavy rainfall and peak floods. The minimum discharge for the year 2010–2011 was recorded (0.67 m^3/s) in the months of November 2010& January 2011 and maximum (4.60 m^3/s) in the month of August 2010. The whole year's average discharge was recorded as 1.74 m^3/s. About 22.48% of annual water discharge was recorded in the month of August 2010 followed by September 2010 (20.15%).

During the winter, i.e. November–December 2010, January–February 2011 monthly water discharge was recorded very low (Tables 5 and 6). The data showed that 65.80% of water discharge was monitored in the rainy season, 13.19% in the winter season and 19.01% in the summer season of 2010–2011, which is higher in percentage in the summer season compared to previoustwo year, water discharge data, due to heavy rainfall (Fig. 8).

Water discharge data correlated with rainfall data showed that maximum rainfall and water discharge were recorded at 22.48% and 20.15% respectively in the month of August 2010& September 2010. The highest rainfall month was August 2010 recorded, at 31.32% rainfall and September 2010 was the second-highest 28.17%

Table 5 Monthly rainfall, water discharge, sediment budget and bed load of Tons watershed (2010–2011)

Month	Rainfall in (mm)	% of rainfall	Water Discharge				Sediment load in (Tons)					
			m^3	%	m^3/s	Discharge (l/s)	Suspended	%	Dissolve	%	Total	%
Jul-10	661.40	21.75	10,654,212	19.43	3.98	3977.83	15561.23	22.63	11564.24	23.74	27125.47	23.09
Aug-10	952.40	31.32	12,325,817	22.48	4.60	4601.93	22131.32	32.19	14856.33	30.50	36987.65	31.49
Sep-10	856.70	28.17	11,051,987	20.15	4.26	4263.88	19987.23	29.07	12784.24	26.24	32771.47	27.90
Oct-10	8.30	0.27	2,052,919	3.74	0.77	766.47	2639.21	3.84	2019.24	4.15	4658.45	3.97
Nov-10	15.40	0.51	1,745,191	3.18	0.67	673.30	1228.63	1.79	1054.24	2.16	2282.87	1.94
Dec-10	34.90	1.15	1,817,871	3.31	0.68	678.72	1163.61	1.69	1054.21	2.16	2217.82	1.89
Jan-11	17.50	0.58	1,795,881	3.27	0.67	670.51	545.63	0.79	487.21	1.00	1032.84	0.88
Feb-11	33.00	1.09	1,873,925	3.42	0.77	774.61	415.32	0.60	401.21	0.82	816.53	0.70
Mar-11	15.00	0.49	3,113,854	5.68	1.16	1162.58	548.62	0.80	545.32	1.12	1093.94	0.93
Apr-11	22.10	0.73	3,049,056	5.56	1.18	1176.33	621.21	0.90	787.32	1.62	1408.53	1.20
May-11	83.50	2.75	1,883,520	3.43	0.70	703.23	757.64	1.10	825.63	1.69	1583.27	1.35
Jun-11	341.00	11.21	3,475,464	6.34	1.34	1340.84	3154.63	4.59	2332.52	4.79	5487.15	4.67
Total	3041.20	100%	54,839,697	100%	20.79	20790.22	68754.28	100%	48711.71	100%	117465.99	100%
Average	253.43	–	4569974.75	–	–	1732.52	5729.52		4059.31		9788.83	23.09

Table 6 Seasonal rainfall, water discharge, sediment budget and bed load of Tons watershed (Jul 2010–Jun 2011)

			Water Discharge				Sediment load in (Tons)						Bed load	Total
Season	Rainfall in (mm)	% of rainfall	m3	%	m^3/s	Discharge (l/s)	Suspended	%	Dissolve	%	Total (D + S)	%	Total	D + S + B
Rainy	2478.8	81.51	36,084,935	65.80	3.40	3395.53	60318.99	87.73	41224.05	84.63	101543.04	86.44	58698.65	160241.69
Winter	100.8	3.31	7,232,868	13.19	0.70	697.61	3353.19	4.88	2996.87	6.15	6350.06	5.41	–	6350.06
Summer	461.6	15.18	11,521,894	21.01	1.09	1093.08	5082.10	7.39	4490.79	9.22	9572.89	8.15	–	9572.89
Total	3041.2	100%	54,839,697	100%	5.19	5186.22	68754.28	100%	48711.71	100%	117465.99	100%	58698.65	176164.64

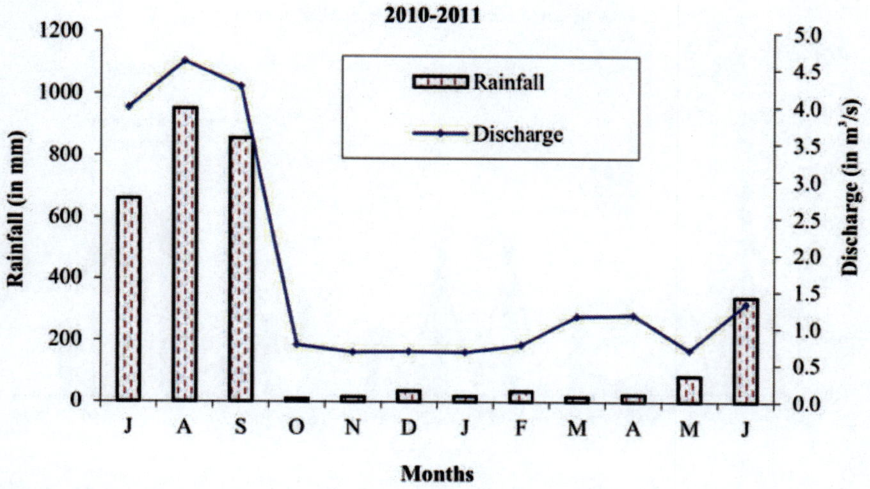

Fig. 8 Average monthly rainfall and water discharge (July 2010–June 2011)

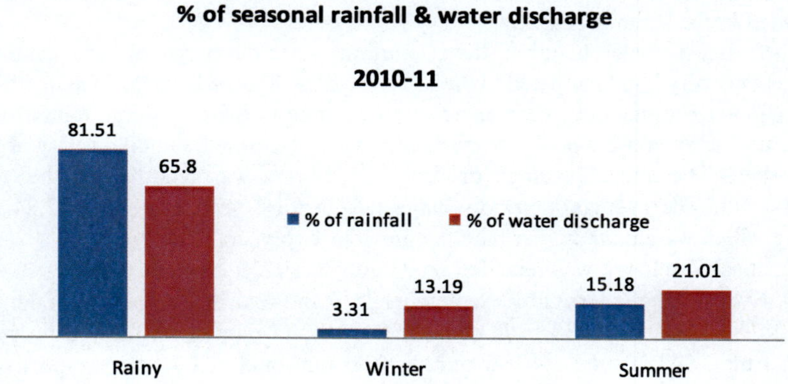

Fig. 9 Average seasonal rainfall and water discharge (2010–2011)

rainfall month in the year. In this year June 2011 month was recorded the highest rainfall compared to 2008–09 & 2009–10 and also record thehighest water discharge. The seasonal variation of data is also shown in Fig. 9.

It has been seen that 65.80% of water discharge is recorded in the rainy season from July to October in the third year. Similarly 81.51% rainfall is also recorded in the rainy season. Minimum 3.31% annual rainfall is recorded in the winter season. Water discharge and rainfall are closely related to each other, Higher the rainfall, higher the amount of water discharge observed in this year. About 65.80% of water discharge in the rainy season was generated. Only 13.19% annual water discharge is recorded in the winter season which received only 3.31% annual rainfall. Similarly

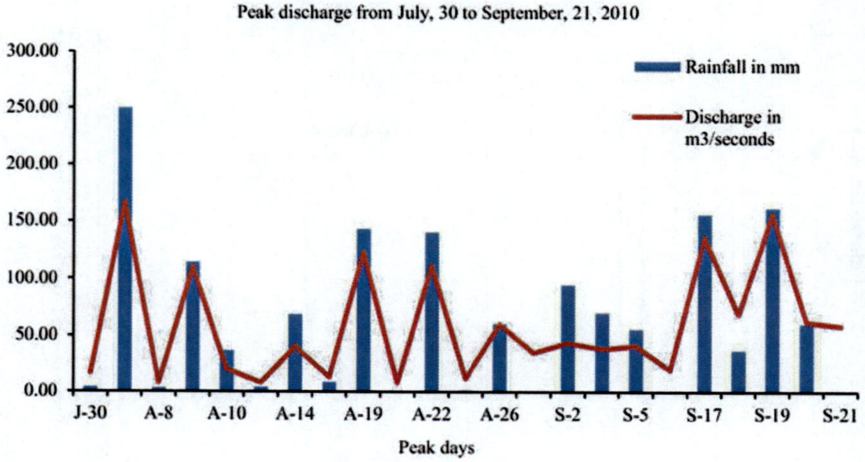

Fig. 10 Peak rainfall and water discharge (30th July, to 21st September, 2010)

81.51% rainfall is also recorded in the rainy season and 15.81% annual rainfall is recorded in the summer season.

There is a notable change in the relation of water discharge and the sediment budget was very high compared to the last two years of recorded data. Where winter and summer seasons have decreasing discharge due to the occasional rainfall and water use for swimming pools for tourists at many places in the middle valley of the watershed. The annual average discharge 1.74 m^3/s was recorded in the year 2010–2011. The average water discharge rate in third year was about 42.24 l/s/km^2/y which were quite higher than the previous two years.

The peak discharge was recorded on 31 July 2010, 19 August 2010, 22 August 2010, 17 September 2010 and 19 September 2010 and peak discharge was 0.66 m^3/s, 0.23 m^3/s, 0.13 m^3/s, 0.36 m^3/s and 10.56 m^3/s and rainfall on these days were 249.80 mm, 143.20 mm, 140.40 mm, 156.00 mm, and 161.40 mm respectively (Fig. 10). The water discharge recorded in 2010–11, was 1.72 times higher than to 2009–10 and 1.56 times to 2008–09, and rainfall was 2.40 times and 1.99 times lower respectively.

2.1.3 Average of Three Years of Water Discharge (2008–2011)

The average rainfall and water discharge data of the three years are as shown in Table 7 and represented in Fig. 11. It reveals that the three years average rainfall is 162.08 mm which shows that the study area comes under semidry (medium rainfall) region. The maximum rainfall month was August 31.01% and followed by July 26.90%. For those years the maximum water discharge was recorded in the month of August 22.72%. But the next higher discharge month was July 21.74%. The minimum percentage of water discharge is recorded in the month of February 2.91% followed by December 3.00% for 2008–2011.

Table 7 Three years average rainfall, water discharge, sediment budget and bedload of Tons watershed (2008–2011)

Month	Rainfall in (mm)	% of rainfall	Water discharge				Sediment load in (Tons)					
			m^3	%	m^3/s	Discharge (l/s)	Suspended	%	Dissolve	%	Total	%
Jan	10.33	0.53	1,317,877	3.24	0.49	492.04	334.50	0.75	333.68	1.05	668.18	0.88
February	43.17	2.22	1,183,948	2.91	0.49	489.40	252.12	0.57	252.46	0.79	504.58	0.66
March	7.40	0.38	1,913,109	4.70	0.71	714.27	283.54	0.64	282.80	0.89	566.34	0.74
April	14.03	0.72	1,880,321	4.62	0.73	725.43	292.47	0.66	347.53	1.09	640.00	0.84
May	55.63	2.86	1,838,541	4.52	0.69	686.43	426.87	0.96	523.03	1.65	949.89	1.24
June	195.47	10.05	2,554,469	6.28	0.99	985.52	2692.26	6.05	1919.14	6.04	4611.40	6.04
July	523.23	26.90	8,844,716	21.74	3.30	3302.24	11506.38	25.85	8471.47	26.65	19977.85	26.18
August	603.13	31.01	9,242,846	22.72	3.45	3450.88	13697.73	30.77	9558.03	30.07	23255.76	30.48
September	405.43	20.85	6,921,264	17.01	2.67	2670.24	11335.23	25.46	7621.80	23.98	18957.03	24.85
October	64.67	3.33	2,492,421	6.13	0.93	930.56	2293.80	5.15	1243.87	3.91	3537.68	4.64
November	10.57	0.54	1,274,711	3.13	0.49	491.79	853.48	1.92	712.73	2.24	1566.21	2.05
December	11.83	0.61	1,219,207	3.00	0.46	455.20	546.52	1.23	515.49	1.62	1062.01	1.39
Total	1944.89	100%	40,683,431	100%	15.39	15394.01	44514.92	100%	31782.03	100%	76296.95	100%
Average	162.07	–	282523.82	–	–	1282.83	3709.58		2648.50		6358.08	

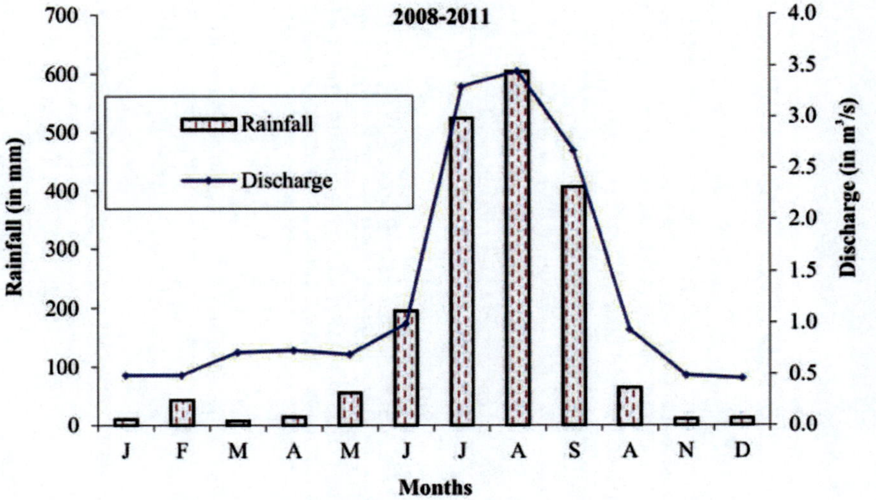

Fig. 11 Average three year's monthly rainfall and water discharge (2008–11)

The preliminary results reveal that the water generating capacity of the watershed varyaccording to the season annually. The average seasonal water discharge has been tabulated and represented in Tables and Figures. It shows that the first-year average flow of discharge was recorded 1.12 m^3/s or 118.08 l/s, second-year water discharge was 1.01 m^3/s and third-year average water discharge was 1.74 m^3/s. The third-year water discharge was 1.6 times higher than the first year and 1.7 times higher than the second year. The three years annual average flow of water discharge recorded 1.28 m^3/s or 1290.06 l/s. Out of that flow 67.60% water discharge is monitored in the rainy season, 12.28% in the winter season and 20.12% in summer season Table 8 and Fig. 12. The average seasonal water discharge data is shown in Fig. 12. It reveals that 82.08% of rainfall is recorded during the rainy season while 14.01% and 3.90% rainfall is recorded during summer and winter seasons respectively.

The seasonal discharge regime has largely remained within historicalranges, but the frequency and magnitude of water level fluctuations have increased considerably in the rainy season depending on the intensity of rainfall. The summer rain is recorded approximately four times higher than winter rain. Itis because of that the pre-monsoon storm rainfall in the month of June hasincreased the overall amount of rainfall in the summer season. The localexperiences indicate that the duration of winter rain is decreasing every coming year Fig. 12.

If the water discharge data is correlated with rainfall data then it is foundthat higher rainfall 82.08% generates higher discharge 67.60% in the rainy seasonand rainfall in winter season 3.90% generate the same discharge 4.46% Fig. 12,which is approximately same in percentage. But in the case of summer seasons therainfall percentages are not approximately the same, due to uncertainty of monsoon, water

Table 8 Three years average (seasonal) rainfall, water discharge, sediment budget and bedload of Tons watershed (2008–2011)

			Water Discharge				Sediment load in (Tons)						Bedload	Total
Season	Rainfall In (mm)	% of rainfall	m^3	%	m^3/s	Discharge (l/s)	Suspended	%	Dissolve	%	Total (D + S)	%	Total	D + S + B
Rainy	1596.46	82.08	27,501,247	67.60	2.59	2587.82	38833.15	87.24	26895.17	84.62	65728.32	86.15	36934.18	102662.50
Winter	75.90	3.90	4,995,742	12.28	0.48	481.84	1986.63	4.46	1814.37	5.71	3800.99	4.98	–	3800.99
Summer	272.53	14.01	8,186,441	20.12	0.78	776.64	3695.14	8.30	3072.49	9.67	6767.63	8.87	–	6767.63
Total	1944.89	100%	40,683,431	100%	3.85	3846.30	44514.92	100%	31782.03	100%	76296.94	100%	36934.18	113231.12

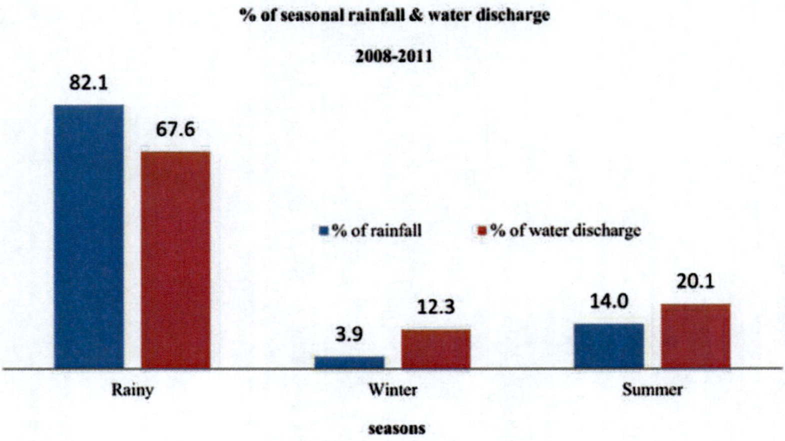

Fig. 12 Average seasonal rainfall and water discharge (2008–11)

Table 9 Rate of water discharge, yearly and average

Year	2008–09	2009–10	2010–11	Average of (three years)
Total discharge in m^3	35,259,708	31,950,887	54,839,697	40,683,431
Water Discharge m^3/s	1.11	1.01	1.73	1.28
Rate $m^3/s/km^2/y$	0.03	0.02	0.04	0.03

used for irrigation and construction purpose, weather conditions, slope and anthropogenic activities in the basin. Thus, after analyzing three years'water discharge and rainfall data of the Tons watershed. It can be concludedthat 1.28 m^3/s water discharges is generated by 162.07 mm rainfall (per month) in the watershed.

2.1.4 Rate of Water Discharge

A correlation of the water discharge for three years in the Tons watershedcan be seen in Table 9. If the three years the annual average rate of waterdischarge in the catchment was 1.28 m^3/s. The highest water discharge rate wascalculated in third year 1.73 m^3/s and the lowest was found in the second year (1.01 m^3/s). While the first year discharge was 1.12 m^3/s, the rate of waterdischarge in $m^3/s/km^2/y$ is tabulated in Table 9.

2.2 Water Balance

Water plays a significant role in natural resources & ecosystems. It is the initial medium for dissolved and suspended solids, and it determinesthe rate at which these

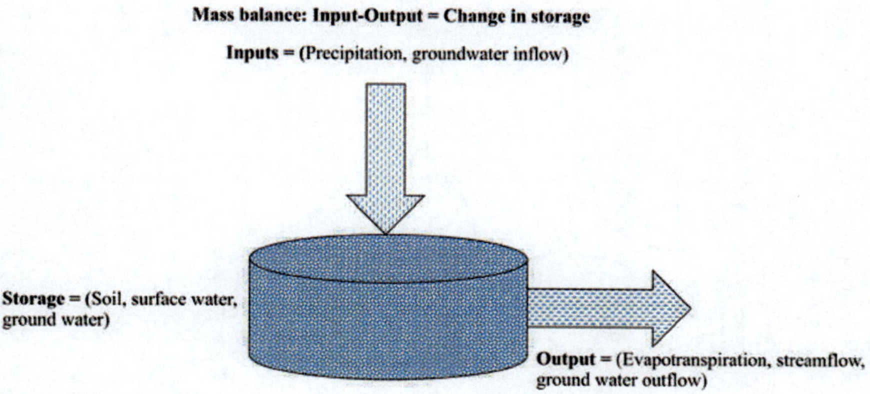

Fig. 13 At a specific duration of time and at a reference temperature and density, from the continuity equation change of storage within a specific command area can be determined as (*e.g.1*)

solids are removed from the system, traditionally defined as the output flux. If one is to gain an understanding of the biological, chemical and physical processes that operate within aphysical system, a detailedunderstanding of the hydrological characteristics is needed.

The catchment, or drainage basin, is the basic and comprehensive unit of study in hydrology, because it represents an area with an easily definable topographic boundary, which as a first approximation, also defines the watershed boundary we canunderstand in simple way to see the water balance example 1 & 2 in (Figs. 13 and 14). Water balance or budget is not an easy task, for a lake, pond, ocean, andwatershed, and not to calculate exactly inflow and outflow of the water; withinan area.

Water balance models were inducted to measure the importance of differenthydrologic components under a variety of hydrologic conditions but itspresentapplications are the most common studies in the field of water resources management. In spite of the simple concept of water balance equation, specific considerationsare need to proper application. With numerous affecting factors on hydrologicprocesses, the caution characteristic of water balance equation.

Figure 13 shows the water Balance sketch, inputs variables are precipitation, ground water and moisture in the soil, where as the output for the water balance are evapotranspiration and stream flow and groundwater flow (*e.g.1*).

2.2.1 Water Balance in Watersheds

Long-term water storage changes in watersheds, including surface waterand ground water, are expressed in the form of residuals (accumulated orscattered water) in water balance equation (snow and ice amounts can beremoved) (Berezovskaya et al. 2005) (Eq. 2):

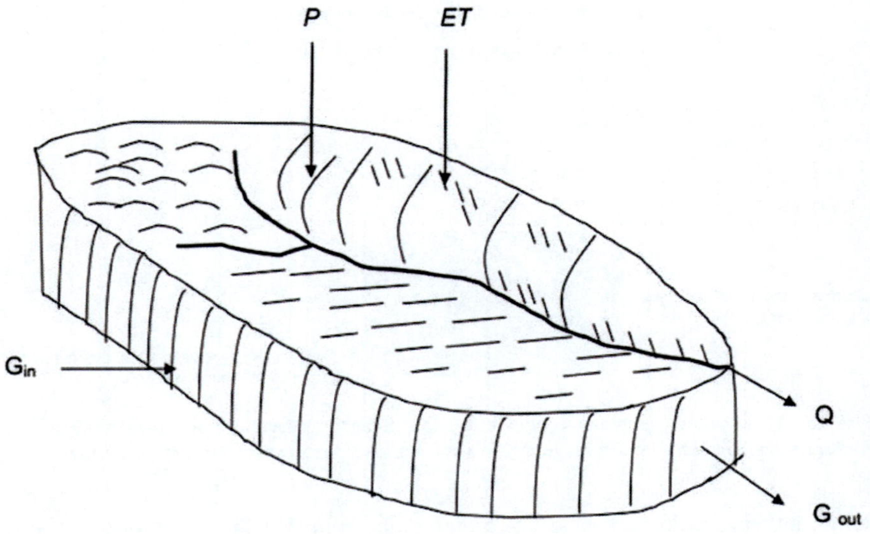

Where:
- P = Precipitation
- ET = Evapotranspiration
- Q = Stream outflow
- G_{in} = Ground water in
- G_{out} = Ground water out

Fig. 14 Water Balance (*e.g.2*) (Modified after Dingman, 2015, physical hydrology)

$$\frac{dS}{dt} = \text{P-Q-ET} \tag{2}$$

Where, dS/dt is total water change in the watershed, P is average precipitation, ETis evapotranspiration and Q is the surface water discharge at the main drain of thebasin. This simple expression of water balance is valid where the groundwateroutput and its withdrawals are negligible. Classification of watersheds accordingto homogeneity characteristics (single or multiparametric methods), climateconditions and physiographical situations (closed or open watersheds), can beevaluated as a basis for using the same equations for similar catchments (comparison methods). In closed watersheds, usually, the level or volume of thelakes or reservoirs is a controlling factor for evaluating water balance equations [24]. So in the present study of water balance in Tonswatershed was evaluated on this method. Efforts have been made for themeasure the water balance, on the basis of available literature and referencesespecially on [21].

2.2.2 Parameters Required for Water Balance

Precipitation

The precipitation and streamflow were measured routinely in the Tonswatershed in mm and cubic meter per second or liter per second. The threeyears (2008–2011) average precipitation was measured as 1544.90 mm/y andyearly were measured 1528.30 mm/y in the first year (2008–09), 1265.20 mm/y in thesecond year (2009–10) and third-year (2010–11) was 2041.20 mm/y precipitationin the basin. And Table 9 and streamflow water discharge 1.28 m^3/s werecalculated for water balance studies. Precipitation data are more widely available than streamflow data and are less affected by land-use changes. Precipitation is the main input to the land-based component of the hydrologiccycle. Accurate calculation of precipitation is an essential input to the waterbalance, and careful attention must be paid to two areas in particular, so wehave correlated rainfall data of two metrological stations established near orwithin the watershed, one is nearby Gucchupani i.e. Forest Research Institute, Dehradun station and another at Mussoorie within the watershed area, accuracy of point measurements of precipitation over the watershed area.

Stream Outflow

After converted streamflow to depth (or water yield) by dividing thevolume by the watershed area and converted in metric units from (m^3/s) to (mm/y). This approach using water yield expresses inputs and outputs per unitarea and allows for direct comparison of flows across Tons watershed. Stream flow (Q) was measured in 2008–09 as 857.91 mm/y, in 2009–10 as 773.65 mm/y and maximum in the year 2010–11 as 1332.83 mm/y with three years an average ofstream discharge was measured 988.13 mm/y.

Evaporation and Evapotranspiration

Evaporation/evapotranspiration is a collective term for all the processesby which water in the liquid or solid phase, at or near the earth's surfacebecomes atmospheric water vapor. "The term includes evaporation of liquidwater from rivers and lakes, bare soil and vegetative cover. Evapotranspiration is the loss of water from soil and plant surfaces. The later isone of the least understood aspects of the hydrologic cycle" [5]. The approach also used to calculate Evapotranspiration yearly andaverage basis in present study, ET was measured to be 670.39 mm/y in 2008–09, 491.55 mm/y in 2009–10 and 1708.37 mm/y in 2010–11. The average of threeyears 2008–11 ET was measured to be 956.77 mm/y (Table 10).

Table 10 Simplified water budget on Tons watershed

Year	P In (mm/y)	Q In (mm/y)	ET In (mm/y)	Q/P
2008–09	1528.30	857.91	670.39	0.56
2009–10	1265.20	773.65	491.55	0.39
2010–11	3041.20	1332.83	1708.37	0.56
Average	1944.90	988.13	956.77	0.50

Ground-Water Inflow & Ground-Water Outflow (Gin, Gout)

Long term difference between groundwater inputs and outputs are smallcompared to other terms, and in such condition the water balance can besimplified by assuming the difference between Gin and Gout is essentially zero (Physical hydrology) [5, 24]. In the present study in long term waterbalance that starts and ends at the same time of the year, the net change instorage is often negligible compound to other terms, and the same approachwas used and applied the two assumptions in Tons watershed, the water balanceequation (Eq. 3) simplified:

$$\text{P-(ET-Q)} = 0 \tag{3}$$

Water storage in the mountainous basin including the soil moisture anddeep seepage is practically unmeasurable component of the water balance equation. Different orientation and steepness of the slopes created specificconditions for vegetation growth and consequently contributed to the development ofthe characteristic soil profiles in the mountainous areas [25]. Thewater storage practically depends on a short-term recharge only, although thelong-term changes require much better understanding of the particularhydrogeological structure of the watershed and are usually neglected. The studyconcluded that water balance components show their significance and at thesame time close mutual relationships. Error in the measured terms ofprecipitation (P) and water discharge (Q) in unmeasured term, and the doublecorrection factor of the evapotranspiration (ET) and depends on the sunshineintensity and duration, for the correct calculation (error-free) of water balance, hydrologist to be proved to correct of the component i.e. total runoff, waterstorage changes, and actual evapotranspiration.

3 Conclusion

A sediment budget and rates of denudation for the middle order streamsand fourth order streams of the Tons watershed of Lesser Himalaya shows thataverage sediment transported is about 113231.12 ton every year. The averagerate of erosion is 1.02 mm/y. Out of the total sediment budget, 32.62% bed loadwas measured, which was the most contribution in denudation for load type and 39.31% suspended load and only 28.07% was dissolved load generates thedenudation rate in Tons watershed. The generation and flow of sediment load in theHimalaya region may be the result of rainfall variability, land use, slopes, naturalhazards, neotectonics and anthropogenic activities within thecatchment. Degradation may be the result of enhanced sediment delivery to thestreams in the Himalaya.

The relationship between land use, erosion and sedimentation is yet notclear, despite many decades of research. The only substantive progress madesince the 1980s is the identification of the high Himalaya as the likely dominantsource of sediment. The present study shows that more than 1/3 of bed load andsuspended load is produced by anthropogenic activities but the contribution ofthis source to stream load is unclear [26]. Although three years denudation and sedimentation results are givenbut this short period and effort made for this study is not sufficient to reach thefinal conclusions. This approach using water yield expresses inputs and outputs per unit area and allows for direct comparison of flows across Tons watershed. Stream flow (Q) was measured in 2008–09 as 857.91 mm/y, in 2009–10 as 773.65 mm/yand maximum in the year 2010–11 as 1332.83 mm/y with three years an average of stream discharge was measured 988.13 mm/y.The study area is very sensitive and mountainous andlandslides occurred often. So research work needs to be done by moderntechniques and instruments with ten years and above database. The studyneeds a holistic approach with Geomorphological studies combined withhydrological aspects over periods of at least ten years.

Estimated water balance values were quantified from available metrological parameters in the Tons watershed. Three years average value of water balance was quantified 0.50, whereas minimum value in (2009–10) was estimated 0.39 and maximum and same values in (2008–09 and 2010–11) were estimated 0.56. These values show that the positive trend of water balances in the Tons watershed. The results indicate that the study area has received the sufficient amount of preciptaion and have good recharge zones in the study area.

4 Recommendations

Present study recommends that the results will be worthwhile in planning toreduce the inflow of silt into the reservoir, conserve soil and minimize run-off in the extreme rainfallyears. Results suggest to planners and engineers for formulating strategies to control erosion and landslides for watershed management on precursory basis. The outcome of this study has proved to recommend that useful in understanding the geographic conditionsand hydrological behavior (discharge and sediment flux) as well as the water balance and availability of the recharge zone in of the basin in the context of setting up small hydropower plants in the region. The studyalso recommends that agency should act as a valuable reference for hydropowerstation management and governmental oversight.

References

1. Asl-Rousta B, Mousavi SJ, Ehtiat M (2018) SWAT-based hydrological modelling using model selection criteria. Water Resour Manag 32:2181–2197
2. Carson B (1985) Erosion and sedimentation processes in The Nepalese Himalaya. Occasional Paper No. 1. Kathmandu, Nepal, International Centre for Integrated Mountain Development
3. Dinpashoh Y, Jahanbakhshasl S, Rasouli AA, Foroughi M, Singh VP (2019) Impact of climate change onpotential evapotranspiration (case study: west and NW of Iran). Theor Appl Climatol 136:185–201
4. Petts GE (1996) Water allocation to protect river ecosystems. Regul Rivers: Res Manage 12: 353–365
5. Bosch JR, Hewlett JD (1982) A review of catchment experiments to determine the effect of vegetation change on water yield and evatranspiration. J Hydrol 254:124–144
6. Singh S (1978) Physiographic regions landforms and erosion surface of The Ranchi Plateau. National Geographer 13(1):43–65
7. Wu J, Miao C, Zhang X, Yang T, Duan Q (2017) Detecting the quantitative hydrological response to changes in climate and human activities. Sci Total Environ 586:328–337
8. Cressey GB (1963) Asian land and peoples, 3rd edn. McGraw Hill, New York
9. Brunjnzeel LA (1996) Predicting the hydrological impacts of land cover transformation in the humid tropics: the need for integrated research. In: Gosh JHC, Nobre CA, Rovert JM, Victoria RL (eds) Amazonian reforestation and climate. Institute of Hydrology, Wallingford, pp 17–55
10. Ghandhari A, Ziaii (2009) Lakes the best indicators for climate phenomena. Research Project. Azad University Branch of Torbat Heidarie (In Persian)
11. Gregory KG, Walling DE (1979) Drainage basin form and processes: a geomorphological approach. Edward Arnold, London, p 200
12. Molnar L et al (1990) Problems of the water balance components determination in a mountainous watershed. IAHS Publ. no. 190
13. Rawat JS, Geeta R, Rai SP (2000) Impact of human activities on geomorphic processes in the Almora Region, Central Himalaya, India, Geomorphology, human activity and global environmental change. Wiley, New York, pp. 285–299
14. Clark C (1987) Deforestation and floods. Environ Conserv 14:67–69
15. Yan R, Zhang XP, Yan SJ, Zhang JJ, Chen H (2017) Spatial patterns of hydrological responses to land use/cover change in a catchment on the Loess Plateau, China. Ecol Indic
16. Sahin V, Hall MJ (1996) The effect of afforestation and deforestation water yields. J Hydrol 178(1–4):293–309
17. Hofer T (1993) Himalayan deforestation, changing river discharge and increasing floods: myth or reality? Mt Res Dev 13:213–233
18. Gentry AH, Lopez-Parodi J (1980) Deforestation and flooding of the upper Amazon. Science 210:1354–1356
19. Lamparter G, Nobrega R, Kovacs K, Amorim RS, Gerold G (2018) Modelling hydrological impacts of agricultural expansion in two macro catchments in Southern Amazonia. Brazil Reg Environ Chang 18:91103
20. Rawat MS (1996) Surface hydrologic responses of a pine forested micro watershed, Kumaun Lesser Himalaya. Final Technical Report Submitted to CSIR, Dept. of Geograpy, Kumuan University, Almora
21. Milliman JD, Meade RH (1983) World-wide delivery of river sediments to the oceans. J Geol 91:l–21
22. Rawat JS, Rawat MS (1994) Accelerated erosion and denudation in the Nana Kosi Watershed, Central Himalayan, India. Part I: Sediment load. Mt Res Dev 14(1):25–38

23. Mallakpour I, Villarini GA (2016) simulation study to examine the sensitivity of the Pettitt test to detect abruptchanges in mean. Int Assoc Sci Hydrol Bull 61:245–254
24. Datt D (1991) Land system landuse and natural hazard in the lower bino basin (Lesser Himalaya) India. Mt Res Dev 11(3)
25. Hamilton LS (1987) What are The Impacts of Himalayan Deforestation on the Ganges Bharmputra lowlands and Delta? Assumption and facts. Mt Res Dev 7:256–263
26. Unni KS (2003) Ecology of Indian River, International Book Distributors, Dehradun, 9/3 Rajpur Road, pp 2–55

Water Quality Status of the Narmada River Across Its Basin: Predicting Water Quality Using Artificial Neural Network

Satanand Mishra and Rahi Jain

Abstract Rivers play a critical role in the functioning of any ecosystem. Narmada is a major perennial river of India, which drains through the Deccan plateau and is used for bathing, drinking, irrigation, and industrial purposes. Its utilization had a detrimental effect on its water quality and is a cause of concern. However, data constraints could make it difficult to determine the water quality and take appropriate water quality management measures. This study is performed to evaluate the Artificial Neural Network (ANN) model for predicting Narmada river water quality. The study involved secondary data collection from different river water quality monitoring stations across the basin followed with the data analysis of data. The study used four different types of parameters namely physical, chemical, biological and toxic compounds. The study performed three stages of analysis. In the first stage, the analysis was performed to determine water quality parameters that are not meeting the recommended water quality across different water quality stations. The second stage of the study involved cluster analysis of the parameters to identify the correlated parameters. The third stage of the study involved the use of ANN to predict the water quality status based on other data. The study can predict the Chemical Oxygen Demand (COD) category i.e., high COD level and low COD level with 85.71% accuracy. The study showcases the ability of the ANN to perform the prediction modeling for Narmada River.

Keywords Artificial neural network · Narmada River · Water quality prediction · Chemical oxygen demand

S. Mishra (✉)
Advanced Materials and Processes Research Institute (AMPRI), Bhopal, India

R. Jain
Centre for Technology Alternatives for Rural Areas (CTARA), Indian Institute of Technology Bombay (IITB), Mumbai, India

S. Yadav et al. (eds.), *Water Quality, Assessment and Management in India*, Earth and Environmental Sciences Library, https://doi.org/10.1007/978-3-030-95687-5_8

1 Introduction

Surface water is an important freshwater source, which is affected by water pollution, especially in developing countries. A strong water quality management system plays a critical role in pollution control and river basin planning. Good management requires water quality data for planning and decision-making [1, 2]. However, in data constraint countries like India, the water quality management is a challenge owing to the lack of water quality data for many rivers and its tributaries [3]. Thus, there is a need to develop new mechanisms that could help in predicting the water quality status based on the limited data environment.

Machine Learning is a well-developed approach, which is known to play a role in predicting water quality parameters [4, 5]. Machine learning incorporates the use of various techniques such as tree-based approaches (decision tree, random forest), perceptron based approaches (Artificial Neural Network), statistical-based approaches (Bayesian Networks), instance-based learning (k-Nearest Neighbours) and support vector machines [6]. For example, Dissolved Oxygen levels are measured using Artificial Neural Network for River Yamuna [7].

In India, Narmada is one of the rivers for which water quality prediction model is not available. Narmada is the major perennial river of central India that receives huge quantities of waste from domestic and municipal sewage discharge, industrial effluents and agricultural run-off. The rapid urbanization and industrialization will only increase the pollution rate of Narmada [8]. This makes the need to model and predict the water quality of the river increasingly important. Accordingly, the current study focuses on assessing and predicting the water quality of Narmada River using Artificial Neural Network (ANN).

2 Artificial Neural Networks (ANNs)

Artificial Neural Network (ANN) is a machine learning technique, which is inspired by the brain and nervous system of biological organisms. Similar to the nervous system in the biological systems. The basic unit of ANN is a neuron. There are three types of neurons, namely input neuron, hidden neuron, and output neuron (Fig. 1). The input neurons take the data and feed into the ANN. This input data is processed at the hidden neurons and the processed data is sent to output neuron(s) as the outcome of ANN. The hidden neurons can exist in single or multiple layers. In the case of multiple layers, the processed output of one hidden layer acts as the input for the next hidden layer [7].

In the processing step at the hidden neuron, two actions are performed. The first action is to assign a weight to each input neuron. At each hidden neuron, an output is created as weighted sum of individual neurons. This output is the linear combination of input neurons. However, many complex problems could not be solved using only the linear transformation of the data. The second action is using an activation

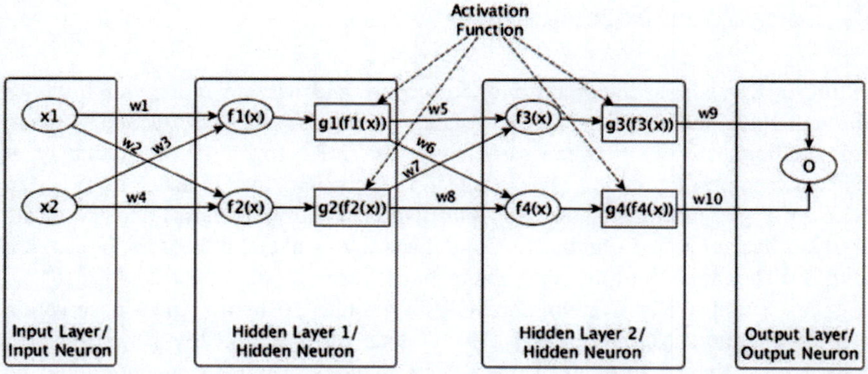

Fig. 1 Artificial Neural Network Structure for multi-layer. x1, x2 represents input neurons containing individual parameter data, w1-w10 are the weights assigned to different input neurons, f(x) represent linear equation for each hidden layer, g(f(x)) represents the output of hidden layer after activation function based transformation and o represent the output neuron containing final output

function, which helps in transforming the linear form of hidden neuron output into the non-linear form (Fig. 1) [6]. Some of the commonly used activation functions are Hyberbolic tangent function (TanH), Sigmoid or Logistic function, and Piecewise linear activation units (PLU) [9, 10].

The ANN is a supervised machine learning technique. It needs to be trained on data with known output before it can be used to predict the output of unknown data. In the training process, it tries to identify the most appropriate weights for each neuron such that the difference between the output generated from ANN and the actual output from the data is minimum. ANN training is an iterative process, once the output is generated with a given set of weights, the weights are altered and ANN training is done to generate new output. The process of altering weights continues till the difference between ANN output and actual output from data is reduced below a threshold level [6]. Different approaches are used to alter the weights in the ANN training. Some of them are ART [11], ADALINE [12], Rosenblatt's perceptron [13], Hopfield's network [14], Error-back propagation [15] and Self-organizing network [16].

Once the ANN is trained, it is tested with a dataset that is not used to train ANN (unseen dataset) to determine its prediction performance. If the prediction performance is just above the threshold, the model is used for the intended use. The advantage of using ANN for problem-solving are: (1) no need to have prior knowledge of relationships between various aspects under investigation, (2) unlike, standard optimization or statistical tool that requires complete information to give optimal solution, ANN always converges to provide an optimal (or, sub-optimal) solution, and (3) constraints and prior solution is neither strictly enforced nor always assumed in the ANN development [7].

3 Study Area: Narmada River

Narmada River is an important river for central and western India, which passes through three Indian states namely Madhya Pradesh, Maharashtra and Gujarat. Longitudinally, it lies between east longitudes 72°32′ to 81°45′. Latitudinally, it lies between north latitudes 21°20′ to 23°45′. It covers a total distance of 1312 km with an origin at Amarkantak (Annupur district, Madhya Pradesh) and enters the Arabian Sea at Gulf of Cambay in Gujarat and drains a total area of 98,796 sq. km (Fig. 2). The majority of the river and its basin lies in Madhya Pradesh [17, 18].

The Central Water Commission (CWC) through 26 hydrological observation stations monitors Narmada River [19]. A total 35 water quality parameters are measured at these hydrological observation stations. These parameters could be classified into physical, chemical, biological, trace, and toxic chemicals and chemical indices. The parameters measured under each category are given in Table 1.

4 Methodology

The methodology adopted for this study is shown in Fig. 3. The data on the water quality parameter is collected for two hydrological monitoring stations such that one is at downstream and another one is at upstream of the Narmada river basin. Accordingly, Dindori and Mandaleshwar sites are selected. The data for different years regarding water quality parameters are collected for both sites. The water quality parameters used in the study are mentioned in Table 1. The data is collected from the Central Water Commission. The data is available for financial years

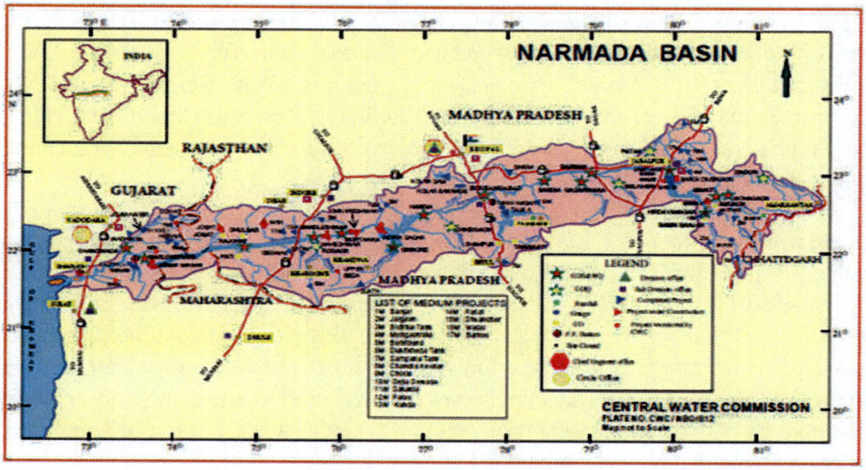

Fig. 2 Map of Narmada River Basin (Source: Central Water Commission)

Table 1 Water Quality parameters commonly measured at hydrological monitoring stations

Code	Parameter Category	Parameter Name	Code	Parameter Category	Parameter Name
1	Physical	Q (cumec)	18	Chemical	Alk-Phen (mgCaCO3/L)[a]
2		EC_GEN (Âµmho/cm)[a]	19		ALK-TOT (mgCaCO3/L)[a]
3		pH_FLD (pH units)	20		B (mg/L)
4		pH_GEN (pH units)[a]	21		Ca (mg/L)[a]
5		TDS (mg/L)[a]	22		Cl (mg/L)[a]
6		Temp (deg C)[a]	23		CO3 (mg/L)[a]
7		Turb (NTU)[a]	24		F (mg/L)[a]
8	Biological	BOD3–27 (mg/L)	25		Fe (mg/L)
9		COD (mg/L)[a]	26		HCO3 (mg/L)[a]
10		DO (mg/L)	27		K (mg/L)[a]
11		DO_SAT% (%)	28		Mg (mg/L)[a]
12	Trace and toxic	Al (mg/L)	29		Na (mg/L)[a]
13	Chemical indices	HAR_Ca (mgCaCO3/L)[a]	30		NH3-N (mg N/L)
14		HAR_Total (mgCaCO3/L)[a]	31		NO2 + NO3 (mg N/L)
15		Na% (%)[a]	32		NO2-N (mg N/L)[a]
16		RSC (–)[a]	33		NO3-N (mgN/L)[a]
17		SAR (–)[a]	34		o-PO4-P (mg P/L)[a]
			35		SiO2 (mg/L)[a]
			36		SO4 (mg/L)[a]

[a]These parameters were finally used for analysis, as data were available for all years

1993–95, 1997–98, 2001–03 and 2004–13. The parameter for which data is not available is removed from further study. Accordingly, the number of parameters reduced from 36 to 26. The 26 parameters used in the study are given in Table 1 and its values are given in Table 2.

4.1 Evaluating the Water Quality Parameters

The water quality parameters were evaluated to determine the water suitability of the Narmada River. The parameter values are compared with the standard water quality values given in the literature. The threshold values for different parameters are obtained from different sources (Table 2) [20–24]. The threshold value for some parameters is not available.

The second preliminary evaluation is done to perform correlational hierarchical clustering. The Pearson correlation coefficient between all the water quality parameters except the COD parameter is estimated. The correlation coefficient matrix

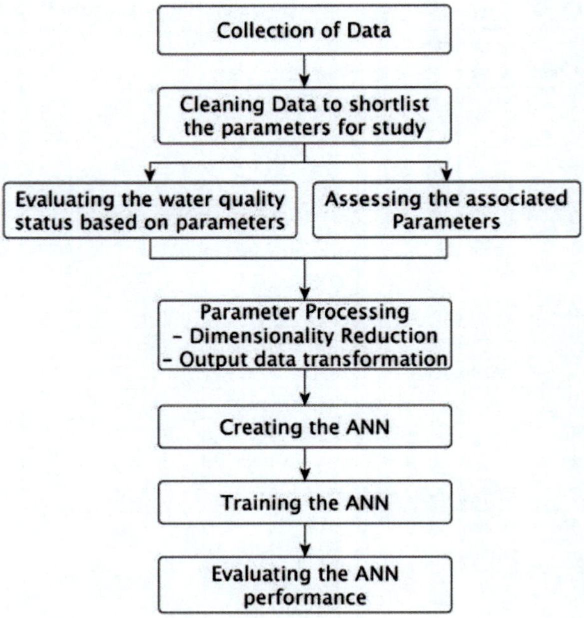

Fig. 3 Methodology adopted in the study

obtained is used to determine the hierarchical clustering of the parameters to identify which water quality parameters are more closely associated with each other. The agglomerative hierarchical clustering is performed [25].

4.2 Predicting the COD

Chemical Oxygen Demand (COD) is an important parameter that determines water quality. Among the 26 water quality parameters, the study used COD as the output parameter, which needs to be predicted from data of the other 25 water quality parameters. In order to predict the COD, the ANN technique is used. The COD data is continuous in nature, which is divided into two categories, namely high COD level and Low COD level. This division is performed using K-mean clustering [26].

The data is randomly divided into training and test set in the ratio of 3:1. The training dataset is used to train and develop the ANN model. During the development of the ANN model, its various hyper-parameters like the number of hidden neurons per layer, activation function are optimized (Appendix A). The performance of the ANN model is evaluated using its ability to predict the COD values of the test data set. The metric used to evaluate the performance of the ANN model is 'Accuracy', which measures the number of instances accurately measured [27].

ANN performance is compared with the baseline model known as Zero Rule (ZeroR) classifier [28]. In this baseline model, the classifier only focuses on the

Table 2 Threshold values used for evaluating the water qaulity at two stations

Parameterers Code			2	4	5	6	7	18	19	21	22	23	24	26	27
Acceptable limit			1000	6.5–8.5	500	None	1	None	None	75	250	None	1	None	None
#	Location	Year	Conductivity	pH	TDS	Temp	Turb	Alk_Phe	Alk_Tot	Ca	Cl	CO3	F	HCO3	K
1	**1**	1	290	8.3	192	23.5	58.8	5.5	147	31	7.0	6.6	0.2	166.0	1.9
2	**1**	2	268	8.4	188	20.7	45.8	7.4	123	27	5.4	9.0	0.2	132.0	1.6
3	**1**	5	274	8.2	165	22.0	81.3	5.4	119	29	6.6	6.5	0.2	132.0	1.1
4	**1**	9	258	8.2	170	21.3	32.6	8.0	151	28	6.2	7.2	0.2	166.0	1.3
5	**1**	10	253	8.2	159	22.0	37.6	6.0	130	26	7.0	7.9	0.2	149.0	1.8
6	**1**	12	267	8.2	172	22.5	3.7	4.3	192	29	8.2	5.2	0.2	219.0	1.0
7	**1**	13	269	8.2	166	23.7	33.2	3.0	162	30	8.6	3.6	0.4	153.0	1.8
8	**1**	14	244	8.2	156	23.8	81.7	2.6	295	29	10.6	3.1	0.5	177.0	2.0
9	**1**	15	292	8.2	193	21.7	38.6	5.2	329	27	11.3	6.2	0.4	213.0	1.9
10	**1**	16	287	8.0	185	22.7	32.0	1.9	154	27	12.8	2.1	0.4	178.0	2.1
11	**1**	17	263	7.9	164	23.2	25.9	0.0	164	30	12.8	0.0	0.3	200.0	2.3
12	**1**	18	287	7.7	178	22.0	38.5	0.6	143	26	10.6	0.7	0.3	173.0	1.4
13	**1**	19	259	8.1	164	22.2	50.5	1.5	151	29	9.7	1.8	0.3	181.0	1.4
14	**1**	20	283	7.9	166	26.2	17.5	0.8	133	27	8.8	1.0	0.2	160.0	1.8
15	**2**	1	292	8.4	191	27.5	11.4	8.3	157	27	6.7	9.9	0.3	171.0	2.4
16	**2**	2	302	8.4	196	25.6	22.3	7.5	140	30	7.1	9.0	0.2	153.0	2.1
17	**2**	5	291	8.4	182	26.4	35.0	11.2	142	30	9.2	13.5	0.2	146.0	1.3
18	**2**	9	250	8.3	177	25.3	24.2	8.7	148	25	7.0	11.3	0.2	162.0	2.1
19	**2**	10	251	8.4	163	26.4	47.5	11.7	143	28	7.4	13.9	0.3	149.0	2.2
20	**2**	12	264	8.3	171	26.1	0.60	5.7	189	31	8.3	6.8	0.3	217.0	1.5
21	**2**	13	254	8.2	158	25.9	18.4	1.9	192	28	8.1	2.3	0.4	153.0	1.9
22	**2**	14	281	8.2	172	26.5	24.8	3.4	292	30	12.1	4.2	0.4	188.0	2.1
23	**2**	15	290	8.3	188	26.1	4.7	4.6	172	28	11.1	5.5	0.4	199.0	2.3
24	**2**	16	273	8.2	178	25.3	5.2	2.2	138	27	9.3	2.7	0.3	163.0	2.1
25	**2**	17	239	8.2	149	25.8	24.7	2.2	137	27	8.5	2.7	0.3	162.0	1.8
26	**2**	18	288	8.2	176	26.1	5.7	1.2	146	28	9.6	1.4	0.4	175.0	2.0
27	**2**	19	251	8.2	159	25.2	9.9	0.4	148	27	8.8	0.5	0.2	180.0	1.9
28	**2**	20	259	8.1	162	27.3	15.0	1.7	125	26	8.3	2.0	0.2	149.0	2.5

(continued)

Table 2 (continued)

Parameterers Code			28	29	32	33	34	35	36	13	14	15	16	17	9
Acceptable limit			100	200	0.5	1	None	None	250	None	200	60	None	26	3
#	Location	Year	Mg	Na	NO2_N	NO3_N	PO4	SIO2	SO4	HAR_Ca	HAR_T	Na_Percent	RSC	SAR	COD
1	1	1	9.8	7.2	0.01	0.33	0.021	19.4	11.7	77	118	11	0.6	0.3	39.3
2	**1**	2	10.5	10.3	0.01	0.35	0.022	12.5	7.8	68	112	16	0.3	0.4	23.6
3	**1**	5	8.4	9.2	0.01	0.24	0.017	24.6	4.4	73	108	15	0.2	0.4	22.5
4	**1**	9	12.5	7.9	0.03	0.26	0.025	21.2	4.6	69	121	12	0.6	0.3	28.2
5	**1**	10	14.3	8.1	0.09	0.45	0.057	23.1	9.1	66	126	12	0.2	0.3	26.6
6	**1**	12	14.4	8.3	0.01	0.19	0.014	24.5	7.7	74	134	12	1.1	0.3	35.6
7	**1**	13	13.8	8.1	0.04	1.43	0.126	27.2	5.1	74	131	12	0.2	0.3	37.6
8	**1**	14	10.0	7.8	0.08	1.09	0.086	27.2	7.5	72	113	13	0.8	0.3	30.5
9	**1**	15	11.3	8.0	0.07	1.29	0.077	26.6	7.0	67	114	13	1.4	0.3	29.9
10	**1**	16	13.6	9.8	0.08	1.03	0.057	25.7	6.0	68	125	14	0.5	0.4	31.7
11	**1**	17	12.7	11.9	0.12	0.6	0.147	27.9	5.1	76	129	17	0.8	0.5	39.2
12	**1**	18	12.4	11.0	0.03	0.28	0.245	26.2	21.9	66	117	17	0.5	0.4	29.3
13	**1**	19	12.7	8.3	0.06	0.81	0.074	25.3	4.9	74	127	12	0.5	0.3	31.0
14	**1**	20	14.3	10.4	0.16	0.70	0.012	26.5	8.9	67	127	15	0.2	0.4	24.7
15	**2**	1	10.8	14.8	0.01	0.44	0.032	15.6	7.3	67	111	22	0.9	0.6	42.2
16	**2**	2	12.8	16.0	0.01	0.39	0.016	14.9	5.9	75	129	21	0.3	0.6	25.2
17	**2**	5	10.2	13.7	0.01	0.46	0.013	25.3	4.2	74	117	20	0.5	0.5	20.2
18	**2**	9	12.7	14.1	0.03	0.34	0.038	20.5	5.7	64	117	20	0.7	0.6	26.0
19	**2**	10	14.5	10.8	0.01	0.39	0.015	21.1	7.2	69	130	15	0.3	0.4	26.5
20	**2**	12	11.0	10.5	0.01	0.27	0.011	23.2	10.9	76	122	16	1.4	0.4	37.4
21	**2**	13	8.6	8.7	0.02	1.42	0.205	23.8	5.6	69	105	15	0.5	0.4	37.8
22	**2**	14	10.4	10.4	0.11	1.55	0.052	23.7	5.8	76	119	16	0.8	0.4	33.0
23	**2**	15	7.9	8.7	0.14	1.28	0.060	26.7	8.5	70	103	15	1.4	0.4	34.3
24	**2**	16	9.3	9.5	0.00	0.68	0.029	22.0	10.2	67	106	16	0.7	0.4	36.1
25	**2**	17	9.1	9.0	0.02	0.66	0.142	29.1	9.5	67	105	16	0.7	0.4	30.6
26	**2**	18	10.6	14.9	0.03	1.13	0.276	29.4	11.7	70	114	21	0.6	0.6	37.3
27	**2**	19	10.1	9.1	0.02	1.07	0.081	23.1	5.9	68	110	15	0.8	0.4	30.1
28	**2**	20	9.6	12.8	0.02	1.04	0.028	23.7	18.2	65	105	20	0.4	0.5	21.5

Location 1 is Dindori and Location 2 is Mandaleshwar. The year from 1 to 20 represent years from 1993–94 to 2012–13 output parameter and ignores the input parameters. Accordingly, irrespective of the value of the input parameter it gives the output as the most frequent category of the output parameter.

5 Results and Discussion

5.1 *Narmada Water Quality*

The water quality status of Narmada River is as shown in Table 3. The results indicate that the water quality did not change between upstream and downstream monitoring points.

The recommended threshold water quality values are available for 17 out of the 26 water quality parameters (Table 2). The assessment of Narmada river water quality for these thresholds suggest that except turbidity, nitrate, and chemical oxygen demand (COD), all other parameters' values are meeting the recommended threshold for all years. Thus, Narmada River needs to focus on its turbidity, nitrate and chemical oxygen demand (COD) for recommended water quality status.

5.2 *Correlation Clustering*

The study performed Pearson correlation analysis of all input parameters as well as location. Using the correlation data, the hierarchical clustering is performed. The results are shown in Fig. 4 The study found that these 25 input parameters and location parameter could be clustered into three groups. This suggests that input parameters have an association with each other.

5.3 *Categorization of Chemical Oxygen Demand (COD) and Input Parameter Dimensionality Reduction*

The continuous values of COD are converted into two categories namely high COD and low COD values using K-mean clustering. The high COD category with 11 samples has a mean value of 37.25 mg/l with a range of 33–42.2 mg/l. The low COD category with 17 samples has a mean value of 26.95 mg/l with a range of 20.2–31.7 mg/l.

Principal Component Analysis (PCA) is performed to reduce the input data dimensions of 26 parameters (25 input parameters and one location). The use of PCA can explain 95% variance of 26 parameters using only 11 principal components.

Table 3 Descriptive Statistics of Water Quality Parameters used in the study

Code	Water Quality Parameter	Mean (Std. Deviation)			Range			Median		
		O[a]	D[a]	M[a]	O[a]	D[a]	M[a]	O[a]	D[a]	M[a]
1	Conductivity	270.68 (17.45)	271 (15)	270.36 (20.19)	239–302	244–292	239–302	268.50	268.50	268.50
2	pH	8.2 (0.16)	8.12 (0.18)	8.27 (0.1)	7.7–8.4	7.7–8.4	8.1–8.4	8.20	8.20	8.25
3	TDS	172.86 (12.77)	172.71 (12.32)	173 (13.68)	149–196	156–193	149–196	171.50	168.00	174.00
4	Temp	24.39 (2.04)	22.68 (1.36)	26.11 (0.69)	20.7–27.5	20.7–26.2	25.2–27.5	25.25	22.35	26.10
5	Turb	29.54 (21.25)	41.26 (21.69)	17.81 (13.07)	0.6–81.7	3.7–81.7	0.6–47.5	25.35	38.05	16.70
6	Alk_Phe	4.39 (3.27)	3.73 (2.6)	5.05 (3.81)	0–11.7	0–8	0.4–11.7	3.85	3.65	4.00
7	Alk_Tot	166.5 (52.76)	170.93 (62.94)	162.07 (42.17)	119–329	119–329	125–292	148.00	151.00	147.00
8	Ca	28.11 (1.62)	28.21 (1.58)	28 (1.71)	25–31	26–31	25–31	28.00	28.50	28.00
9	Cl	8.83 (1.98)	8.97 (2.4)	8.68 (1.53)	5.4–12.8	5.4–12.8	6.7–12.1	8.55	8.70	8.40
10	CO3	5.24 (3.91)	4.35 (2.95)	6.12 (4.62)	0–13.9	0–9	0.5–13.9	4.70	4.40	4.85
11	F	0.29 (0.09)	0.29 (0.1)	0.29 (0.08)	0.2–0.5	0.2–0.5	0.2–0.4	0.30	0.25	0.30
12	HCO3	170.21 (23.46)	171.36 (26.48)	169.07 (20.94)	132–219	132–219	146–217	166.00	169.50	162.50
13	K	1.84 (0.39)	1.67 (0.38)	2.01 (0.33)	1–2.5	1–2.3	1.3–2.5	1.90	1.80	2.10
14	Mg	11.37 (2)	12.19 (1.91)	10.54 (1.78)	7.9–14.5	8.4–14.4	7.9–14.5	10.90	12.60	10.30
15	Na	10.33 (2.47)	9.02 (1.42)	11.64 (2.63)	7.2–16	7.2–11.9	8.7–16	9.65	8.30	10.65
16	NO2_N	0.04 (0.04)	0.06 (0.05)	0.03 (0.04)	0–0.16	0.01–0.16	0–0.14	0.03	0.05	0.02
17	NO3_N	0.72 (0.43)	0.65 (0.42)	0.79 (0.44)	0.19–1.55	0.19–1.43	0.27–1.55	0.63	0.53	0.67
18	PO4	0.07 (0.07)	0.07 (0.07)	0.07 (0.08)	0.01–0.28	0.01–0.25	0.01–0.28	0.05	0.06	0.04
19	SIO2	23.57 (4.13)	24.14 (4.12)	23.01 (4.21)	12.5–29.4	12.5–27.9	14.9–29.4	24.15	25.50	23.45
20	SO4	8.15 (4.03)	7.98 (4.52)	8.33 (3.64)	4.2–21.9	4.4–21.9	4.2–18.2	7.25	7.25	7.25
21	HAR_Ca	70.29 (3.9)	70.79 (3.89)	69.79 (3.98)	64–77	66–77	64–76	69.00	70.50	69.00
22	HAR_T	117.68 (9.22)	121.57 (7.99)	113.79 (8.95)	103–134	108–134	103–130	117.00	123.00	112.50
23	Na_Percent	15.68 (3.14)	13.64 (2.02)	17.71 (2.73)	11–22	11–17	15–22	15.00	13.00	16.00
24	RSC	0.64 (0.36)	0.56 (0.36)	0.71 (0.34)	0.2–1.4	0.2–1.4	0.3–1.4	0.60	0.50	0.70
25	SAR	0.41 (0.1)	0.35 (0.07)	0.47 (0.09)	0.3–0.6	0.3–0.5	0.4–0.6	0.40	0.30	0.40
26	COD	31 (6.02)	30.69 (5.53)	31.3 (6.67)	20.2–42.2	22.5–39.3	20.2–42.2	30.55	30.20	31.80

[a]O = overall (n = 28), D = Dindori (n = 14) and M = Mandaleshwar (M = 14)

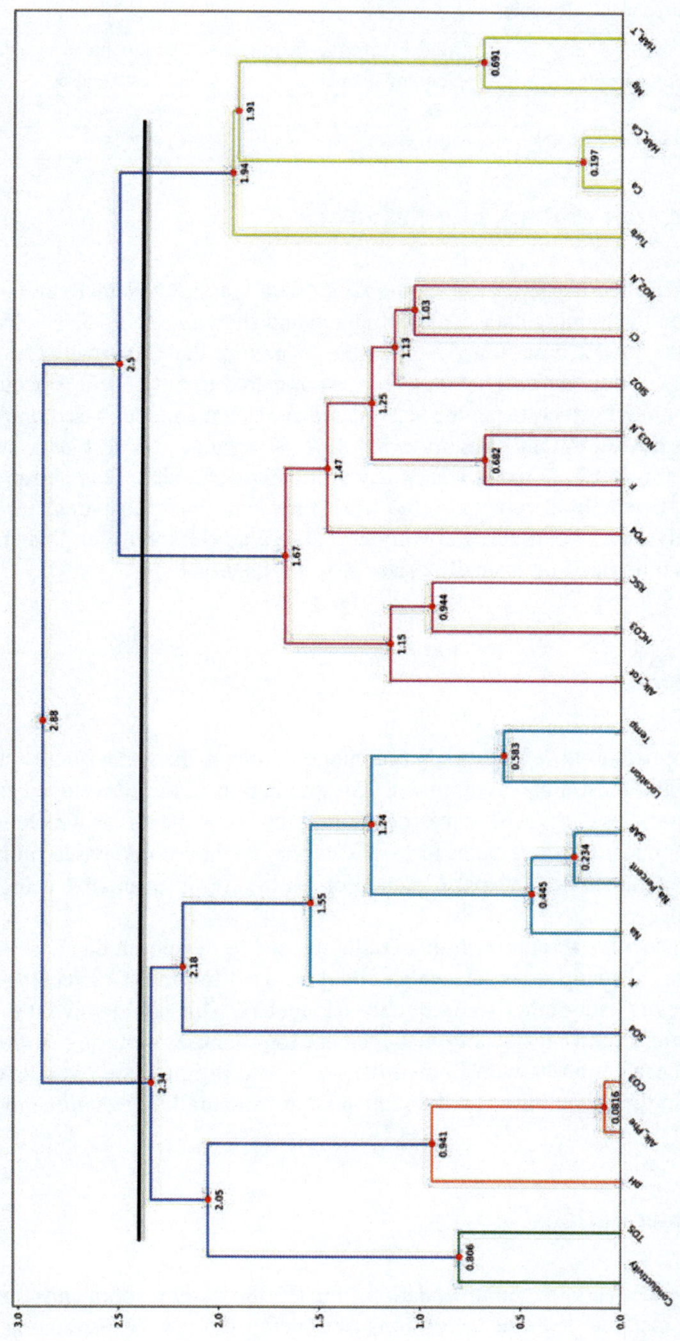

Fig. 4 Correlation based Hierarchical Clustering Dendrogram. The black horizontal line denotes the point at which Dendrogram is cut from the top. The cut creates three clusters

Table 4 Performance of ANN models in terms of accuracy for training (n = 21) and test (n = 7) dataset as compared to ZeroR model

	Model Performance in terms of prediction accuracy	
Machine Learning Model	Training dataset	Test Dataset
ZeroR	60.71%	60.71%
Artificial neural network (ANN)	100.00%	85.71%

5.4 Application of ANN and Results

The study performed classification of the COD data using ANN in Python. The results based on the training dataset are computed and shown in Table 4. The results showed that the model based on ANN is able to predict the COD category with 100% accuracy in training data set (n = 21) as compared to only 60.71% accuracy from ZeroR model. Further, using the test dataset (n = 7), the model based on ANN is able to predict the COD category with 85.71% accuracy in test data set as compared to only 60.71% overall accuracy from ZeroR model. This shows that the ANN model is able to categorize the COD based on the input parameter data more accurately than the base model without input parameters. Further, this shows that ANN can train itself on a small dataset of water quality.

6 Conclusion

This study proposes to develop a prediction model to assess the water quality status of rivers in data constraint environment. The prediction model developed using Artificial Neural Network (ANN) was developed for Narmada River which could predict the Chemical Oxygen Demand level category in the sample based on other water quality parameters. The performance of the model is evaluated using the accuracy metric.

The study provides an approach that could be used to develop a database of all limited data and develop a model, which could be used to predict different water quality parameters using other water quality parameters. This could solve the very critical problem in the water quality management for countries suffering from data constraints. These countries with limited data set could estimate the overall water quality status of their river based on the sequential and/or parallel prediction models.

7 Recommendations

This study provides certain recommendations. Firstly, the study recommends the use of supervised machine learning algorithms to develop models for estimating and predicting the water quality status of rivers in limited data settings. Secondly, the study recommends use of ANN algorithm-based model to predict the Narmada water quality status.

Acknowledgments We expressed our heartiest thanks to the Central Water Commission, Ministry of Water Resources, New Delhi for providing data. Without data, it was impossible to complete the research work.

Appendix

Hyper-parameter tuning performed in the study

ML technique	Hyper-parameter	Range	Optimal Range
ANN	Hidden layer size	10 to 100	50
ANN	Number of hidden layer	3	3
ANN	Activation function	Identity, logistic, tanh, relu	Tanh
ANN	Solver for weight optimisation	Sgd, lbfgs	Sgd
ANN	Learning rate	Constant, adaptive, invscaling	Adaptive
ANN	L2 penalty (regularization term) parameter	0.0001 to 0.8	0.233

References

1. Cosgrove WJ, Loucks DP (1986) Water management: current and future challenges and research directions. Water Resour Res 51:135–145
2. Jain SK (2019) Water resources management in India – challenges and the way forward. Curr Sci 117:569–576
3. Srikanth R (2009) Challenges of sustainable water quality management in rural India. Curr Sci 97:317–325
4. Chau K (2006) A review on integration of artificial intelligence into water quality modelling. Mar Pollut Bull 52:726–733
5. Najah Ahmed A, Binti Othman F, Abdulmohsin Afan H, Khaleel Ibrahim R, Ming Fai C, Shabbir Hossain M, Ehteram M, Elshafie A (2019) Machine learning methods for better water quality prediction. J Hydrol 578
6. Kotsiantis SB (2007) Supervised machine learning: a review of classification techniques. Informatica 31:249–268. https://doi.org/10.1007/s10462-007-9052-3
7. Sarkar A, Pandey P (2015) River water quality modelling using artificial neural network technique. Aquat Procedia 4:1070–1077
8. Katakwar M (2016) Narmada river water: pollution and its impact on the human health. Int J Chem Stud IJCS 4:66–70
9. Karlik B (2015) Performance analysis of various activation functions in generalized MLP architectures of neural networks. Int J Artif Intell Expert Syst 1:111–122. https://doi.org/10.1007/3-211-29956-4_7

10. Liao Z, Carneiro G (2016) On the importance of normalisation layers in deep learning with piecewise linear activation unit. In: 2016 IEEE winter conference on applications of computer vision, WACV 2016. Newyork, USA, pp. 1–9. https://doi.org/10.1109/WACV.2016.7477624
11. Grossberg ST (1982) Neural principles of learning, perception, development, cognition, and motor control. In: Studies of mind and brain. Springer, Dordrecht
12. Widrow B, Hoff M (1960) Adaptive switching circuits. Western electronic show and convention. Inst Radio Eng 4:96–104
13. Rosenblatt, F., 1961. Perceptrons and the theory of brain mechanics. New York, USA
14. Hopfield JJ, Tank DW (1985) "Neural" computation of decisions in optimization problems. Biol Cybern 52:141–152
15. Rumelhart DE, Hinton GE, Williams RJ (1986) Learning representations by back-propagating errors. Nature 323:533–536
16. Kohonen T (1989) Self-organization and associative memory. Springer, Berlin
17. Malviya A, Diwakar SK, Choubey ON (2010) Chemical assessment of narmada river water at Hoshangabad city and Nemawar as navel of river in Central India. Orient J Chem 26:319–323
18. Shraddha S, Rakesh V, Savita D, Praveen J (2011) Evaluation of water quality of Narmada River with reference to physco-chemical parameters at Hoshangabad city, MP, India. Res J Chem Sci 1:40–48
19. Tiwari DK, Tiwari HL, Mishra S, Nateriya R (2018) A literature review on ANN based hydrological modeling on Narmada River basin. Int J Innov Adv Comput Sci 7:2347–8616
20. Bureau of Indian Standards (2012) Indian standard drinking water-specification (second revision). New Delhi, India
21. Helmer R, Hespanhol I (1997) Water pollution control – a guide to the use of water quality management principles edited, United Nations environment Programme water. Water Supply & Sanitation Collaborative Council, World Health Organisation. London, UK. https://doi.org/10.4324/9780203477540
22. Raychaudhuri M, Raychaudhuri S, Jena SK, Kumar A, Srivastava RC (2014) WQI to monitor water quality for irrigation and potable use. Bhubaneswar, India
23. Vaghasiya AR, Patel NR (2016) Physico-chemical study of Narmada river water at Gujarat, India 2005–2010
24. World Health Organization (WHO) (2011) Guidelines for drinking-water quality: Fourth Edition. Geneva, Switzerland. https://doi.org/10.1016/S1462-0758(00)00006-6
25. Johnson SC (1967) Hierarchical clustering schemes. Psychometrika 32:241–254
26. Likas A, Vlassis N, Verbeek J (2003) The global k-means clustering algorithm. Pattern Recogn 36:451–461
27. Hossin M, Sulaiman MN (2015) A review of evaluation metrics for data classification evaluations. Int J Data Min Knowl Manag Process 5:1–11. https://doi.org/10.5121/ijdkp.2015.5201
28. Nasa C, Suman S (2012) Evaluation of different classification techniques for WEB data. Int J Comput Appl 52:34–40. https://doi.org/10.5120/8233-1389

Rain Water Harvesting Methods in Rajasthan

Supriya Singh, Pratibha, Vanshika Singh, and Sudesh Kumar

Abstract Rainwater harvesting is a useful and smart way to conserve the water falling freely on our roofs. The water from roofs is flowed into the catchment area through pipes. The storage area may be above the ground or maybe underground also. In the coming times, of increased population and decreased demand for water, rainwater harvesting can firmly supply clean and fresh water to the community.

Keywords Rainwater harvesting · Catchment area · Water storage units · Advantages · Disadvantages

1 Introduction

Earth when watched from the outer space appears to be a unique and splendid planet, "a blue globe" [1] with red areas from dessert and land and white mass of cloud. Water is indispensable for life and it plays a major role in earth's climate. It is a priceless natural gift but it is not an ultimate constant. Though there are nearly 480 million cubic kilometres of water on our planet, nearly three million is available in the ground form. By modifying land use, the proportion of the different pathways, evaporation, percolation, and run off has changed. Water is a vital element for all human beings [2]. Our body consists mostly of the Water. We need water for drinking, cooking, washing, agriculture and to run our industries. We usually take it for granted because of its availability; but when in scarcity it becomes our most precious resource [3].

Water will determine whether if India remains poor or become wealthy. The key to good management lies in the wise use of water and not simply making more and more dams [4]. Management involves teaching every person around the world the value of water and the high demand to maintain the water level. Water management is all about using the existing technologies to maximise the use of water and technologies to share water with all [4].

S. Singh (✉) · Pratibha · V. Singh · S. Kumar
Department of Chemistry, Banasthali Vidyapith, Rajasthan, India

S. Yadav et al. (eds.), *Water Quality, Assessment and Management in India*, Earth and Environmental Sciences Library, https://doi.org/10.1007/978-3-030-95687-5_9

There is a lot to learn from people of old days. Their water math was simple. They all worked on the principle of rain water harvesting. Since they knew, they could get all their water needs fulfilled from the first cloudburst. The only solution was to capture the rain water and to use it to recharge the groundwater level [5].

Urbanisation has led to digging of more and more bore wells and underground tanks, which has drastically resulted in the reduced ground water level [6, 7]. As we all know water constitutes majority of earth crust, but out of this entire amount the usable form is very small. Human population depends totally on water including all kind of needs the household and commercial. The rainwater holds so much importance due to its purity and softness. The industries and agriculture over exploit water for commercial use. Even the farmers overexploit the water resources [8] which are adding to non-sustainable farming practices. Because of these non-sustainable practices, the future generations have to suffer a lot of water scarcity.

Water is a limited natural resource [9], and it plays an important role in maintaining a stable environment, but the industrialisation has led to depletion of the natural resources and contaminating it with industrial effluents. Water, air and soil are affected with industrial pollution. As per the data after industrialisation water quality problem is said to affect around 44 million people in the whole world [10].

2 Rain Water Harvesting: Overview

With the increasing population, it has become essential to meet the demands of water. Surface and groundwater resources are used faster than their recharge [11]. These water sources could be recharged through rainwater harvesting. Rain water harvesting is one of the environmental sound techniques that could be carried out at individual level to meet the demands of falling ground water level [12]. Rainwater is the ultimate source of fresh water.

Rainwater is valued for its purity and softness. The rain waters chemical analysis also approves its quality [13]. The water has a neutral pH and it is free from other dissolved impurities. Rain water harvesting is wise technique to store the rain water in the underground tanks so that it could be used further for different purposes [14, 15]. "Rain water harvesting is usually treated as an umbrella which describes the whole method of collection and runoff forms" [16].

Rainwater storage can meet the needs of day-to-day life or it could be used to improve the ground water level.

The center of science and environment at Delhi defines rain water harvesting as "CATCH WHERE IT FALLS" [17]. The process of rainwater to be described in simpler terms includes catching the falling rainwater directing it into the storage tanks, filtering it in order to make it free from any sort of contamination and in the end using it to recharge the ground water level. The water harvested could be used to meet the needs of people residing in cities. Roof top rainwater could be made fit for drinking taking advantages of the available technologies. The water from the first rain is usually discarded [18] (Fig. 1).

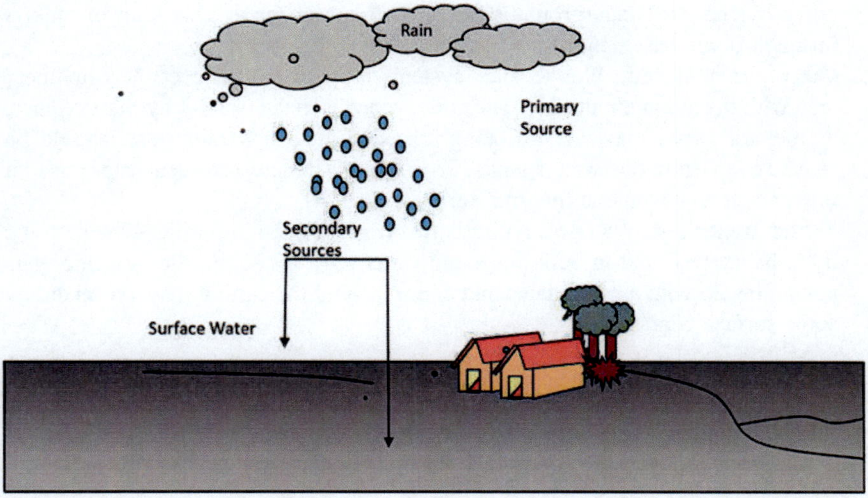

Fig. 1 Where does water come from? Figure describing the various sources of water. Reprinted with permission from Ref [19]

2.1 Components of Rain Water Harvesting

The two main system of rain water harvesting system are the rain water collection system and storage and supply system respectively. The components of collection system include guttering, roof, and piping which aims at get the best out of rainwater collection. On the other hand, storage and supply system facilitates the supply of rainwater so that the needs are met [20].

There are six basic types of components for a rainwater harvesting system:

1. **Catchment area:** Catchment area is the surface which receives water at first i.e. rain water directly hits the catchment area. This surface could be tiled, woody open ground, or the rooftop [21]. Generally, the catchment area preferred is artificial because they reduce the risk of contamination compared to the natural surfaces like lawns and grounds. It is also essential to keep in mind the nature of roofing material. The roofs coated with lead are considered unsuitable for the harvesting system.
2. **Downspouts:** Most of the conveyance systems are linked directly to the natural recharge systems so that the water is used soon for recharging purpose.
3. **Roof washers and screeners:** Leaves and other debris may end up entering the system; to prevent this wired mesh is installed on both the sides of drainpipe.

   A first flush device is also installed. It is a valve that ensures the rainwater of the first rain is not entering the system.
4. **Storage tanks:** storage tanks could be constructed both above and below the ground. The materials that could be used are fiberglass, stainless steel,

polyethylene, etc. underground tanks are often cemented. This kind of tank is installed in the basement of the building [22].

5. **Conveyance system:** Conveyance systems are used to connect the catchment area with the storage unit [23]. The conveyance is made usually by plastic pipes, as metallic pipes may end up being corroded. The first rain water should be moved away from the storage tanks. To make this happen sediment trap is used it prevents debris from entering the storage units [24].
6. **Water treatment:** Water from rain is free from mineral impurities which on the other hand are found in bulk in ground water [24]. However, the rainwater gets contaminated with air pollutants and after reaching the earth it may get mixed by some surface contaminants.

Before determining the treatment techniques, firstly the water needs to be tested in a laboratory to determine whether it is potable or non-potable.

There are various types of treatment that could be used [25]:

- Filtration
- disinfection
- pH control

Dirt and sediments mostly find a way to enter the system. So in order to restrict their path the water is treated. The most common kind of treatment method used is filtration (Figs. 2 and 3).

3 History

India being a historic country, it has its deep roots in the historical cultural tradition. We still have retained the essence of our traditions. Some traditions evolved have a deep impact on our day-to-day lives. In the earlier times, people used to go to banks to get water, gradually the lakes and streams have started drying up, so the human settlement started looking for a more trusted source of water. Indus valley civilization was the first major human civilization of India. Archaeological evidence has shown the existence of drinking water supply from dams and wells [26].

Rain water harvesting is not a new concept. It existed back to 1700 BC. Archaeologists have found the rain water collecting dams on the islands of Crete. Romans were considered to be masters with the development of rain water harvesting technologies [27]. In the ancient times, the underground tanks were left opened. This served a two way purpose by providing water for the household needs and maintaining the climate around as it regulated the temperature by providing a cooler effect [28]. However, gradually as the population increased and the pits and dams needed to be covered so that the water could be prevented from contamination. Therefore, the covering of dams also had an advantage over the uncovered dams. The covered dams prevented the loss of water through evaporation [29].

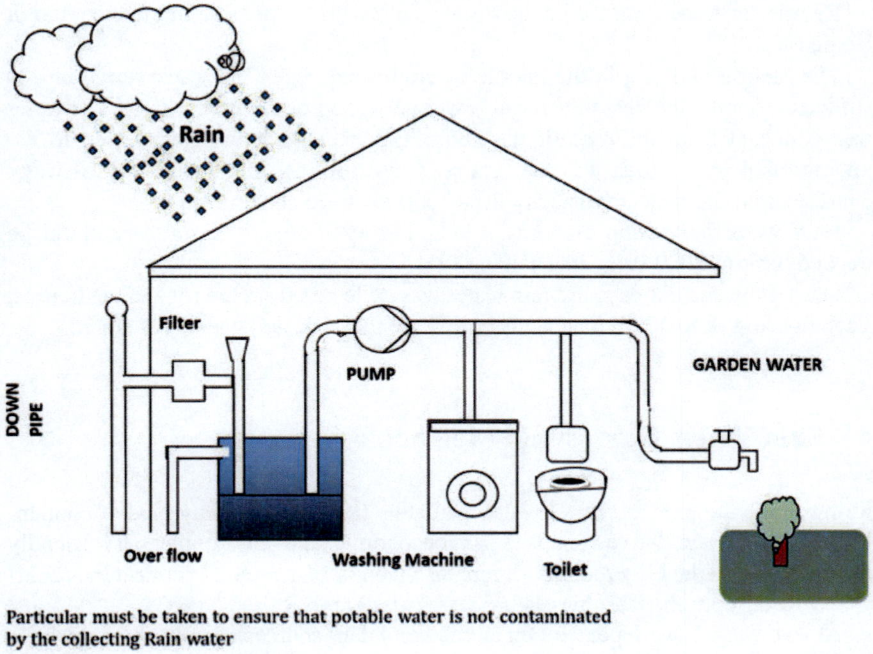

Fig. 2 Particular must be taken to ensure that the collecting Rainwater does not contaminate potable water. Reprinted with permission from Ref [19]

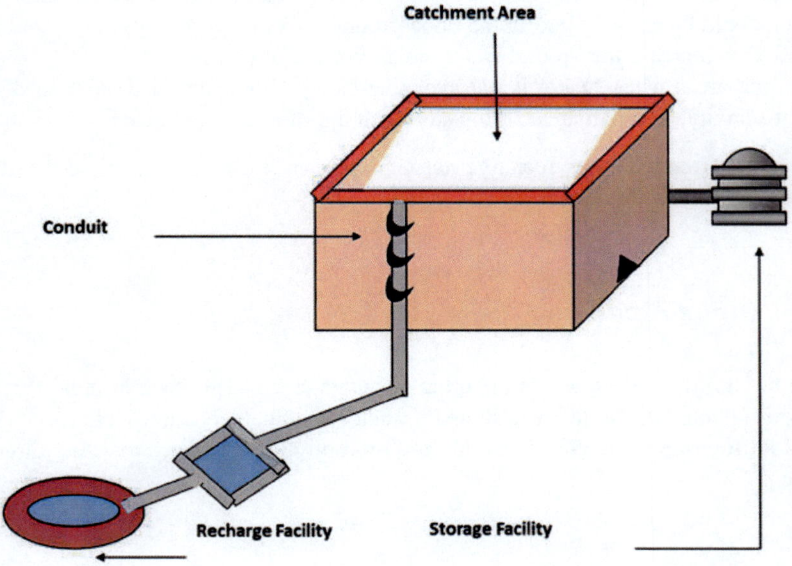

Fig. 3 Components of a rainwater harvesting system. Reprinted with permission from Ref [19]

Rain water harvesting technique disappeared with an increase in urbanization in Rome [30].

The first pit was dug in the middle of the Bronze Age. The water stored in it is sufficient to meet the demands for at least one season of the water crisis. The water was collected from the roof of the house and later on it was transferred to the underground pits. Africa was the first country to follow this technology of water conservation; the tanks of capacity 200–2000 m^3 were constructed.

Rain water harvesting also has a long history in Asia. The history could be tracked back to 2000 years in Thailand [31].

During the ancient days the rain water was collected from the roof of the houses. Each house was found to have a cistern where the collected water was stored.

4 Rain Water Harvesting: Methodology

Rainwater is the purest form of water available. It is free from any type of contamination. Rainwater harvesting is a socioeconomic and environmentally friendly practice [32]. In the hilly regions, diversion channels were made to collect the runoff water. Roof top rain water harvesting was prevalent in Rajasthan [33].

Nowadays, we are dependent on all the secondary sources of water such as rivers and lakes. We have stopped paying attention to the fact that rainwater is the primary source of water and the purest one. To keep the water bodies recharged, we should start conserving rain water [34].

Rain water harvesting as described earlier is the technique of capturing rain water which could be either stored in the underground tanks for further use or it could be directed to increase the ground water table by percolation [34].

In the areas where water is not available easily, this is an ideal solution of the water scarcity, in urban areas; rainwater from the rooftop is collected.

Methods Ground water recharge through rainwater harvesting could be done mainly by following methods:

4.1 Urban Areas

Due to less available space in the urban areas its essential to keep in mind that the underground tanks should be designed in such a way that it occupies less space [35].

Roof top rain water could be collected by rooftop rain water harvesting through [17]:

4.1.1 Recharge Pits

This technique is suitable for alluvial areas with permeable rocks exposed on the land surface and with a roof area of about 100 sq. meters they should be provided with a mesh to prevent some leaves and other debris to come along with the roof runoff water [36].

The pits should be in the graded form – boulders at base, middle area filled with gravels and coarse sand at the top.

4.1.2 Trench

Buildings having area of about 200–300 sq. m are suitable for building a trench, and a permeable top surface is demanded at shallow heights. The trench may be wide up to 1 m. Presence of mesh is essential to prevent the flow of debris along with rainwater and a collection chamber should also be designed to arrest the flow of some finer particles. "The top layer of sand should be cleaned periodically to maintain the recharge rate" [36] (Fig. 4).

4.1.3 Rooftop Rainwater Harvesting Through Existing Tube Wells

This technique could be practiced in the regions where the shallow aquifers have dried up and the demand is fulfilled by deep tube wells [17]. The roof runoff is directed to tanks by connecting the roof drains through PVC pipe of diameter 10 cm. the first runoff is discarded through the drain pipe allowing the washing of roof. The size of the filter varies according to the size of roof's (15 cm if area area is less than 15 sq. m and 20 cm if more than 150 sq. m).

4.2 *Rural Areas*

4.2.1 Rain Water Harvesting Through Dug Wells Recharge

The dug wells abandoned for a long time, once cleaned and desilted may be used as harvesting structures. The recharge water is guided to the tank below the water level to avoid water bubbles. Periodic chlorination is an essential step for maintenance [37].

4.2.2 Rain Water Harvesting Through Recharge Shaft

It is the most easy and convenient method. A shaft may be dug manually of a diameter of more than 2 m for a depth till it reaches the impermeable strata and it

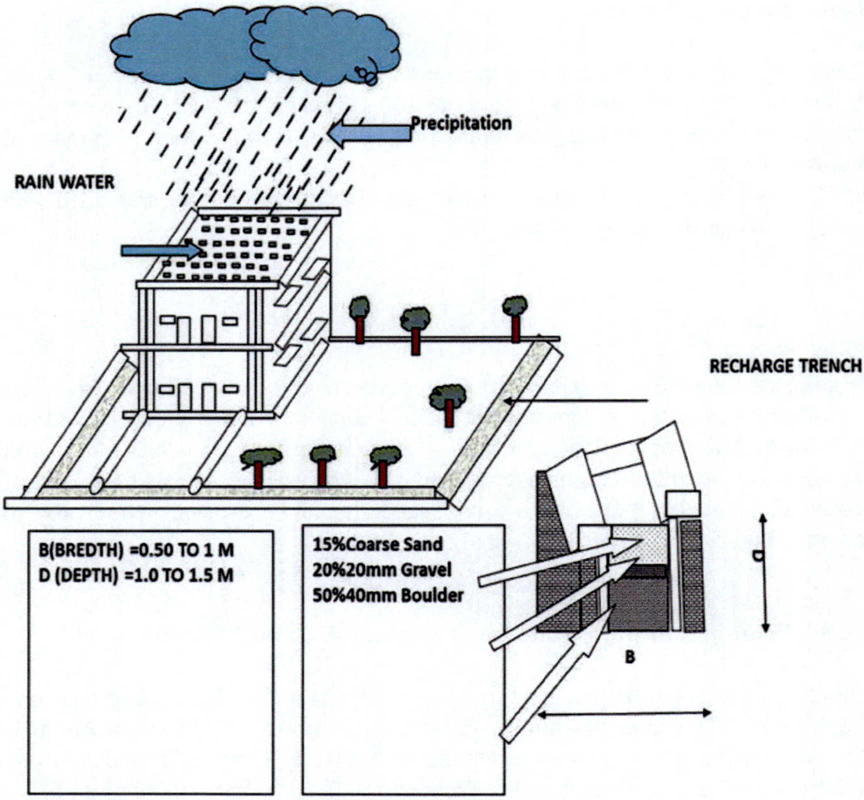

Fig. 4 Rooftop rain water harvesting through a trench. Reprinted with permission from Ref [19]

may not touch the water table. The water from these tanks does not percolate down due to siltation. The amount of surplus water can be recharged to ground water level through the shaft tanks [38] (Fig. 5).

4.2.3 Rain Water Harvesting Through Percolation Tanks

Percolation tank are created artificially in order to capture the surface runoff water. These are earthen tanks and used only for recharge of ground water level [36] (Fig. 6).

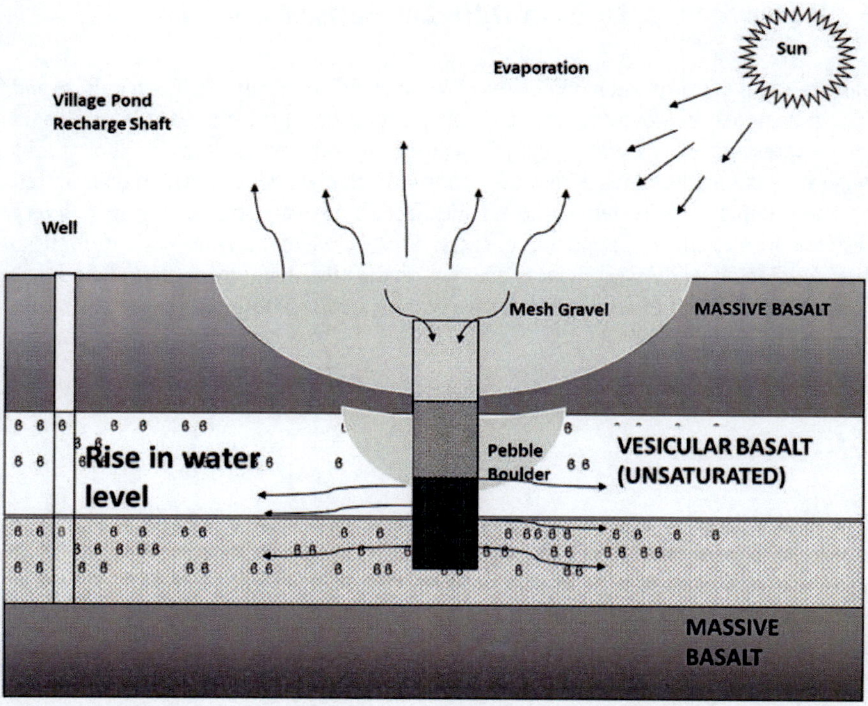

Fig. 5 Rain water harvesting through recharge shaft. Reprinted with permission from Ref [19]

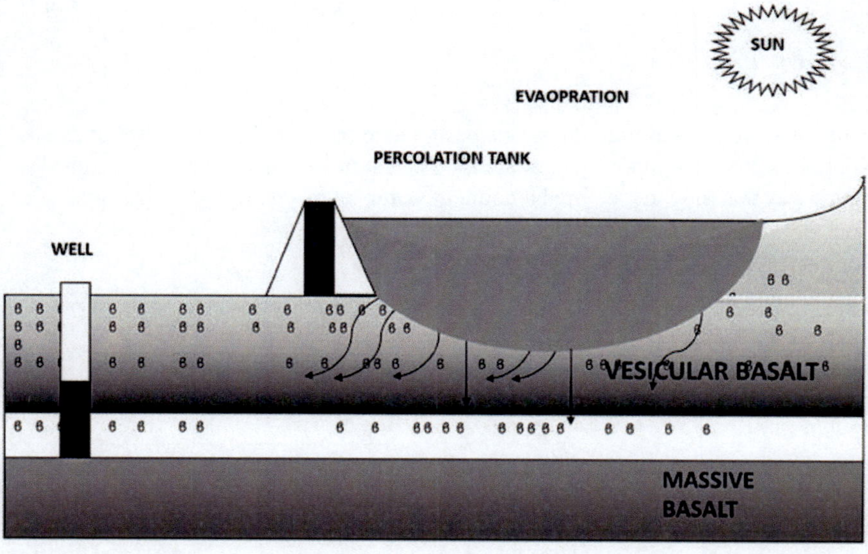

Fig. 6 Rainwater harvesting through percolation tanks. Reprinted with permission from Ref [19]

5 Water Storage Units in Different Parts of Country

India is a vast country and it comprises of various different cultures. Each culture and tribe reflects its own culture and geographical peculiarities [39]. As it is often said India represents *unity in diversity*; [40] it could be seen when studying the various water harvesting structures across the country. Each state has its own annual rainfall rate and demand for water, based on the needs rain water harvesting or delivery systems are developed across the country. Some areas experience high rainfall so they develop water delivery system whereas on the other hand some states face water scarcity o they need to develop water harvesting structure to recharge ground water table [41] (Table 1).

5.1 *Jack Wells*

These structures are found in the area of **Andaman and Nicobar islands**. These are found in the southern part of the islands, where the topography is very rugged. Pits are made from the logs of hard wood and then the water is collected in them. A full length bamboo is cut longitudinally and then placed in a slanting manner with lower end placed in the pits. They serve as conduits of rainwater harvesting. The rainwater is collected drop by drop in the pits called **jack wells.** A series of jack wells are connected to one another through these bamboo splits. The subsequent jack well, being bigger than the previous one and preventing overflow [43].

5.2 *Korambus*

Found in districts of Kerala. These are designed to increase the water level of canals. Korambus are the temporary dam across the mouth of canal. They are designed in such a way that only the required amount of water is flowed through fields and rest of

Table 1 Water harvesting structures in the different states [42]

Name of water harvesting structure	State
• Baudi • Nawn • Khatris • Nalas	Himachal Pradesh
Nala	Uttarakhand
Dongs	Assam
Virdas	Assam
Apatani	Arunachal Pradesh
Johad	Rajasthan

the water flows through diversion channels. The height of Korambus are so maintained that the fields remain upstream are not submerged [44].

5.3 *Bamboo Drip Irrigation*

Practiced in **Meghalaya**. It is a 200 year old system followed by tribal farmers of jayntia hills. The bamboo channels are used to divert the springs at the hill tops to the places owner in gravity. The water entering the bamboo channel is approx. 20 L and transported all over drop by drop at the rate of 20–80 drops per min [45].

5.4 *Zabo*

Practiced in **Nagaland**. Though in this area drinking water is a problem but it receives high rainfall. The zabo combines the water runoff through fields, forests hills etc. All the water is finally collected in the pond like structure designed at the foot of the hill [46].

5.5 *Virdas*

These could be spotted in **Gujarat**. They are the wells dug in the *jheels*, chiefly they are shallow. Developed by the nomadic tribes' *maladharis.* They observed the topography of the region, rain pattern, identified the depressions and made **virdas** in certain places so that rain water could be collected in it. These structures followed a brilliant technology of separating potable drinking water from the non-potable saline water based on their densities [47].

5.6 *Johads*

Johads are found in **Rajasthan.** They are the small earthen tanks that are made to raise the GWT. These are simple mud and stone barriers. These are made to capture the rain run-off water. Dr. Rajendra Singh referred to as **water man of India** played a catalysing role by initiating the construction of 8600 Johads spread at the area of 6500 sq. km. once the water percolates the moisture content of soil increases [33].

5.7 Kuhls

These are the traditional irrigation structures found in the area of **Jammu and Kashmir and Himachal Pradesh.** These are used to transport water from glaciers to villages. Kuhls were prevalent since the British rule in India. These were also used to irrigate the fields. Apart from irrigation these structures were also used to run flour mills. The wheel used here was the wooden one. If the terrain got muddy, stones were used to prevent clogging of the wheel [48].

6 Rajasthan

Rajasthan is the north western state of India [49]. It is the second largest state in terms of area. Thar Desert falls in the arid region with a population of about 30 billion [50]. The situation of groundwater level in Thar dessert is alarming [51]. It is one of the states whose economic growth is dependent on water [52]. There is scarcity of safe drinking water. The available ground water is either brackish or saline. The expected annual rainfall rate is approx. 16 inches during June to September [53]. Rajasthan is well known for its efficient water management techniques. This region involves water crisis due to low annual rainfall. Low annual rainfall affects the crop production and livelihood of the residents.

The dominant structures for holding water are JOHADS [54]. The Johads are crescent shaped dams made up of rock and sand [55]. These communal water conservation structures not only providesa water for the day-to-day usage but also holds water to be used during the dry weather, in the absence of rainfall by playing an important role in recharge of aquifers.

Many human practices have led to a tremendous decrease in the ground water table. Gradually after 1940s deforestation paced up and Johads were replaced by tube wells [56]. Tube wells were one time attraction as they led to adepletion of ground water level. The water was continuously driven out and no recharge was taking place. The fallen ground water was also a major cause of soil erosion [57].

Analyzing the rainfall data of the year 1991–1999 depicts drought once in every 2 years. The increase in population and their needs have made the situation more alarming. Sweet water is very costly and is collected by private water tanks which are further sold at higher prices.

6.1 Water Storage Units in Rajasthan

There are five types of rainwater harvesting structures in Rajasthan. These are [33]:

6.1.1 Beri

Its underground reservoir that can hold 400,000 cu L and the small one can hold 100,000 cu L of water respectively. This capacity could easily meet the demands of 4–5 family for a few months. This is found in the area with impermeable rocks lined with clay [47]. These are found in the western Rajasthan. Beri's are 10–12 m deep. The mouth is made narrow to prevent evaporation, while they are broad at the bottom. Wooden planks to prevent contamination cover these.

6.1.2 Tanka

It is designed to collect the water obtained from monsoon rain. Tanka is an underground tank. It could be as large as a room. This storage unit is constructed from cement. Tankas are a part of roof top rainwater harvesting. These are connected too slopes of roof through pipe. Rainwater could be stored in for over a year in tankas. Since these are covered, they are contamination free [58].

6.1.3 Johads

These are check dams which are used for percolation and ground water recharge. These storage units are spread to over 650 villages in Alwar district. Some rivers were dry and have come again alive today [54].

6.1.4 Baori

Baori's are community well. Used for drinking water. Local nomadic tribes constructed most of them. Water is retained for a long as evaporation is absent [59].

7 How Rain Water Harvesting Works

As studied above, rain water harvesting means to collect and store rainwater to meet the needs of people throughout the year. Rainwater is the source of the most pure form of water available naturally when air is free of pollutants and it is the primary source of water. Rain water harvesting technology aims at recharging the ground water level [29]. Either the rainwater collected can be used directly to meet the needs or it could be used to recharge the ground water level. Rivers and lakes are the secondary sources of water, but presently all of us have started depending on the secondary sources and have forgotten the need to save rainwater, which holds crucial importance. The cities waste a lot of water every year by not collecting them, if

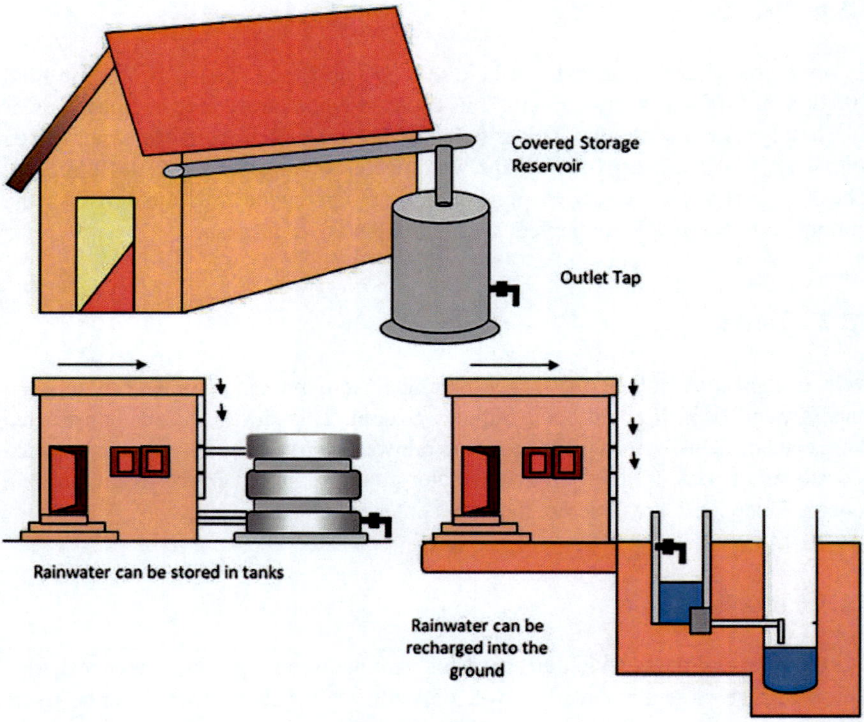

Fig. 7 Harvesting water from Rooftop. Reprinted with permission from Ref [19]

harvested properly it would have met the needs in the water scarce regions. During the last thirty years, different researchers have classified the RWHT on the base of specific factors such as size, location, design [60].

Rain water could be harvested from the following sources [13]:

- Rooftops: to collect water from the roofs the first thing needed is impervious rooftops connected with tanks or recharge pits.
- Drains: cities with low annual rainfall should develop a neat setup of their drains as they provide a large catchment area (Fig. 7).

7.1 Rain Water Harvesting Potential

To understand rainwater harvesting potential we first need to understand the term rainwater endowment. Rainwater endowment refers to the total amount of water that a place receives in the form of rain. Now out of the total received water the amount of water hat is harvested is called rain harvesting potential. There are various factors that influence rainwater harvesting potential. Out of all the characteristics that were most important are climatic condition and the area of catchment [61].

7.1.1 Rainfall

When talking of this characteristic its essential to keep in mind the two factors, i.e. **quantity** and **pattern**. To understand the quantity of rainfall, it is essential to have a deep knowledge of the rainfall over the years. At least a data of 10 years must be needed to understand how much rainfall occurs. The pattern of rainfall on the other hand, is essential to be understood to construct the dams as for example if a certain place is having more of the drier days then large dams needed to be constructed.

7.1.2 Catchment Area Characteristics

It's very essential to have better perspective of sizing of tank which is carried out by keeping in mind the size of area, rainfall data around the year, type of surface and the demand of potable water [62].

The catchment area characteristics are used for determining the storage conditions. Water harvesting potential could be calculated by the formula given below [35]:

$$\text{Water harvesting potential} = \text{Rainfall X Collection efficiency}$$

When calculating collection efficiency, the wastage of water through the natural process is taken into account. The water spilled after he first showers, water evaporated and all the water that cannot be harvested are taken into knowledge.

7.1.3 Runoff Coefficient

The water that runs away from the surface and does not gets trapped in the catchment area falls under the category of **runoff water.** The amount of runoff water depends on the nature of the surface rain is falling, if the surface is tiled then there will be more runoff water compared to rainwater falling on a muddy surface. Runoff coefficient works on the fact that all water falling on the surface can't be collected. The following table shows data of runoff coefficient from various surfaces [35] (Fig. 8 and Table 2).

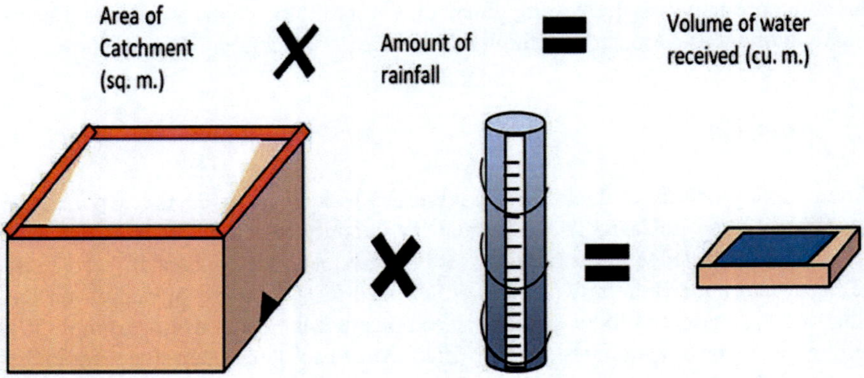

Fig. 8 Water Harvesting Potential. Reprinted with permission from Ref [19]

Table 2 Runoff coefficient from various surfaces [19]

Type of catchment	Coefficient
Roof catchment	
• Tiles	0.8–0.9
• Metal sheets	0.7–0.9
Ground surface	
• Concrete	0.6–0.8
• Brick pavement	0.5–0.6
Untreated ground catchment	
• Soil	0.0–0.3
• Rocky surface	0.2–0.5
• Green area	0.05–0.10

8 Needs and Advantages of Rain Water Harvesting

8.1 Needs of Rain Water Harvesting System

1. The ground water is getting depleted
2. To meet the needs of day-to-day life.
3. Enhance the ground water level.
4. Improve the ecology.
5. To check soil erosion due to runoff.
6. To overcome the problem of health hazards due to polluted water.

8.2 Advantages of Rain Water Harvesting [63–65]

1. The facility to get water without having to walk long distances is the prime most feature.

2. Rainwater is a relatively pure source of water and it is free from contamination in case air contaminant-free.
3. Rainwater is beneficial for plants as it is free from mineral impurities.
4. It lowers the cost of water supply.
5. Trusted backup for water supply.
6. Reduces top soil erosion from runoff.
7. Prevalent in the water scarce regions.
8. The water is soft and free from hardness.
9. It is used to recharge the ground water level.
10. The technology used is simple and easy to use.
11. Reduces flood flows.
12. Ground water is not wide-open to pollution.
13. Solves the problem faced during drought.
14. Avoid the cost of accessing the public water systems.
15. Reduction in water bills.
16. Relief during the non-rainy days.
17. Better quality of water in the coastal areas.
18. Obtainability of water throughout the year

8.3 Disadvantages of Rain Water Harvesting [66]

The following are the disadvantages of rainwater harvesting [66].

8.3.1 Extra Expenditure

The rainwater needed to be treated before making it fit for consumption. The treatment involves extra charges. To avoid these extra charges we need to get water supplied from local council because then it is already treated.

8.3.2 Enormous Labours and Assets Required

Construction of dams and underground tanks is not an easy task. Before harvesting water there are a lot of pre requirements.

8.3.3 Rainfall Dependency

This process is exclusively rainfall dependent so during the days of scanty rainfall, there is no alternate source.

8.3.4 Restricted Storage

The areas with prolonged rainfall, the storage tanks become limited to sore the water. Not all water could be harvested.

8.3.5 Acid Rain

Harvesting water from acid rain could pose a threat to all the living forms of life as this water would be used for irrigation and consumed by humans as well. When irrigated through this water, the pH of soil increases leading to stunted growth of plants [67].

9 Projects Encouraging Rain Water Harvesting in Rajasthan

9.1 Johad: Water-Shed in Alwar District

The semi-arid landscape of Alwar in Rajasthan has undergone a major transformation in the past decade [56]. The fall in ground water level was an alarming situation. The legislations enforced in order to maintain the ground water level were also not so effective. In the Alwar district of Rajasthan, the ground water level is below the critical level [68].

Johad is one such technology to restore the ecological balance of the region. This initiative was considered to be rewarding and conventional watershed process. It began in 1984 when **Tarun Bharat Sangh** an organisation in the **Bhikampura** of Alwar district [68]. Johads have transformed lives. In the recent years deforestation coupled with mining have led to destruction of landscape and massive soil erosion. 4 blocks were declared dark zone by Government of Rajasthan. Water is below the recoupable level.

Identification of johad to bring change was the first important step for **TBS.** The next thing to ensure was the active involvement and participation in repairing the damaged structures along with constructing new ones. There were very few available to cooperate because most of the people have started migrating due to shortage of water. TBS approached adults and insisted to cooperate. Today more than 2500 water-harvesting structures are present. TBS is also responding immediate needs [69].

Johads are much more than mere water harvesting structure for the community. Johads act as fundamental part of the socio-cultural milieu.

It was also an area of concern to understand the willingness to pay for johad initiative, which fulfils three chief requirements:

Table 3 Ground water level before and after Johad [71]

S. No.	Depth of well (in feet) 1988	Water level before Johad	Water level of well after johad (in feet) 1994
1	81	Dry completely	44.5
2	73	Dry completely	37
3	67	3 feet	40.5
4	55.5	4 feet	27
5	81	10	66
6	69	20	50
7	43	15	35
8	83	20	58
9	80.5	19	55
10	66.5	Dry completely	25

i. Livestock gets water to drink
ii. Residents get water for irrigation and drinking purpose.
iii. Ecological stability by increasing ground water level.

Johads changed the socio-economic status of the region [70]. Nobody could imagine that a single johad could bring about such a huge change. Water supply could now be ensured for the entire year. Apart from meeting the daily needs of water for drinking cooking and cleaning purpose, there has been a noticeable change in the biomass productivity, and two seasonal rivers were converted to perennial; Aravri and Ruparel (Table 3).

9.2 *Rain Water Harvesting and Groundwater Recharge*

Barmer is the second largest district, located in the west Rajasthan. It also constitutes the part of Thar Desert. The temperature experienced here is 46–51 °C. This district falls under the arid climatic zone. Barmer experiences situations of drought on a frequent basis. The agriculture here is mainly rain fed. Since there is no assurance of rain the area follows dry land farming, but this is also very risky so as an alternate to the livelihood, livestock rearing is carried out. Approx. 277 mm rainfall per year is experienced for 10–15 days.

TechnoService is a non-profit organisation for the economic development which was founded in 1968. The agency is improving the natural resource conservation by constructing and repairing structures at individual and community level. This main objective is to construct runoff structures for the prevention of water loss. As the data states this service has constructed 157 rain water harvesting known as **khadin**. During the kharif season 2015 these tanks were able to harvest 20,47,000 rain water which was later able to recharge natural aquifers by percolation [72].

9.2.1 Objective

1. Khadin is a well-planned water structure for the preservation of rain water from runoff and then utilising it for agricultural purposes. This rain water harvesting structure is built chiefly for utilisation of water for agricultural purposes [73]. The structures are built on agricultural field and the water is used subsequently for crop production. Redesign of Nadi Structure Nadi, a customary Water Harvesting Structure in the Western Rajasthan is uncovered or embanked for collecting pitiful precipitation, to moderate the shortage of savoring water the lean time frame i.e. late-spring and amid summer. Rain water gathered in the Naadi from the overflow are made accessible for periods beginning four months to a year after rain, contingent upon the catchment qualities, the measure of precipitation got and its force. This is an old practice and the Nadis are the most vital water wellsprings of the district utilized for drinking purposes by individuals and additionally animals. In any case, at the appointed time of time, these Nadis are silted and henceforth the water stockpiling limit is decreased to a substantial expand. Henceforth, these Nadis become scarce amid summer. Again because of absence of legitimate administration rehearse, absence of upkeep, the group endured a considerable measure to get the water for drinking. With due demand from the group and Panchayats, the undertaking taken up the redesign work of such two Nadis with dynamic investment and commitment from group. The procedure received in building up a Nadi is in association with Community. In the wake of getting the No Objection Certificate from the Panchayat, the remodel work started with dynamic cooperation, supervision and commitment with dynamic group investment. Amid the year 2015, two such structures are redesigned and finished before the rain amid June, 2015 and permitted the water passage and reaped in the structure. Amid July,2015, it is evaluated that 1,90,000 of water is put away in the two Nadi structures over a capacity zone of 5 hectares and furthermore contributing towards ground water revive.
2. Development of Community Khadin Structures Community Khadin structures are comparable Rainwater Harvesting Structures as for the expansive catchment region (more than 100 hectares) in a bigger zone of capacity. This is an overflow reaping structure, intended to gather surface spill over water for farming and ground water revive. It is an earthen bank worked over the general slant which preserves the most extreme conceivable water overflow inside the rural field. The Khadin framework depends on the rule of collecting water on farmland and resulting utilization of this water-soaked land for trim generation. Two substantial group Khadin structures for two gatherings of agriculturists (74 Farmers) are developed in the Bhadkha Gram Panchayat of Barmer Block. With dynamic support of the ranchers, the structures are finished amid June, 2015 and are prepared before rain for gathering and putting away of water amid blustery season. If it is not too much trouble, discover a portion of the photos demonstrating the measure of water put away amid rain which contributed towards ground water energize.

9.3 Mukhyamantri Jal Swavlamban Abhiyan (MJSA)

It is a multi-stakeholder project which aims at making even the most remote village water sufficient [74].

Shockingly the real test for the task, which was propelled in January this year, was not rare precipitation but rather keeping a monstrous yearly wastage of water.

Covering 295 squares in 33 areas with a precipitation over 450 mm, every one of that was required was to guarantee that precipitation is caught, however the arrangement had an inborn issue—water does not know any limit.

"Rain water streams all around, be it backwoods arrive, government land or private fields. We expected to apply propelled water reaping techniques, however for that, many individuals needed to go ahead board and that was a test," said Sriram Vedire, Chairperson of Rajasthan River Basin and Water Resources Planning Authority [75]. As per Vedire, getting all partners to concur was a "test" however "transforming it into a mass development did the trap."

Gaining by Andhra Pradesh's effective explore different avenues regarding the 'Four Water' idea—rain water to surface water to soil dampness and at last to ground water—the program, which has been separated into various stages, concentrated on raising the ground water table. Phase 1 saw development of almost 93,000 ease water gather structures. These incorporate town lakes, field bundings, anicuts, minor water system tanks (MIT), minor permeation tanks (MPT) and minor stockpiling tanks (MST).

Great as it is for an administration plan to accomplish so much, the way that a ton of it was managed without dislodging individuals or securing land is the thing that prevailed upon the general population.

"For a MPT, we did not have to go in for arrive procurement since water is put away in these tanks for only fourteen days or months, and from there on, it permeates into the ground. Ranchers advantage since they are getting clammy soil which will help support efficiency," said Vedire.

Maybe the most intriguing piece of the story lies in the amalgamation of strategies and nearby knowledge about water gathering. Geo-labelling each work site helped screen every single development other than guaranteeing specialized attainability and monetary practicality.

10 Conclusions

With the increase in population and the growth in industrialisation, the demand for water is increasing day by day. To make the resources available to our future generations we need to follow the concept of sustainable development. Several methods and techniques have been implemented in order to overcome the needs of people but technology is not easy for all people to use, it does not go hand in hand. The technological advancements and the changed lifestyle has also been one of the reasons in overexploitation of water resources. Engaging in the use of new technologies and forgetting the old and traditional water conservation practices could not bring a change in water problems. Water problem is shared equally across the globe, so it has become a global problem. In order to meet the demands for water we need to start practicing rainwater harvesting. The amount of water harvested varies from place to place. It depends on the amount of annual rainfall data. Rainwater is found to be the most pure form of water as it is free from impurities and contaminants. Rainwater has many potentials to increase the ground water level. Rainwater harvesting is the need of the hour. A useful technique should be developed in a country like India.

Acknowledgements All these development have been carried out with the financial support of **Cairn CSR.**

References

1. Ishaku HT, Majid MR, Johar F (2012) Rainwater harvesting: an alternative to safe water supply in Nigerian rural communities. Water Resour Manag 26(2):295–305
2. Niemczynowicz J (1999) Urban hydrology and water management – present and future challenges. Urban Water 1(1):1–14
3. Aslan D, Selçuk SA (2018) A biomimetic approach to rainwater harvesting strategies through the use of buildings. Eurasian J Civil Eng Architect 2(1):27–39
4. Shah T, Raju KV (2002) Rethinking rehabilitation: socio-ecology of tanks in Rajasthan, north-west India. Water Policy 3(6):521–536
5. Wurthmann K (2020) Rainwater harvesting system installations required to offset new water demand created by growing populations in Broward and Palm Beach Counties: a dataset for decision making based on numbers of installations, costs, and water and energy savings. Data Brief 28:105016
6. Foster SSD, Chilton PJ (2003) Groundwater: the processes and global significance of aquifer degradation. Philos Trans R Soc Lond B Biol Sci 358(1440):1957–1972
7. Li Y et al (2018) Multi-objective optimization integrated with life cycle assessment for rainwater harvesting systems. J Hydrol 558:659–666
8. Singh RB (2000) Environmental consequences of agricultural development: a case study from the Green Revolution state of Haryana, India. Agric Ecosyst Environ 82(1):97–103
9. Oki T, Kanae S (2006) Global hydrological cycles and world water resources. Science 313(5790):1068
10. Fletcher TD, Andrieu H, Hamel P (2013) Understanding, management and modelling of urban hydrology and its consequences for receiving waters: a state of the art. Adv Water Resour 51: 261–279
11. Chatterjee R et al (2018) Mapping and management of aquifers suffering from over-exploitation of groundwater resources in Baswa-Bandikui watershed, Rajasthan, India. Environ Earth Sci 77(5):157
12. Villarreal EL, Dixon A (2005) Analysis of a rainwater collection system for domestic water supply in Ringdansen, Norrköping, Sweden. Build Environ 40(9):1174–1184
13. Farreny R et al (2011) Roof selection for rainwater harvesting: quantity and quality assessments in Spain. Water Res 45(10):3245–3254
14. Joshua R, Vasu V (2013) Characteristics of stored rain water and its treatment technology using moringa seeds. Int J Life Sci Pharma Rev 2:155–174

15. Alamdari N et al (2018) Assessing climate change impacts on the reliability of rainwater harvesting systems. Resour Conserv Recycl 132:178–189
16. Stahn H, Tomini A (2010) A drop of rainwater against a drop of groundwater: does rainwater harvesting really allow us to spare. Groundwater
17. Glendenning C et al (2012) Balancing watershed and local scale impacts of rain water harvesting in India—a review. Agric Water Manag 107:1–13
18. Vicente-Serrano SM et al (2010) A complete daily precipitation database for northeast Spain: reconstruction, quality control, and homogeneity. Int J Climatol 30(8):1146–1163
19. BlueDrop Series Rain water harvesting and Utilisation. Un-habitat
20. Abas PE, Mahlia T (2019) Techno-economic and sensitivity analysis of rainwater harvesting system as alternative water source. Sustainability 11(8):2365
21. (2010) Rainwater harvesting: model-based design evaluation. Water Sci Technol 61(1):85–96
22. Novo AV et al (2010) Review of seasonal heat storage in large basins: water tanks and gravel–water pits. Appl Energy 87(2):390–397
23. Basinger M, Montalto F, Lall U (2010) A rainwater harvesting system reliability model based on nonparametric stochastic rainfall generator. J Hydrol 392(3–4):105–118
24. Campisano A et al (2017) Urban rainwater harvesting systems: research, implementation and future perspectives. Water Res 115:195–209
25. Li Z, Boyle F, Reynolds A (2010) Rainwater harvesting and greywater treatment systems for domestic application in Ireland. Desalination 260(1–3):1–8
26. Krishna R, Mishra J, Ighalo JO (2020) Rising demand for rain water harvesting system in the world: a case study of Joda Town, India. World Scientific News 146:47–59
27. Mahmoud WH et al (2014) Rainfall conditions and rainwater harvesting potential in the urban area of Khartoum. Resour Conserv Recycl 91:89–99
28. Angelakis A, Spyridakis D (2010) A brief history of water supply and wastewater management in ancient Greece. Water Sci Technol Water Supply 10(4):618–628
29. Rockström J et al (2010) Managing water in rainfed agriculture—The need for a paradigm shift. Agric Water Manag 97(4):543–550
30. Braud I et al (2013) Evidence of the impact of urbanization on the hydrological regime of a medium-sized periurban catchment in France. J Hydrol 485:5–23
31. Cech TV (2010) Principles of water resources: history, development, management, and policy. Wiley
32. Nijhof S et al (2010) Rainwater harvesting in challenging environments: towards institutional frameworks for sustainable domestic water supply. Waterlines 29(3):209–219
33. Glendenning C, Vervoort R (2011) Hydrological impacts of rainwater harvesting (RWH) in a case study catchment: The Arvari River, Rajasthan, India: Part 2. Catchment-scale impacts. Agric Water Manage 98(4):715–730
34. Rockstrom J (2013) Balancing water for humans and nature: the new approach in ecohydrology. Routledge
35. Khastagir A, Jayasuriya N (2010) Optimal sizing of rain water tanks for domestic water conservation. J Hydrol 381(3–4):181–188
36. Bhattacharya AK (2010) Artificial ground water recharge with a special reference to India. Int J Res Rev Appl Sci 4(2):214–221
37. Kumar MD et al (2006) Rainwater harvesting in India: some critical issues for basin planning and research. Land Use Water Resour Res 6(1):1–17
38. Kumar MD et al (2008) Chasing a mirage: water harvesting and artificial recharge in naturally water-scarce regions. Econ Polit Wkly:61–71
39. Pareek A, Trivedi P (2011) Cultural values and indigenous knowledge of climate change and disaster prediction in Rajasthan, India
40. Ragin CC, Amoroso LM (2010) Constructing social research: The unity and diversity of method. Pine Forge Press
41. Raya RK, Gupta R (2020) Rural community water management through directional tunnelling: visual modelling of rainwater harvesting system. Water Pract Technol 15(3):734–747

42. Ammar A et al (2016) Identification of suitable sites for rainwater harvesting structures in arid and semi-arid regions: a review. Int Soil Water Conserv Res 4(2):108–120
43. Kumar R (2015) An economic study on impact of rainwater harvesting structures for agriculture in semi-arid areas of Rajasthan. Division of Agricultural Economics ICAR-Indian Agricultural Research Institute New Delhi-1
44. Singh DR (2011) An economic analysis of water users associations in canal irrigated area in Tamil Nadu. IARI, Division of Agricultural Economics
45. Pattanaaik S, et al (2012) Traditional system of water management in watersheds of Arunachal Pradesh
46. Sharma BR, et al (2010) Water poverty in the northeastern hill region (India): potential alleviation through multiple-use water systems: cross-learnings from Nepal Hills. Vol. 1. IWMI
47. Krishan S (2011) Water harvesting traditions and the social milieu in India: a second look. Econ Polit Wkly:87–95
48. Singh HP et al (2010) Sustainability of traditional drinking water sources in Himachal Pradesh. Nat Environ Pollut Technol 9(3):587–592
49. Kumar V, Jain SK, Singh Y (2010) Analysis of long-term rainfall trends in India. Hydrol Sci J 55(4):484–496
50. Chaudhry P, Gupta RK (2010) Urban greenery and its sustainable extension strategies in hot arid region of India. Int J Sustain Soc 2(2):146–155
51. Poonia S, Rao A (2013) Climate change and its impact on Thar desert ecosystem. J Agric Phys 13(1):71–79
52. Everard M (2015) Community-based groundwater and ecosystem restoration in semi-arid north Rajasthan (1): socio-economic progress and lessons for groundwater-dependent areas. Ecosyst Serv 16:125–135
53. Amit D, Jethoo A, Poonia M (2012) Impact of drought on urban water supply: a case study of Jaipur city. Int J Eng Innov Technol 1:170–174
54. Gupta S (2011) Demystifying 'tradition': the politics of rainwater harvesting in rural Rajasthan, India. Water Alternat. 4(3)
55. Glendenning C, Vervoort R (2010) Hydrological impacts of rainwater harvesting (RWH) in a case study catchment: The Arvari River, Rajasthan, India. Part 1: Field-scale impacts. Agric Water Manag 98(2):331–342
56. Dai A (2011) Drought under global warming: a review. Wiley Interdiscip Rev Clim Chang 2(1): 45–65
57. Qureshi AS et al (2010) Challenges and prospects of sustainable groundwater management in the Indus Basin, Pakistan. Water Resour Manage 24(8):1551–1569
58. Goyal R (2010) Rainwater harvesting: A key to survival in hot arid zone of Rajasthan. In Proceedings of national workshop-cum-brain storming on rainwater harvesting and reuse through farm ponds: Experiences, issues and strategies
59. Saxena D. Water conservation: traditional rain water harvesting systems in Rajasthan
60. Tamagnone P, Comino E, Rosso M (2020) Rainwater harvesting techniques as an adaptation strategy for flood mitigation. J Hydrol 586:124880
61. Aladenola OO, Adeboye OB (2010) Assessing the potential for rainwater harvesting. Water Resour Manag 24(10):2129–2137
62. Teston A et al (2018) Rainwater harvesting in buildings in Brazil: a literature review. Water 10(4):471
63. Kadam AK et al (2012) Identifying potential rainwater harvesting sites of a semi-arid, basaltic region of Western India, using SCS-CN method. Water Resour Manag 26(9):2537–2554
64. Misra AK (2019) Rainwater harvesting and artificial recharge of groundwater. In: Groundwater development and management. Springer, pp 421–439
65. Abdulla F, Abdulla C, Eslamian S (2021) Concept and technology of rainwater harvesting, in handbook of water. Harvest Conserv:1–16
66. De Kwaadsteniet M et al (2013) Domestic rainwater harvesting: microbial and chemical water quality and point-of-use treatment systems. Water Air Soil Pollut 224(7):1629

67. Gwenzi W et al (2015) Water quality and public health risks associated with roof rainwater harvesting systems for potable supply: review and perspectives. Sustain Water Qual Ecol 6: 107–118
68. Torri MC (2010) Decentralising governance of natural resources in India: lessons from the case study of Thanagazi Block, Alwar, Rajasthan, India. Law Env't Dev J 6:228
69. Thanju JP, Shrestha BD (2010) Miracle in Rajasthan: traditional practice of rainwater harvesting. Hydro Nepal: J Water Energy Environ 7:20–22
70. Vij S, Narain V (2016) Land, water & power: the demise of common property resources in periurban Gurgaon, India. Land Use Policy 50:59–66
71. Samantaray R (1998) Johad – watershed in alwar district
72. Sharma R et al (2014) Emergence of dengue problem in India—a public health challenge. J Commun Dis 46(2):17–45
73. Bhattacharya S (2015) Traditional water harvesting structures and sustainable water management in India: a socio-hydrological review. Int Lett Nat Sci 37
74. Dheeraj M (2017) Social mobility and change in a Rajasthan Village Ranawaton ki Sadri reviewed
75. Everard M, et al (2018) Report of the three-day workshop on 'Regeneration of the Banas-Bisalpur Socio-ecological Complex'

Groundwater Exploration Using Remote Sensing and GIS Techniques Coupled with Vertical Electrical Soundings from Hard Rock Terrain: A Case Study in Salem District, Southern India

S. Sankaran and S. Siva Rama Krishnan

Abstract In South India, Salem District can be geologically classified as a hard rock terrain. More than 90% of the district is underline by hard rock of Archaean. The tectonic activities like weathering, fracturing or faulting aid in development of secondary porosity. The weathering and fracturing in hard rocks are highly variable within a very short distance which challenges the location of potential sites. Under such a scenario, a baseline information on various groundwater controlling factors namely Geology, Geomorphology, Topography, Lineament, Drainage and Soil is imperative. Remote Sensing (RS) and Geographical Information System (GIS) techniques are the major tools in deciphering the surface and sub-surface conditions. The study gains importance since an integrated approach using RS and GIS for mapping the structural features have been field checked with collateral data including Vertical Electrical Sounding (VES) while targeting high yielding bore well locations. The entire study are was mapped on 1: 25,000 scale using IRS-P6, LISS-IV satellite data and Shuttle Radar Topography Mission (SRTM 90 m) elevation data. Under the GIS environment, drainage and lineament densities and slope were calculated. Suitable weights were assigned for each layer parameters. Based on the above data analysis, ground water potential maps were generated in the hard rock terrain of Salem District (Southern India). The potential map was divided into six prospective zones (i.e., excellent, very good, good, moderate to poor, Poor and very poor). The above systematic analysis helped to locate 21 VES points wherein the shallow aquifers and deeper fracture zones could be delineated. In majority of these locations, High yielding (755–49 liters/minute) bore wells were drilled with good correlation of litho logs which greatly helped in

S. Sankaran (✉)
Environmental Geophysics Group, National Geophysical Research Institute, Council of Scientific & Industrial Research, Hyderabad, India

S. Siva Rama Krishnan
School of Information Technology and Engineering, VIT, Vellore, India
e-mail: siva.s@vit.ac.in

S. Yadav et al. (eds.), *Water Quality, Assessment and Management in India*, Earth and Environmental Sciences Library, https://doi.org/10.1007/978-3-030-95687-5_10

providing the much needed rural water supply schemes in this water-scarce area of Salem District.

Keywords Groundwater exploration · Remote sensing · Geographic information system · Vertical electrical sounding · Groundwater prospect map · Hard rock · South India

1 Introduction

The Integrated Groundwater Exploration using Satellite Imagery, RS-GIS, and Hydro-geomorphological maps play a significant role in targeting potential shallow & deep aquifers in hard rock areas. Such an approach helps us to identify the Major Lineaments, Faults, and Deeper Fracture zones due to the tectonic activities and degree of weathering in such hard rock terrains [1]. These tectonic activities induce secondary porosities which vary significantly within a short distance and hence the groundwater potential in hard rock areas. In India, most of the hard rock regions are devoid of primary porosity and the low rainfall coupled with high surface run-off restrict the scope of groundwater movement and storage.

A systematic scientific approach is to design a proper data collection procedure through various available information of the study area and it's subsurface. Such information leads to important clues in targeting potential aquifer zones. RS is a useful tool in collecting information regarding Geology, Geomorphology, lineaments and slope from large and inaccessible areas. GIS platform is ideal to merge the interpretation of RS with discrete and continuous data from primary and secondary sources [2]. Such integration of data can converge the Hydrogeological investigations to identify groundwater potential zones in a more economical and rapid method [3]. Integration of Multi-Thematic maps using RS-GIS were found useful for exploration, exploitation and management of Groundwater Resources [1, 4, 5]. Mapping of lineaments and its verification through Ground Geophysical Investigation [6], Hydrogeological and Geophysical investigations in arid and semi-arid regions of India for estimating groundwater potential zones fetched 80% success [7] in crystalline areas. The combination of GIS and Resistivity surveys in targeting potential aquifer zones has been reported by various authors [8–10]. Studies in collaboration with National Remote Sensing Agency (NRSA) India, [11, 12] were successful in identifying potential Groundwater zones. Similarly, many authors [13–21] have deployed RS-GIS Techniques coupled with Geophysical Investigations for identifying Geomorphic units, mapping of Fractures and identifying artificial recharge sites for Groundwater prospecting.

In the present study, the RS-GIS techniques is used for generation of various Thematic maps depicting Geology, Geomorphology, drainage, slope and soil cover. This helped in deciphering Groundwater Potential zones. Vertical Electrical Sounding (VES) in such focused areas were successfully demonstrated by exploratory drilling in targeting high yielding wells.

2 Study Area

The study area is located in a part of the tectonic element of Moyar–Bhavani–Salem–Attur Shear Zone (MBSZ) “Granulite terrain” in the peninsular shield. It covers an area of 384 km^2, comprising three Panchayat unions, namely Konganapuram, Magudanchavadi (MC Donald choltry) and Veerapandy, in Salem district of Tamil Nadu, India (Fig. 1). The MBSZ is 3–4 km wide and its strike is E–W [14]. Kanjamali hill is inferred as doubly plunging syncline structural hill well known on account of the occurrence of magnetite iron ore deposits located at a distance of 10 km west of Salem town.

It has an elevation of 986 m above MSL and a total length of 7.2 km from east to west with width of about 3.2 km. The researchers in [22] identified few fault zones trending NNE-SSW directions on the eastern and western margin of Kanjamalai hill. The MBSZ is trending E-W direction and its related parallel lineaments are playing significant role in the groundwater movement and occurrence. It lies between the latitudes 11°30′ 0″ to 11°40′0″ North and longitudes 77°50′0″ to 78°7′50″ East. The main geological formation in the study area are Peninsular Gneiss and Charnockites, which are tectonically disturbed giving rise to Folds, Faults, Lineaments, Shear Zones and Fractures. The major drainage of the study area is towards the eastern part, contributed by Tirimanimuttar river flowing NE-SW. The topography of the

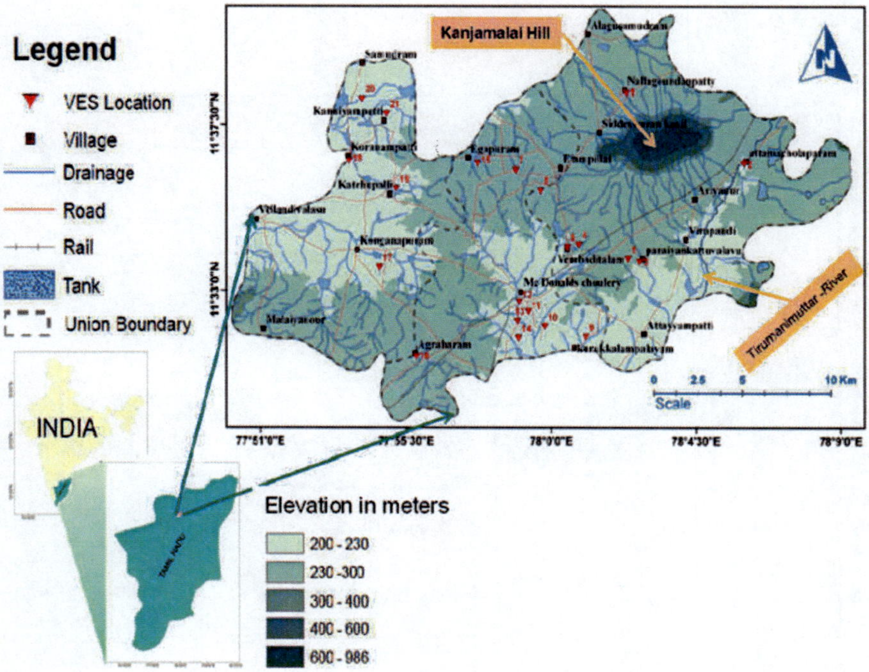

Fig. 1 Location map of the study area

study area varies with 200 to 986 m elevation and steep slopes along the hill ranges and most of the area is gentle and undulating slopes towards the south and south-western sides. Salem district falls in the Rain shadow region hence receives poor rainfall (504.6–920.8 mm/annum). The Groundwater fluctuation varies from 02 m to 13.5 m with lowest levels during summer (March–June). The Fractures and Joints in both the Gniessic and Charnockites play a significant role in Groundwater storage and these layer thickness vary from 5–8 m to 55 m. At the Geomorphic contacts between Gneiss and Charnockites, the potential zone thickness can reach up to 90 m.

3 Materials and Methods

The Flow Chart (Fig. 2) explains the systematic approach and analysis which are divided into three main parts i.e., Generation of Spatial database, Analysis of Spatial data and Data modeling. The materials used for study involves the SOI Toposheet (58I/2 and 58E/14) at 1:50,000 scales to prepare base and drainage maps. The Remote Sensing (RS) and GIS-based image interpretation keys like tone, texture, shape, size, pattern, and association elements are useful for identification and interpretation of the various thematic maps. The IRS-P6-LISS-IV and SRTM data have been digitally enhanced with the help of ENVI digital image processing

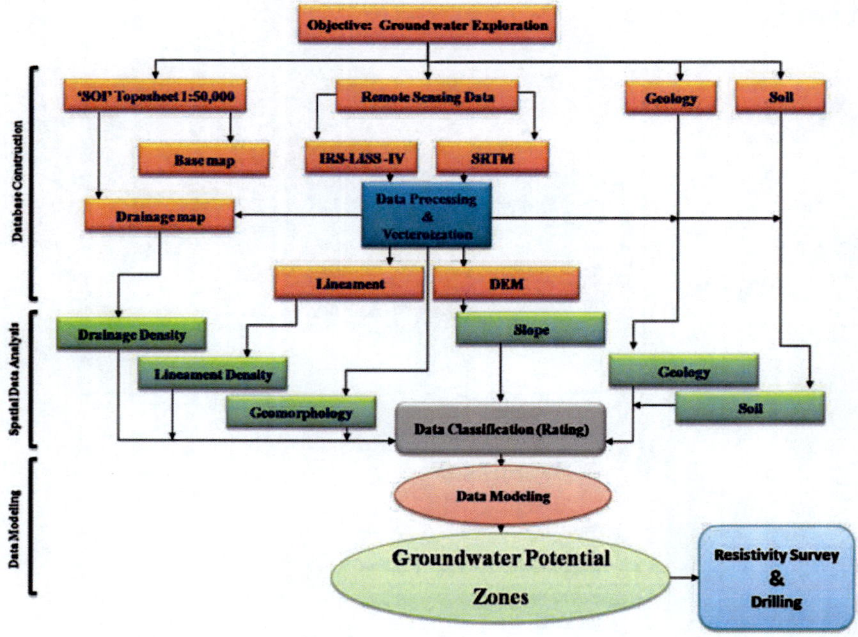

Fig. 2 A flow chart on the methods of groundwater exploration in Salem district, Tamil Nadu

software (Arc GIS v.10) for vectorization of the various thematic layers. The band ratios, principal component analysis, edge enhancement techniques were performed for the deriving lineament and different geomorphologic units. The topography and slope maps have been generated from the SRTM elevation data. The geology map was prepared from the GSI maps at 1:250,000 scale and 34th OCG material from GSI 2010–2011 and updated with RS data. The soil map was prepared from the soil map of India and it was also updated from the satellite data.

3.1 *Spatial Database Construction*

The Arc catalog tool was used for tables, geometric networks, and other items inside the database. The designed maps such as lineament, geomorphology, geology, soil, drainage and slope maps (Fig. 3) were vectorized under GIS environment and processed for topological errors such as dangles, pseudo-nodes and attributes to these maps have been added. In any coverage, attributes need to be added to available features to distinguish them. The drainage, geology, soil maps were further updated with the digital image processing of the individual satellite data coupled with field verification.

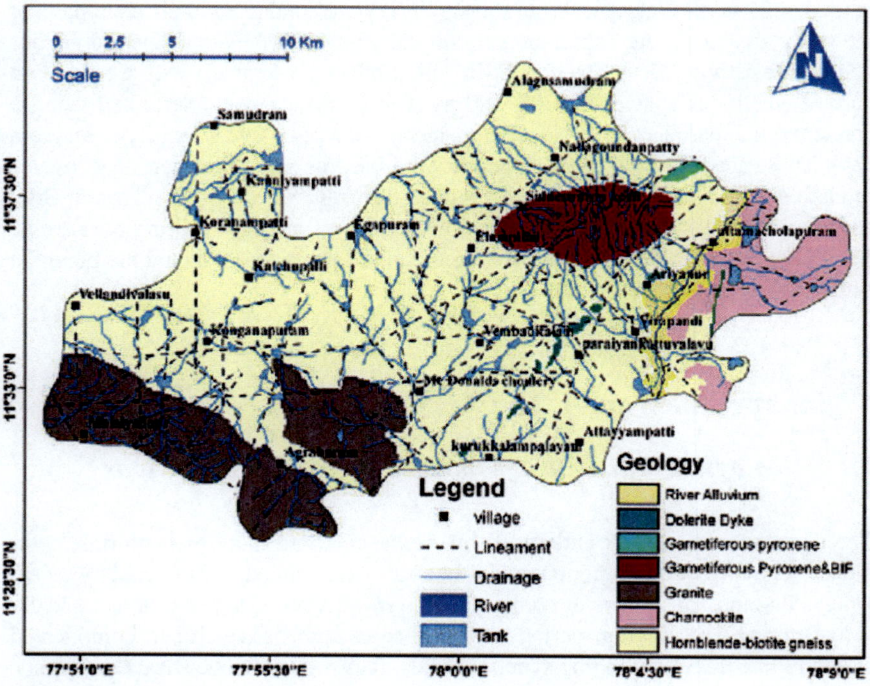

Fig. 3 Geology map of the study area

3.2 Spatial Data Analysis

The various thematic layers representing Geology, lineament, geomorphology, slope, drainage and soil, are converted into the raster formats using 25 m × 25 m grid cell size. Each Layer maps were reclassified with suitable weightages based on the works carried out by other researchers [15, 18, 23].

3.3 Data Modeling

The thematic layer maps were converted to raster data with 25 m × 25 m grid size. The deduced raster maps of the individual parameters were assigned a theme weightage and rating. The two were multiplied to get the overlaid Raster thematic layers which were analyzed under Arc-GIS tool to generate Groundwater prospective maps.

3.4 Validation of Groundwater Prospective Map

A total of 21 vertical electrical soundings (VES) were conducted in different parts of the study area using the Schlumberger configuration with the aid of the indigenous DDR-2 Resistivity Meter (NGRI, CSIR lab, product Hyderabad) with a maximum current electrodes spacing of 300–500 m. The field data was interpreted using a conventional partial curve matching technique [24] and auxiliary diagram [25]. These interpreted layer parameters were used as initial estimates for inverse modeling based on optimization techniques using RESIST v. 1.0 software. Site-specific inferences were drawn for recommending 13 sites for drilling bore wells. The yields of the drilled bore wells were measured using V-notch and the borehole Litho-logs were also observed.

4 Interpretation and Discussion

4.1 Geological and Hydrogeomorphological Investigations

The study area is entirely underlined by Archaean "crystalline metamorphic complexes. The rocks of this group are highly weathered, jointed and covered by recent valley fills and soil covers at some places" [26]. The study area was tectonically active during Precambrian period and its manifestation gave rise to complicated structure like doubly plunging synclinal folds, faults, lineaments, shears and joints. The geology of the area comprises of Hornblende-biotite gneiss, granites,

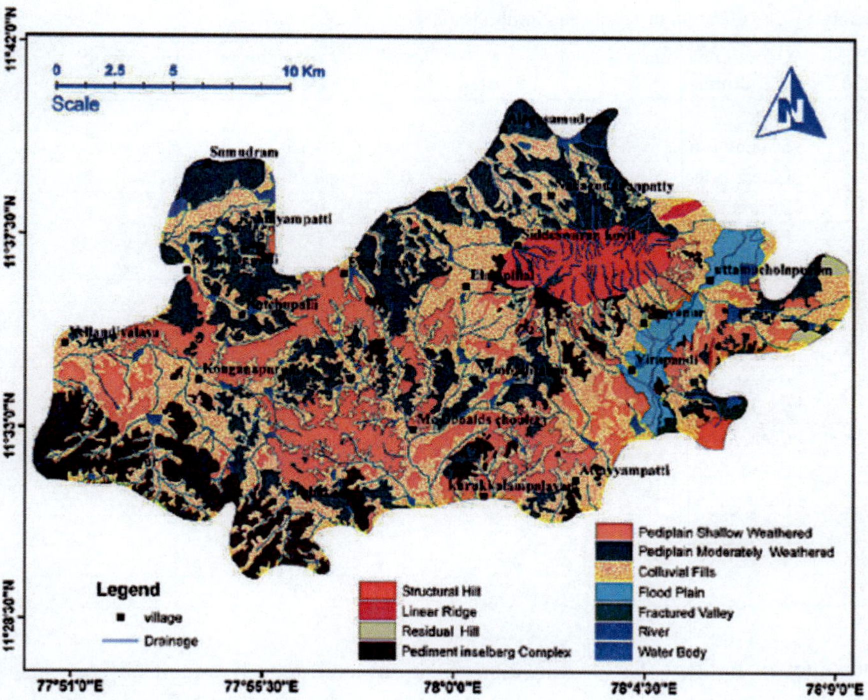

Fig. 4 Geomorphology map of the study area

charnokites, and garnetiferous pyroxene with banded iron ore formation of Kanjamali hills [22]. The recent alluvial deposits are found along the river channel in the eastern part. The values assigned to the geology layer take into account the hydrogeological significance of the rock types for groundwater occurrences. The lithology was characterized based on rock type, thickness of weathering, fracture density and occurrence of dykes.

Geomorphology map (Fig. 4) was prepared using the IRS-P6-LISS-IV coupled with the SRTM data, which in turn are related to surface runoff and infiltration. Geomorphological studies involve the identification and characterization of the fundamental units of landscape. The underlying lithology, slope and the type of existing drainage patterns influence the genesis and processes of different geomorphic units. The satellite images were examined based on visual interpretation and the various geomorphic units in the study area were categorized (i.e., structural hills, linear ridge, residual hill, and pediment inselberg complex) based on their image characteristics which form run-off zones and (i.e., pediplain moderately weathered, pediplain shallow weathered, colluvial fills, flood plain, and fractured valley) form infiltration zones (Table 1). These runoff zones are not suitable for groundwater storage, hence a low weightage (1) were assigned for the preparation of groundwater potential zones. The flood plain, which is in infiltration areas, was assigned

Table 1 Classification of runoff and infiltration zones

Sl. No.	Geomorphic unit (Landforms)	Symbol	Area (km^2)	Area covered (%)	Weightage assigned
Run off zones					
1	Structural hills	SH	18.23	4.75	1
2	Linear ridge	LR	0.95	0.25	1
3	Residual hill	RH	1.34	0.35	1
4	Pediment inselberg complex	PIC	43.99	11.45	1
Infiltration zones					
5	Pediplain moderately weathered	PPM	80.52	20.97	3
6	Pediplain shallow weathered	PPS	71.77	18.69	2
7	Colluvial fills	CF	145.89	37.99	4
8	Flood plain	FP	11.81	3.07	5
9	Fractured valley	FV	1.33	0.35	2
10	River	R	0.34	0.09	–
11	Water body	WB	7.88	2.05	–
Total			384.05	100%	

maximum weightage (5). Identification of various landforms in term of surface water recharge was the next significant step to assign weightage between 2 to 4 (Table 1). In the study area, the weathered, fractured rocks and the recent colluvial deposits constitute the main aquifers. In addition, the presence of Fault/Fracture zones passing through these units which act as conduits for movement and occurrence of groundwater were also considered while giving more weightage. Out of total study area, 83.3% is covered by infiltration zones and 16.7% by run-off zones. Such large areal extent of the infiltration zones indicates fairly good availability/storage of groundwater.

4.2 Elevation and Slope

The slope is the rate of change of elevation and considered as the principal factor of the superficial water flow since it determines the gravity effect on the water movement [27]. Elevation and slope both were considered for identifying the potential zones. The elevation was divided into two zones; plain land with elevation (200–300 m) and hilly terrain with elevation (300–986 m). The elevation ranges against their ratings are given in Table 2. The slope is directly proportional to runoff and groundwater recharge will be lesser in the areas with steep slope. In areas of gentle and undulating plains, the surface water contributes significantly to recharge the weathered and fractured aquifer. The slope was calculated from the Digital Elevation Model (DEM), which was obtained from the SRTM digital elevation data.

Table 2 Assigned weightages and ratings for preparation prospective map

Factor	Weight (a)	Classes	Rating (b)	Score (a*b)
Geology	3	River Alluvium	5	15
		Hornblende-biotite gneiss	3	9
		Granite	2	6
		Garnetiferous Pyroxene & BIF	1	3
		Garnetiferous pyroxene	1	3
		Charnockite	1	3
		Dolerite Dyke	1	3
Hydro-geomorphology	5	Structural hill (SH)	1	5
		Linear Ridge (LR)	1	5
		Residual hill (RH)	1	5
		Pediment inselberg complex (PIC)	1	5
		Pediplain Moderately weathered (PPM)	3	15
		Pediplain shallow weathered (PPS)	2	10
		Colluvial Fills (CF)	4	20
		Flood plain (FP)	5	25
		Fractured valley (FV)	2	10
Topography (m)	1	200–230	4	4
		230–300	2	2
		300–400	2	2
		400–600	1	1
		600–986	1	1
Slope (°)	2	Nearly level (0–1)	4	8
		Very gentle (1–3)	3	6
		Gentle (3–5)	2	4
		Moderate (5–10)	1	2
		Moderate to steep (10–15)	1	2
		Steep (>15)	1	2
Lineament-D (km^2)	4	Low density (<0.5)	1	4
		Medium to low (0.5–1)	2	8
		Medium (1–1.5)	3	12
		High density (1.5–2)	4	16
		Very high density (>2)	5	20
Drainage-D (km^2)	3	Low density/coarse texture (0–1)	4	12
		Medium density/medium texture (1–2)	2	6
		High density/fine texture (2–4)	1	3
		Very highdensity/superfine texture (>4)	1	3
Soil	2	Stony waste	1	2
		Sandy soil with rock out crops	2	4
		Colluvial Soil	4	8
		Brown silty clayey Soil	3	6
		Alluvial Soil	5	10

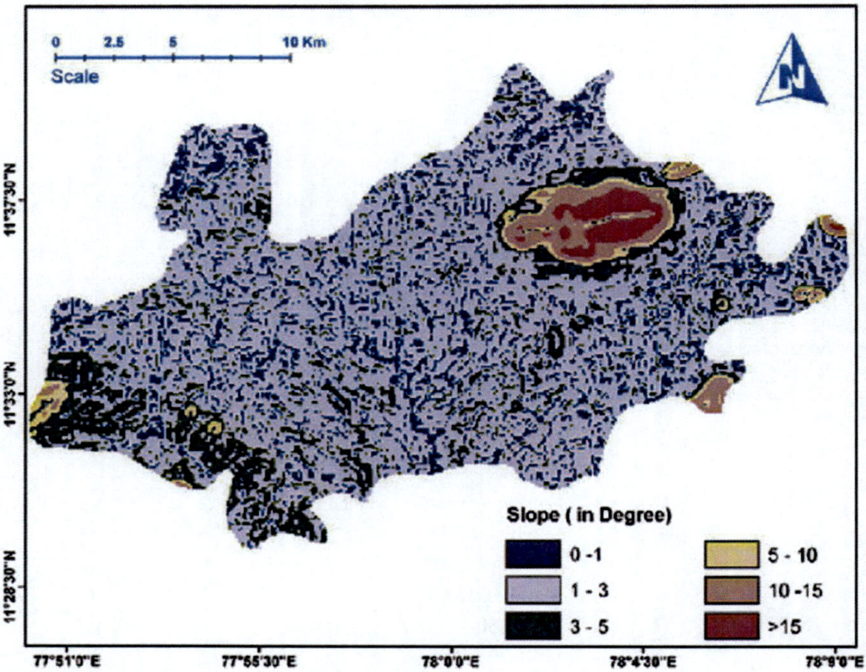

Fig. 5 Slope map of the study area

The identified slope category (in degree) were classified into 6 classes (Fig. 5) which varied from 0–1° (nearly level), 1–3° (very gently sloping), 3–5° (gently sloping), 5–10° (moderately sloping), 10–15° (steep sloping), >15° (very steep sloping).

4.3 Lineament Analysis

Digital image processing systems are ideal for mapping lineaments and working in a digital environment provides many advantages over a traditional manual approach. The authors [26] have suggested that the stronger surface expression and longer lineament is more significant for ground-water exploration. The extension of large lineaments representing a shear zone or a major fault can extend subsurface from hilly terrain to alluvial terrain. It generally forms a productive groundwater reserve. Similarly, intersection of lineaments can also be probable sites of groundwater accumulation [28]. The lineaments are developed by the tectonic activity reflect a general surface manifestation of underground fractures with inherent characteristics of porosity and permeability of the underlying materials [29]. The present study area has under gone intensive tectonic activity and it falls under E-W trending MB shear

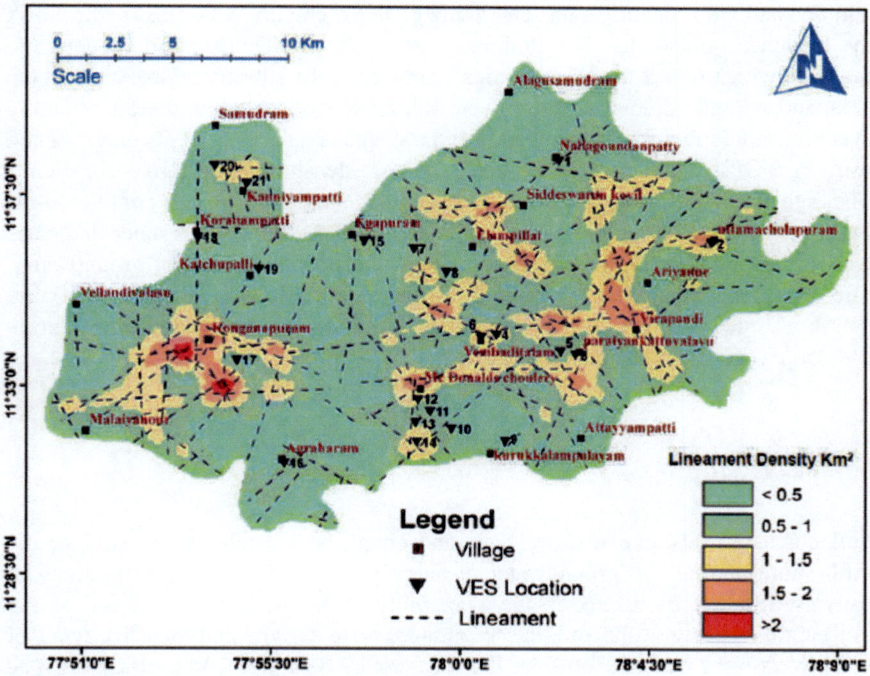

Fig. 6 Lineament density map of the study area

zone. The lineaments are trending NE-SW, NW-SE, NNE-SSW, E-W. The E-W trending lineaments are parallel to the MBSZ acting as major drivers for groundwater movement. The linear NE-SW trending Dolerite dykes are present and they are dissected and shifted, which may also act as good groundwater potential zones. Generally, it is expected that the thickness of weathered/fractured rock is greater along the lineaments and control the movement/availability of groundwater [30]. It was found that normalized transmissivity near the lineaments is high and excellent relationship exists between higher fracture densities and higher well yields [31]. Thus the lineament densities (in km^2) of the study area were delineated into 5 classes (Fig. 6) and they are varied from <0.5 km^2 (very poor), 0.5 to 1 km^2 (poor density), 1 to 1.5 km^2 (moderate density), 1.5 to 2 km^2 (high density) and >2 km^2 (very high density).

4.4 Drainage/Drainage Density

The drainage pattern is one of the important indicators of hydrogeological features, because drainage pattern, texture and density are controlled by the underlying lithology. The drainage map was prepared from the Survey of India (SOI) toposheet and updated with Satellite data. The drainage is structurally controlled and follows the lineament and faults. The drainage pattern shows dendritic and subdendritic drainage which reflect the homogenous character of the subsurface materials in the area, and a Radial drainage pattern (centrifugal) was observed on doubly plunging synclinal structural hill (Kanjamalai hill). Drainage density map has been generated using Arc-GIS 10, Spatial Analyst Tool. Drainage density ranges from 0 to >4 km^2. The high drainage density and lower order drainage indicates the areas of the run-off zones (elevated topography). Higher-order of the drainage channel indicates recharge zones (low lying plain topography) and potential zone for groundwater. The classified drainage densities in the study area are low density/coarse texture 0–1 km^2, medium density/medium texture 1–2 km^2, high density/fine texture 2–4 km^2, very high density/super fine texture >4 km^2.

4.5 Soil Map

Soil characteristics like texture, size, and shape have a significant role in the infiltration of water. In groundwater recharge point of view, the high porosity of soils gives indirect evidence of the availability of the groundwater. Based on the infiltration capacity of the soil, the weightages were derived and total five types of soil covers were mapped form the IRS-P6 satellite data and field verification. The derived soil types are mainly stony waste, sandy soil with rock outcrops, colluvial soil, brown silty clayey soil and alluvial soil.

4.6 Groundwater Potential Zones

The groundwater potential of the study area covering 384 sq.km reveals six distinct zones representing 'excellent', 'good', 'good to moderate', 'moderate to poor', 'poor' and 'very poor' (Fig. 7) and Table 3. The excellent groundwater potential zones mainly encompass colluvial deposits, alluvial deposits and structural controlled areas. It defines the areas where the terrain is most suitable for groundwater storage, and also indicates the availability of water below the ground. The area covered by 'excellent' groundwater potential zone in total study area is about 51.3 km^2 (13.3%). Similarly, the 'good' and 'good to moderate' zones encompass a total area of 198.2 km^2 (51.6%). The 'moderate to poor, poor and very poor' groundwater potential zones in the study area are most likely due to the presence of runoff zones (hillocks, rocky outcrops, and steep slopes). It is encompassing a total area of 134.4 km^2 (35.1%). Table 4 show the Groundwater potential map for the three Panchayat Union viz.: Konganapuram, Mc. Donald. Choultry and Veerapandy.

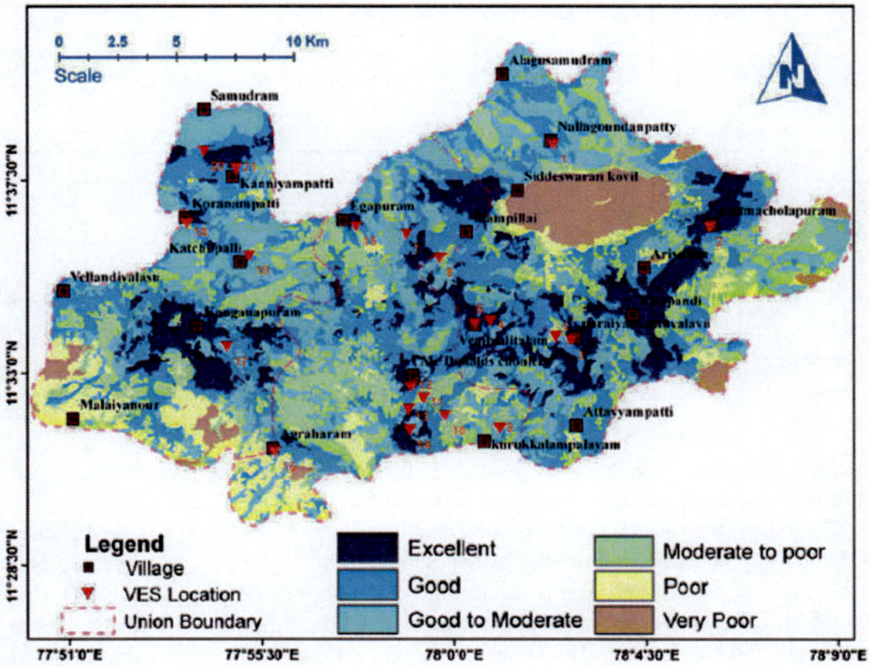

Fig. 7 Groundwater Prospect map

Table 3 Percentage of Groundwater prospective zones

S. No	Potential	Area (km^2)	Area (%)
1	Excellent	51.3	13.3
2	Good	96.7	25.2
3	Good to Moderate	101.5	26.4
4	Moderate to poor	79.3	20.7
5	Poor	31.5	8.2
6	Very poor	23.6	6.2
	Total	384.0	100.0

5 Results

5.1 *Geophysical Exploration*

A total of 187 vertical electrical soundings (VES) were conducted in different parts of the prospective map (Fig. 7) using the Schlumberger configuration with the aid of the indigenous DDR-2 Resistivity Meter with a maximum current electrodes spacing of 300–500 m. The field data was interpreted using a conventional partial curve matching technique [24] using master curve and auxiliary diagram [25]. The

Table 4 Groundwater potential zones Panchayat union wise

S. No.	Potential Zones	Konganapuram		Mc Donalds Choultary		Veerapandy	
		Area (km^2)	Area (%)	Area (km^2)	Area (%)	Area (km^2)	Area (%)
1	Excellent	14.5	13.6	10.4	9.6	26.4	15.5
2	Very good	28.5	26.9	27.6	25.6	40.6	23.9
3	Good	29.2	27.6	29.4	27.2	42.9	25
4	Moderate to poor	18.6	17.6	28.3	26.2	32.4	19.3
5	Poor	11.9	11.2	11.7	10.8	7.9	4.6
6	Very poor	3.2	3	0.6	0.6	19.8	11.7
	Total	106	100	108	100	170	100

interpreted layer parameters were used as initial estimates for interpretation based on inversion and optimization techniques using RESIST v.1.0 software.

The entire study was under National Drinking Water Mission to provide drinking water supply through community water supply scheme by laying pipelines and construction of Overhead Tanks for Public Water supply (PWS) schemes. Out of the 187 VES investigation carried out covering all the villages, 147 location were recommended for drilling. The success rate range from 755 lpm to 1 lpm (12,000 gph to 15 gph). Only at 5 locations the wells drilled were almost dry with only moisture. The success rate was 97% and most of the villages were provided drinking water supply. The 5 villages were the wells turned out to be dry were provided water supply through pipelines from the nearby villages through PWS.

In general Table 5 gives the overall picture of the subsurface layer parameters indicating Multilayer Lithological scenario. The top layer (p1,h1) represents the soil cover, the second layer (p2,h2) is the weathered zone. The fracture zones which occur as third layer sometimes occur between two high resistivity layers. Such deeper fracture zone occurs in areas of Tectonic activities as deduced from the RS-GIS interpretation.

5.2 *Exploratory Wells and Yield Ranges*

Geophysical investigation were verified with Exploratory drilling at 13 selected locations. The total depth of the drilled wells ranges from 60–140 m. The yield of wells ranges from 1.5 to 755 lpm (liters per minute) with only one well having poor yield. It clearly indicates that high yielding wells are located in excellent and good groundwater potential zones (Fig. 8). The yield is ranging from 250 to 755 lpm on excellent category, 109 to 276 lpm on good category, 49 lpm on good to moderate, and 8.6 lpm on moderate to poor category. The Groundwater Prospect Map with 13 selected drilling location falling under various categories of potential zones are

Table 5 Layer parameters of Selected VES in the study area (p1,h1... indicates the layer Resistivity & Thickness)

VES No	Potential zone	Village	ρ1/h1	ρ2/h2	ρ3/h3	ρ4/h4	ρ5/h5	Total (H)
4	Excellent	Edayankadu	27.56/2.2	75.21/6.91	56.02/5.48	592.63/48.4	180.99/18.42	134.55
6	Excellent	Vembaditalam	81.9/1	97.1/9.1	201.9/9.2	High/51.9	108.5/9.2	80.4
12	Excellent	Devarayapuram	39.6/1.1	78.6/7.1	High/56.2	137.7/7.3	High	71.7
1	Good	N.G.patty	52.32/0.51	15.77/2.74	293.91/18.46	High/95.01	88/23	139.72
3	Good	Parayan K.V	7.3/1	338.3/3.1	771.7/19.5	189/25.8	High	49.5
7	Good	Konerippatti	93/1.9	115/10.4	High/77.3	150/10	High	99.6

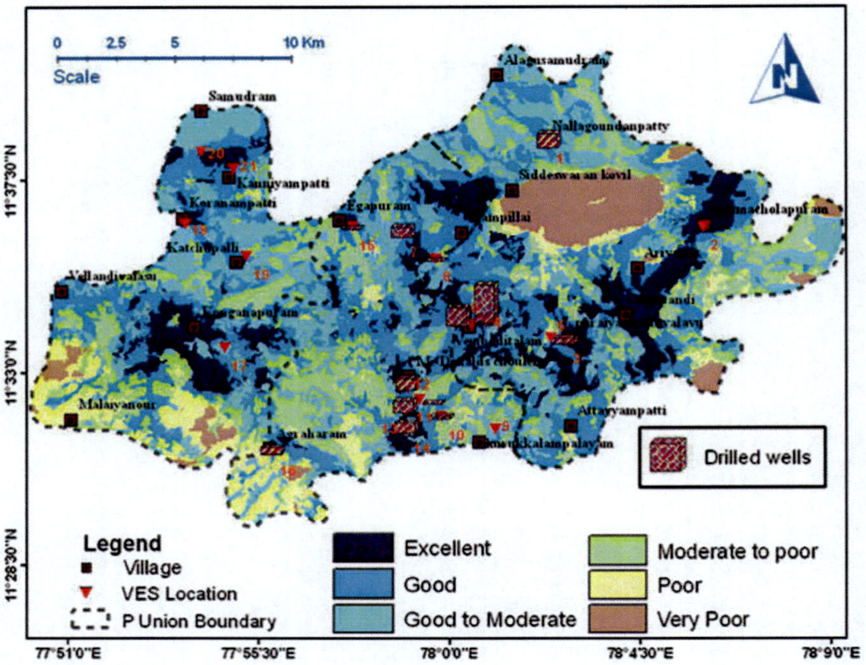

Fig. 8 Exploratory Groundwater Prospect map

shown in Fig. 8. Selected drill time lithologs and its correlation with VES are discussed.

5.3 *Veerapandy Panchayat Union (VPU)*

The sounding location at Edayankadu VES 4 was chosen right on the lineament of E-W trending and adjacent a small tank as inferred from the drainage map and located on the colluvial fills (CF). The 'Excellent' potential zone derived indicates three aquifer zones tapped and the yield has progressively increased right from the depth of 36 m to deeper depth of 134 m wherein the total yield obtained was 755 lpm.

Figure 9 shows the details of drilling Lithology and the corresponding correlation with the VES interpretation. Observed lithology from drilling indicating the resistivities of top soil as 27.5 Ω-m, weathered rock as 56–75 Ω-m, the granite gneiss with minor fractures as 592.6 Ω-m. At 20 to 63 m depth due to presence of the minor fractures in hard rock the water struck at 36 m within this zone. The yield has progressively increased at 68 m depth due to presence of highly fractured gneiss with low resistivity of 180.9 Ω-m. The deeper aquifer is highly fractured and highly

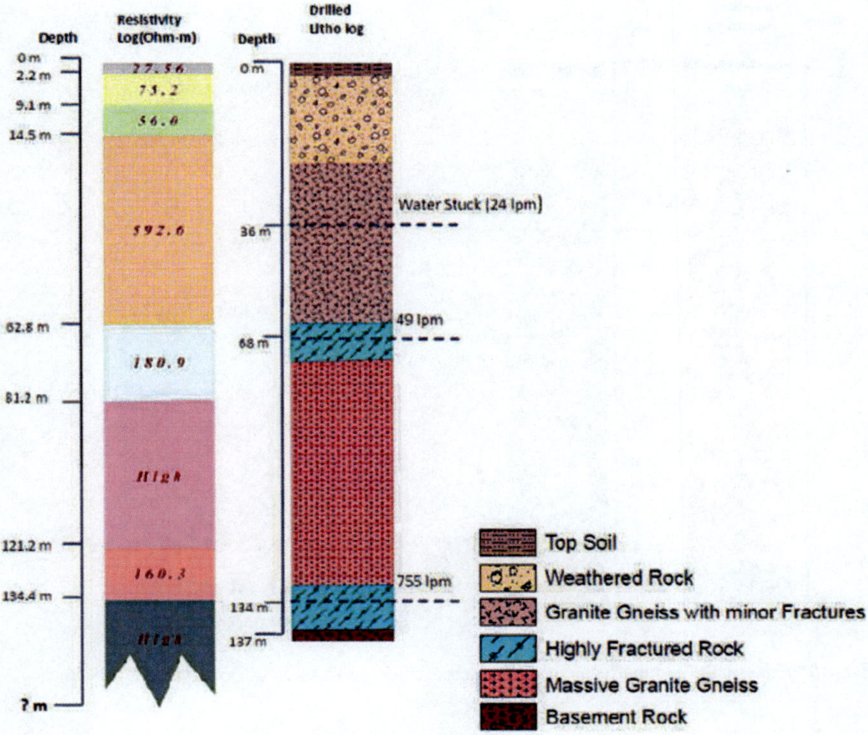

Fig. 9 Edayankadu The highest yield obtained at a depth of 134 m

saturated at 134 m depth with 160.3 Ω-m resistivity and it was sandwiched between two high resistivity layers within the massive granite gneiss. The yield obtained at this depth is 755 lpm and the total depth drilled was 137 m.

5.4 Mc Donalds Choultrey Panchayat Union (MDCPU)

The sounding location at Konerripatti VES 7 was chosen right on the NNE-SSW direction lineament, occupied by pediplain moderately weathered (PPM) and having contact with colluvial fills (CF). It sits in 'Good' potential zone derived from the Groundwater Prospect map. The drilling data indicate three aquifer zones tapped (Fig. 10) and the yield has progressively increased right from the depth of 36.5 m to deeper depth of 94 m wherein the total yield obtained was 198 lpm.

Observed lithology from drilling indicates the resistivities of topsoil as 93 Ω-m and weathered rock as 115 Ω-m. At 36.5 m depth due to presence of the minor fractures in hard rock the initial water struck within this zone. The yield progressively increased at 79 m and 94 m depth due to presence of highly fractured gneiss

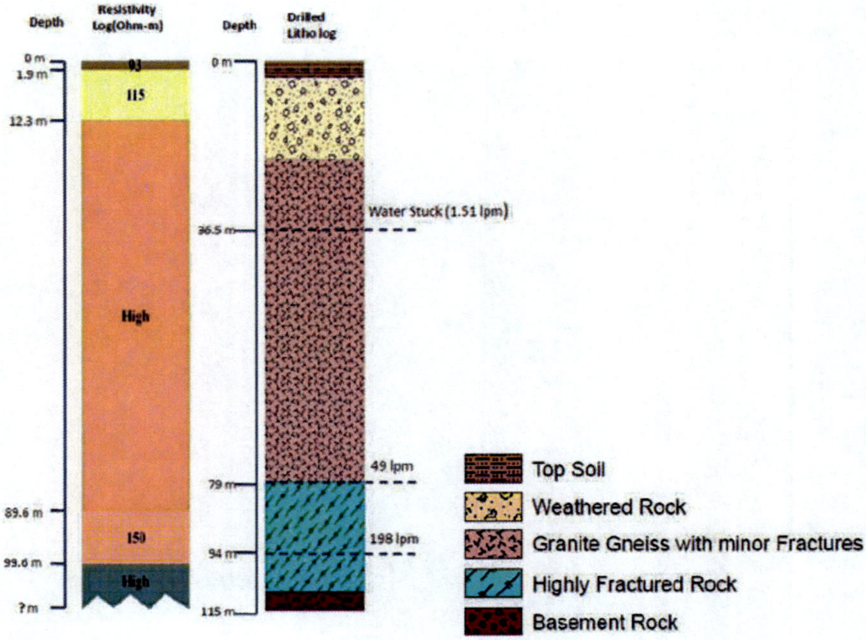

Fig. 10 Koneripatti, comparisons of VES interpretation and Litholog. Total yield is 198 lpm

with low resistivity as 150 Ω-m. The fracture zone after 79 m depth was the potential zone with a total yield of 198 lpm at 94 m depth.

The sounding location at Kunnipalayam VES 14 was chosen at the intersection of N-S direction lineament and NE-SW lineament beset with colluvial fills (CF). This location was chosen on the basis of 'Good' potential zone as well as the VES interpretation indicating fracture zone between 25–39.9 m resistivity ranges. The main aquifer occurs at 25 m depth, and extends down to a depth of 34 m yielding 1.5 to 200 lpm observed lithology from drilling (see Fig. 11). The lithologs obtained reveals coarse sandy soil with boulders upto 2.7 m with resistivity of 50.2 Ω-m, followed by highly weathered granite gneiss with resistivity 113.7 Ω-m, and highly fractured gneiss with resistivity of 166.4 Ω-m values followed by massive granite gneiss with high resistivity values.

5.5 *Konganapuram Panchayat Union (KPU)*

The sounding location at Agraharam VES16 was chosen at the lithological contact on NE-SW lineament, Pediment inselberg complex (PIC) and the Colluvial fills (CF). This location was chosen on the 'Good' potential zone and the total depth drilled was 60 m. The main aquifer was met at 18 m depth and extends down to a

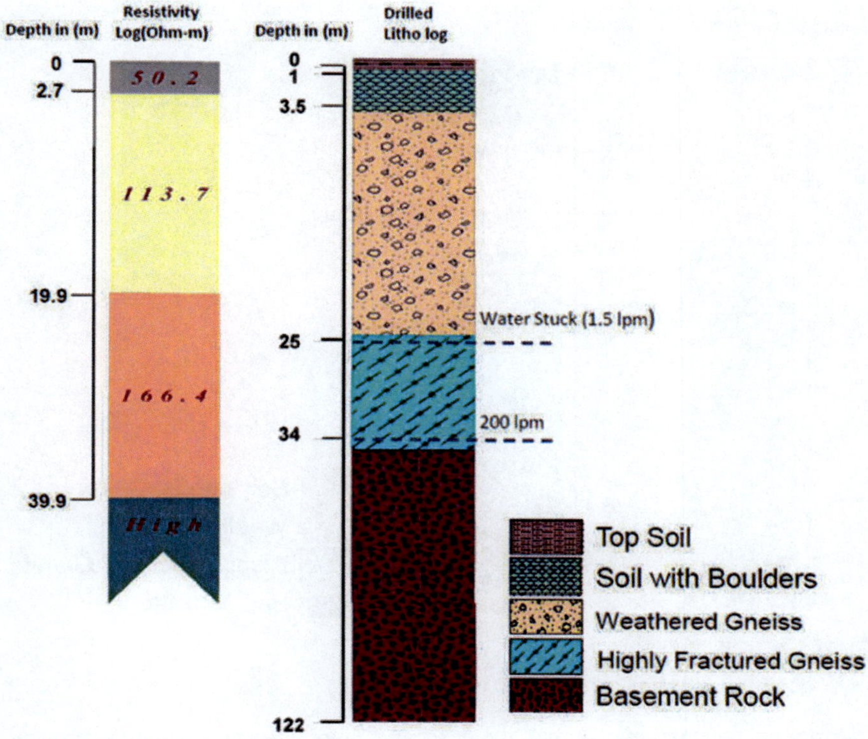

Fig. 11 Kunnipalayam, shallow fracture zone, Total yield 200 lpm

depth of 45 m yielding 82.2 to 264 lpm cumulative drilling discharge. The litho logs show highly fractured granite with a resistivity of 107 Ω-m, followed by massive pink granite with high resistivity values (Fig. 12).

6 Conclusion

The integration of RS and GIS techniques coupled with geophysical data reduces the time and cost in identifying high yielding sites for drilling wells in the complex hard rock terrain. The present study has proved the success rate while adopting such systematic approach. In hard rock areas, it is imperative to use satellite imagery in identifying the structural controls of GW storage & movement. These should be verified through geophysical surveys for locating deeper fracture zones and high yielding wells. RS-GIS plays an important role which help geophysicists in locating potential groundwater zones and geophysical data interpretation will be more accurate in predicting high yielding wells.

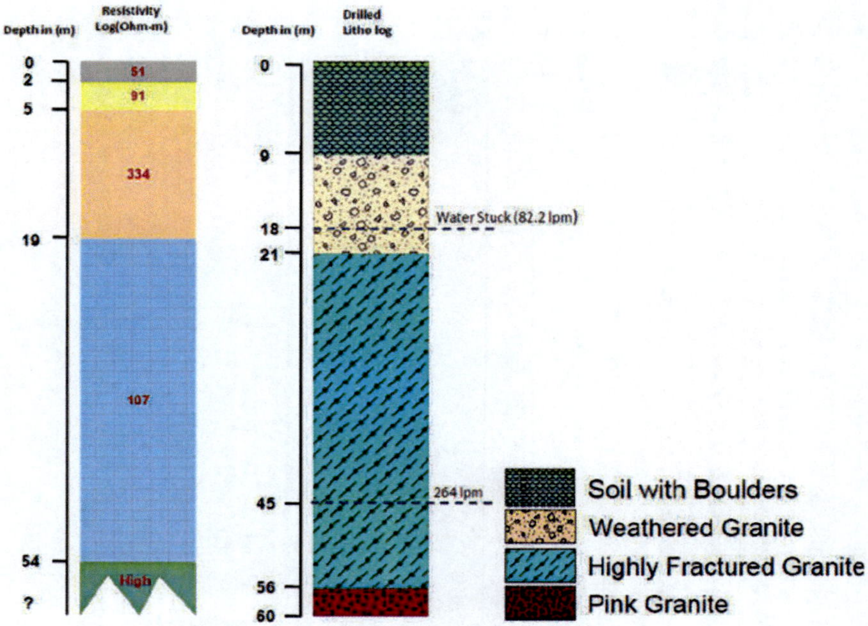

Fig. 12 Agraharam, Total yield is 264 lpm

From the potential map output, the panchayat union wise areal extent have been assessed and most prospective panchayat union falls under the category of excellent, good, and good to moderate that is Konganapuram (68.1%) followed by Verapandy (64.4%) and Mc Donalds Choulery (62.4%). The total depth of the drilled wells ranges from 60–140 m. The yield of wells range from 1.5 to 755 lpm (liters per minute. The entire study was a guiding tool to the National Groundwater Mission in providing the much needed drinking water supply in 143 villages of Salem District, Tamil Nadu.

Acknowledgments This is a part of joint R&D study with National Geophysical Research Institute, Hyderabad and National Remote Sensing Centre, Hyderabad and the support rendered by Director NGRI and Director NRSC is gratefully acknowledged.

References

1. Adinarayana J, Krishna NR (1996) Integration of multi-seasonal remotely-sensed images for improved landuse classification of a hilly watershed using geographical information systems. Int J Remote Sens 17(9):1679–1688
2. Nakagawa A, Ochi S, Shibaski R (1998, November). An application of RS and GIS for hydrological modeling in continental scale. In Proceedings of the 19th Asian conference on remote sensing, Manila, Philippines (pp. 16–20)

3. Prasad RK, Mondal NC, Singh VS (2008) Evaluation of groundwater resource potential using GIS in Kurmapalli watershed of Andhra Pradesh. J Geol Soc India 71(5):661
4. Chowdhury A, Jha MK, Chowdary VM, Mal BC (2009) Integrated remote sensing and GIS-based approach for assessing groundwater potential in West Medinipur district, West Bengal, India. Int J Remote Sen 30(1):231–250
5. Jha MK, Chowdary VM (2007) Challenges of using remote sensing and GIS in developing nations. Hydrogeol J 15(1):197–200
6. Taylor KC, Minor TB, Chesley MM, Matanawi K (1999) Cost enectiveness of well site selection methods in a fractured aquifer. Groundwater 37(2):271–274
7. Rao BV, Briz-Kishore BH (1991) A methodology for locating potential aquifers in a typical semi-arid region in India using resistivity and hydrogeologic parameters. Geoexploration 27(1–2):55–64
8. Dhiman SD, Keshari AK (2002, February) GIS based correlation between groundwater quality parameters and geological units. In Proceedings of the Map India Conference, New Delhi, India (pp. 6–8)
9. Israil M, Al-Hadithi M, Singhal DC (2006) Application of a resistivity survey and geographical information system (GIS) analysis for hydrogeological zoning of a piedmont area, Himalayan foothill region, India. Hydrogeol J 14(5):753–759
10. Sharma AK, Thakur PK (2007) Quantitative assessment of sustainability of proposed watershed development plans for Kharod watershed, western India. J Indian Soc Remote Sens 35(3): 231–241
11. Reddy PR, Kumar KV, Seshadri K (1996) Use of IRS-1C data in groundwater studies. Curr Sci 600–605
12. RGNDWM Project Team (2009) Satellite remote sensing applications. Bull NNRMS 33 (B):7–10
13. Bahuguna IM, Nayak S, Tamilarsan V, Moses J (2003) Groundwater prospective zones in basaltic terrain using remote sensing. J Indian Soc Remote Sens 31(2):101–105
14. Sahu PC, Sahoo H (2006) Targeting ground water in tribal dominated Bonai area of drought-prone Sundargarh District, Orissa, India – a combined geophysical and remote sensing approach. J Hum Ecol 20(2):109–115
15. Krishnamurthy J, Venkatesa Kumar N, Jayaraman V, Manivel M (1996) An approach to demarcate ground water potential zones through remote sensing and a geographical information system. Int J Remote Sens 17(10):1867–1884
16. Reddy GO, Mouli KC, Srivastav SK, Srinivas CV, Maji AK (2000) Evaluation of ground water potential zones using remote sensing data – A case study of Gaimukh watershed, Bhandara District, Maharashtra. J Indian Soc Remote Sens 28(1):19
17. Sankar K (2002) Evaluation of groundwater potential zones using remote sensing data in Upper Vaigai river basin, Tamil Nadu, India. J Indian Soc Remote Sens 30(3):119–129
18. Saraf AK, Choudhury PR (1998) Integrated remote sensing and GIS for groundwater exploration and identification of artificial recharge sites. Int J Remote Sens 19(10):1825–1841
19. Sikdar PK, Chakraborty S, Adhya E, Paul PK (2004) Land use/land cover changes and groundwater potential zoning in and around Raniganj coal mining area, Bardhaman District, West Bengal-a GIS and remote sensing approach. J Spat Hydrol 4(2)
20. Teeuw RM (1995) Groundwater exploration using remote sensing and a low-cost geographical information system. Hydrogeol J 3(3):21–30
21. Toleti BVMR, Chaudhary BS, Kumar KM, Saroha GP, Yadav M, Singh A, ... Singh PK (2000, December) Integrated groundwater resources mapping in Gurgoan District (Haryana) India using remote sensing and GIS techniques. In The 21st Asian Conf. on Remote Sensing, Taipei, Taiwan
22. Rajendran S, Thirunavukkaraasu A, Poovalingaganesh B, Kumar KV, Bhaskaran G (2007) Discrimination of low-grade magnetite ores using remote sensing techniques. J Indian Soc Remote Sens 35(2):153

23. Srinivasa Rao Y, Jugran DK (2003) Delineation of groundwater potential zones and zones of groundwater quality suitable for domestic purposes using remote sensing and GIS. Hydrol Sci J 48(5):821–833
24. Bhattacharya P (2012) Direct current geoelectric sounding: principles and interpretation. Elsevier
25. Orellana E, Mooney HM (1966) Master tables and curves for vertical electrical sounding over layered structures; Tablas Y Curvas Patron Para Sondeos Electricos Verticales Sobre Terrenos Estratificados. Interciencia
26. Murugesan V, Krishnaraj S, Kannusamy V, Selvaraj G, Subramanya S (2011) Groundwater potential zoning in Thirumanimuttar sub-basin Tamilnadu, India—a GIS and remote sensing approach. Geo-spatial Inf Sci 14(1):17–26
27. Preeja KR, Joseph S, Thomas J, Vijith H (2011) Identification of groundwater potential zones of a tropical river basin (Kerala, India) using remote sensing and GIS techniques. J Indian Soc Remote Sens 39(1):83–94
28. Mondal MS, Pandey AC, Garg RD (2008) Groundwater prospects evaluation based on hydrogeomorphological mapping using high resolution satellite images: a case study in Uttarakhand. J Indian Soc Remote Sens 36(1):69
29. Rao NS (2006) Groundwater potential index in a crystalline terrain using remote sensing data. Environ Geol 50(7):1067–1076
30. Mabee SB, Hardcastle KC, Wise DU (1994) A method of collecting and analyzing lineaments for regional-scale fractured-bedrock aquifer studies. Groundwater 32(6):884–894
31. Magowe M, Carr JR (1999) Relationship between lineaments and ground water occurrence in western Botswana. Groundwater 37(2):282–286

Part IV
Water Quality Assessment and Management

Management of Coal Fly Ash Leachates Generated from Disposal Sites Near Thermal Power Plants

Deblina Maiti, Sundararajan Muniyan, and Iqbal Ansari

Abstract Coal fired thermal power plants generate large volumes of fly ash every year. Although, more than 60% of the fly ash produced in India is utilized in various industries yet the left bulk amount of ash is disposed in landfills. Presence of hazardous leachable heavy metals in fly ash have an adverse impact on the water environment due to their potential leaching into the groundwater and surface water. Moreover, many old disposal facilities are unlined which creates a potential risk for groundwater contamination. In this context, various national standards for monitoring fly ash disposal sites have also been made. Added to this the expense of the thermal power industry can hike if the existing or abandoned disposal sites come under regulation. This chapter aims to highlight the characteristics of fly ash and its leachates which can potentially harm the water environment. Various conditions which can potentially affect the leachate quality arising from the dumpsites have been indicated. The management and treatment measures have also been summed up which can be implied by the industries before coal combustion, and for fly ash disposal so as to prevent the adverse effects on the environment. The bottom line from the chapter suggests that sustainable utilization of technologies like, demineralization of coal, devising a suitable leaching method as per the local environment, treatment of fly ash using additives like lime, incorporation of attenuation barriers underneath fly ash ponds, efficient leachate collection system and further treatment of the leachates can ensure best management practices regarding pollution control from coal fly ash.

Keywords Water quality parameters · Heavy metals · Treatment · Groundwater contamination · Additive

D. Maiti · S. Muniyan (✉) · I. Ansari
Natural Resources and Environment Management, CSIR-Central Institute of Mining and Fuel Research, Dhanbad, Jharkhand, India

Department of Mathematics & Computer Science, Mizoram University, Aizawl, Mizoram, India

S. Yadav et al. (eds.), *Water Quality, Assessment and Management in India*, Earth and Environmental Sciences Library, https://doi.org/10.1007/978-3-030-95687-5_11

1 Introduction to Fly Ash Generation and Utilization

Fly ash (FA) is the left-over product after coal combustion in furnaces of thermal power plants (TPP) after electricity generation. Following combustion, the fine particles of FA are carried away along with flue gas, captured in electrostatic precipitators and then either utilized or disposed [1]. The magnitude of FA generation is in millions of tons each year. According to a published article 1 ton of ash is generated after burning 4 tons of coal [2]. In India itself, a major part of the coal extracted every year is used for electricity generation which accounts for 71% of the total electricity. Favorably, FA is also used in various industrial activities, yet the global utilization is only 25%, and the rest of the ash is considered as a waste [3, 4]. FA has been denoted as a Green List waste by the Organization for Economic Cooperation and Development but not regarded as waste by Basel Convention. Comparably, FA utilization in India is higher compared to global utilization as it was 64% in 2016–17. The dry form of FA is utilized (Fig. 1) for cement manufacturing, reclamation, making roads or embankments, mine filling, manufacturing of bricks or tiles, and used as an amendment in agriculture [5]. FA utilization as a backfilling material is a bulk utilization technique, but the strategy is still at a developing stage. Though in agriculture, alkaline type of FA is often used to neutralize acid soils or clayey FA has been used to improve the properties of sandy topsoil yet, FA utilization in agriculture is not always beneficial for living organisms, because the plants can uptake toxic metals from FA and pass it to the higher trophic levels which can show deleterious results in the long run

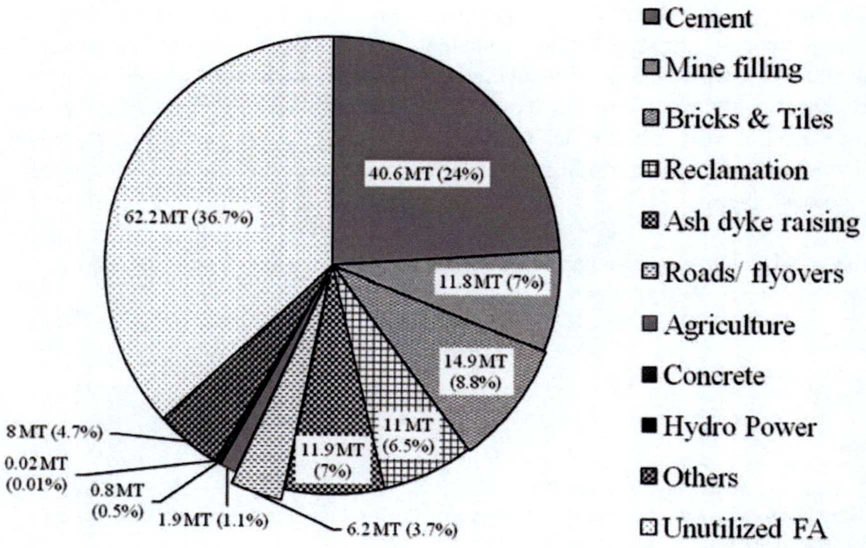

Fig. 1 Utilization of fly ash in India in the year 2016–17. MT signifies million ton of fly ash used in each case with the percentage used in brackets [5]

[6, 7]. FA utilization for biomass production in forestry with economic trees has also been reported with high reclamation costs [1, 8, 9]. Till now 100% utilization of FA is a difficult issue and the bulk amount of unutilized FA is disposed off in terrestrial lands creating huge pollution in the nearby area [9]. Moreover, the projected coal demands in the power sector show that in the near future more FA will be generated which will continue to exceed its consumption [4, 10]. With this background, the present chapter aims to deal with the problem of leachate generation from fly ash disposal sites, environmental impacts, leachate treatment strategies and management measures.

2 Fly Ash Properties

FA properties vary along with the type of coal, conditions of coal combustion, particle size, emission control devices, collector efficiency and handling methods [11, 12]. The properties have been categorized into physical properties, chemical properties and heavy metals as given below:

2.1 Physical Properties

The physical properties of FA have been summarized in Table 1. The colour of FA varies according to the percentage of unburnt carbon and iron oxide present in it. FA is composed of spherical glass-like particles of diameter 0.01–100 μm. The particles are often observed to be hollow from inside and are called cenospheres. Cenospheres are further filled with further smaller crystals called pedospheres [16]. Presence of cenosphere in FA results in lower values of specific gravity (SG) and bulk density (BD), while the small particle size increases the surface area of the material as well as its sorption capacity [13, 14]. The texture of FA is silty loam and is finer if produced from bituminous coal [14].

Table 1 Physical properties of fly ash* [1, 9, 10, 13–15]

Parameters	Worldwide	Indian
Particle diameter (μm)	0.01–100 (20)	–
Texture	Silty loam	Silt or Sandy loam
Specific surface area (m^2 kg^{-1})	300–500	336–428
Specific gravity (g cm^{-3})	1.6–2.6	2.12–2.40
Bulk density (g cm^{-3})	0.54–0.86	0.70–1.05
Water holding capacity (%)	40–60	60–65

2.2 Chemical Properties

Chemical composition of FA is predominated by oxides, carbonates, hydroxides, sulfates and silicates of Ca (calcium), Fe (iron), Al (aluminum) and other trace metals. The particles mainly contain ferro-aluminosilicate compounds of the above elements. The chemical properties and mineralogical compositions of different FAs have been listed in Table 2. Presence of unburnt carbon in FA is analyzed as loss on ignition (LOI) values which range from 0.01–52% [13]. The pH varies from 4–12 and the value is influenced by the content of S (sulphur) and CaO (calcium oxide) in fly ash. Generally, anthracite coals produce acidic ashes and lignite coal produces alkaline ashes [4, 17]. As per American Society for Testing and Materials (ASTM), FA has been categorized into two types based on its Ca content or pozzolonic properties, as well as silica (SiO_2) and alumina (Al_2O_3) content. It has been reported that the pozzolonic activity decreases with increase in carbon percentage in FA [18, 24]. Generally, Indian coal ashes contain 23–73% SiO_2, 12–53% Al_2O_3, and

Table 2 Mineralogical composition and chemical properties of fly ash [13–15, 17–23]

Parameters	Indian ash			Lignite ash	Bituminous ash	Anthracite ash
	Fly ash	Pond ash	Bottom ash			
pH	7.9–12.8			11	4.5–11	4.5
SiO_2 (%)	38–63	38–75	23–73	48	38–63	52–54
Al_2O_3 (%)	27–44	12–53	13–27	30	27–44	22–24
TiO_2 (%)	0.4–1.8	0.2–1.4	0.2–1.8	1.40	0.4–1.8	–
Fe_2O_3 (%)	3.3–6.4	3.5 - 35	4–11	5.40	3.3–6.4	6.1–7.2
MnO (%)	0.5	0.6	0.3	–	–	–
MgO (%)	0.01–0.5	0.1–0.8	0.1–0.7	2.6	0.01–0.5	0.9–1.2
CaO (%)	0.2–8	0.2–0.6	0.1–0.8	7.9	0.2–0.8	0.3–0.5
K_2O (%)	0.04–0.9	0.1–0.7	0.56	0.20	0.04–0.9	2.8–3.0
Na_2O (%)	0.07–0.4	0.05–0.3	0.3	0.4	0.07–0.4	0.2–0.4
LOI (%)	0.2–3.4	0.01–21	0.6–52	5.7	0.2–3.4	9.6–12.9
SO_3 (%)	0.13–3.12			2.80	0.03–0.16	0.05–0.27
EC (dS cm^{-1})	8.6–8.7			–	–	–
CEC (meq 100 g^{-1})	0.52–1.28			–	–	–
TN (%)	200–500			–	–	–
TP (mg kg^{-1})	200–3500			–	–	–

SiO_2 silicon dioxid,. *Al_2O_3* aluminium oxide, *TiO_2* titanium dioxide, *Fe_2O_3* ferric oxide, *MnO* manganese oxide, *MgO* magnesium oxide, *CaO* calcium oxide, *K_2O* potassium oxide, *Na_2O* sodium oxide, *LOI* loss on ignition, *SO_3* sulphur trioxide, *EC* electrical conductivity, *CEC* cation exchange capacity, *TN* total nitrogen, *TP* total phosphorus

0.1–8% CaO. As a thumb rule, combustion of sub-bituminous or lignite coal leads to the production of Class C FA containing >20% CaO, 1–3% lime, 50–70% SiO_2 while class F FA is produced by combustion of bituminous or anthracite coal and contains >70% silica, <10% CaO. Mineralogical composition of a particular FA found through X-ray diffraction analysis shows mainly quartz, feldspars, carbonates, and chlorites [8, 18, 25].

2.3 Heavy Metals

Coal possesses significant amounts of various metals which is often 4–10 times higher in the FA yielded by it (listed in Table 3). During coal combustion, the minerals present in it are volatilized at a higher temperature which get condensed on the FA surfaces [34, 35]. The amorphous oxidized minerals of Fe and Al act as active sites for metals adsorption which can later on leach into the aqueous medium under favourable conditions. Metal composition varies in different FAs; commonly it contains considerable amounts of essential metals like Ca, K, magnesium (Mg), Cu, manganese (Mn), molybdenum (Mo), Zn, boron (B), and Fe. Patra et al. [36] confirmed through X-ray emission spectroscopy that titanium (Ti), K, Ca, and Fe are present significant amounts while Cr, cobalt (Co), Se, Mn, Ni, Zn, Cu, As, and Pb are present in minute amounts. But other toxic metals like cadmium (Cd), Pb, Ni, Se, Hg which are present in FA, may cause harmful effects in living organisms even at low concentration if they enter the food chain [26, 37].

Table 3 Metal concentration (mg kg^{-1}) in fly ash [9, 16, 26–33]

Metals	India	Worldwide	Metals	India	Worldwide	Metals	India	Worldwide
Fe	68	0.31–37	P	10.8	2.1	As	1–4	20.4
Al	4.8–312	0.5–109	Co	21–58	7–26	Cd	0.1–52	0.03–1.3
Ca	0.03–34	1.84–86	Cu	17–170	19–57	Cr	5–330	15–148
Mg	0.02–1	0.02–12	Zn	20–203	39–167	Hg	–	0.18–0.4
K	10.8	24.5	Mo	4–33	3–4.2	Pb	10–115	16–97
B	17–38	0.4–50	Ni	2–296	15–88			
Mn	12–739	100–679	Se	0.6–2.6	8–10			

Fe iron, *Al* aluminium, *Ca* calcium, *Mg* magnesium, *K* potassium, *B* boron, *Mn* manganese, *P* phosphorus, *Co* cobalt, *Cu* copper, *Zn* zinc, *Mo* molybdenum, *Ni* nickel, *Se* selenium, *As* arsenic, *Cd* cadmium, *Cr* chromium, *Hg* mercury, *Pb* lead

3 Environmental Impacts Due to Fly Ash Disposal

During coal combustion in TPPs, both FA and BA are generated which are mainly differentiated by their fine and coarse size particles respectively. After utilization, the left out FA is disposed on low lying areas or ash ponds near TPPs. During disposal both FA and BA are mixed with water into slurry, and transported through pipelines to ponds. The water from the slurry drains out from ponds and the left residue which is locked in the ponds is known as pond ash (PA). The drained water is termed as leachate and contains mainly dissolved metals and solids. Although FA and BA possess different properties yet synonymously for convention the ash which is observed in disposed form is called "fly ash". During landfilling, more space for the further generated ash is made by digging or emptying the landfills. In the process the PA is carried and dumped in other low lying areas which create huge FA dumps. Another form of ash disposal from TPP is in dry form which also creates FA dumps. In this form of disposal, the FA from ESPs are taken to the disposal site in a partially wet and covered condition [6, 8, 33]. Specifically, the FA dumps are very unstable as the ash on the surface is completely dry; the particles get easily dispersed by wind, remain in the atmosphere for a large period and travel to long distances to cause air pollution. The particles ultimately settle down on plant leaves causing harmful effects. They also affect the health of many living organisms through inhalation [5, 38]. In the recent past, Rao et al. [39] estimated 40,000 Ha of usable land covered by FA dumps which in turn is continuously increasing the land polluted by these dumps. During heavy monsoon, the FA erodes into the nearby water sources and soil system as slurry to contaminate them. When FA reaches the sub-soil, it causes siltation, and clogs the natural drainage system. Leaching is also a huge problem from fly ash disposal sites as described below.

4 Leaching from Fly Ash and its Management

The land covered by FA is expanding critically and is a menace to the environment, more because it contains potentially toxic heavy metals; namely Cr (chromium), As (arsenic), Cd (cadmium), Pb (lead), Hg (mercury), Se (selenium) etc. [13]. After FA slurry disposal in ponds, a huge amount of water is drained out in the form of leachates which contains various heavy metals and dissolved solids (Fig. 2). Further weathered FA in ponds and dumps contain high levels of soluble salts which can contaminate the groundwater below migrate from non-lined FA ponds [40, 41]. Leaching of Zn (zinc), Cu (copper) and Pb was found in higher concentrations in tube well located near ash pond in Orissa [42–44]. Choi et al. [45] indicated the influence of the ash leachates on the nearby groundwater composition through comparison of the chemical characteristics of leachates, FA slurry and groundwater. In the year 2008, a TPP in Kingston USA incurred massive destruction by spilling of million gallons of ash slurry into the nearby landscape from its damaged 40-acre ash

Fig. 2 View of fly ash ponds also containing the drained water after disposal in slurry form

pond. The slurry covered 300 acres of the nearby land and contaminated a river with high levels of As above the allowable concentration in drinking water standard. Additionally high levels of Pb, thallium, barium (Ba), Cd, Cr, Hg and Ni were also reported [46]. Government has imposed various stringent rules for FA disposal which restricts land availability for disposal and also levies various duties like continuous monitoring of pollution caused by the dumps, in view of the environmental impacts caused by the disposal sites [9]. Management of leachate generated from the dumpsites is one of the factors which should also be considered during management of such wastes. Below paragraphs will elaborate on the chemistry of leachate generation, properties, treatment and management strategies.

4.1 Characteristics of Fly Ash Leachates

Leachate is a liquid which is produced when a solution or water percolates through a permeable material. FA leachates contain various ions like Ca, Mg, SO_4, as well as toxic metals such as Fe, Mn, Cd, Cr, Ni, Cu, Pb, Zn, As, Ba, Hg, Se, B (shown in Table 4). Some of them are often found above the desirable limits prescribed in drinking water standards [47–52, 66]. These metals may also leach into the groundwater and contaminate the drinking water source. Therefore, it is essential to devise efficient treatment and management procedures for these leachates to prevent the contamination of water bodies. The various methods and approaches in this aspect are described below.

4.2 Management and Treatment of Fly Ash Leachates

Demineralization of coal is the appropriate concept which can solve the problem of harmful ions present in the fly ash leachates.

Table 4 Characteristics of fly ash leachates. Parameters are in mg L^{-1} except for pH, turbidity, electrical conductivity [47–65].

Parameter	Value	Parameter	Value
pH	4.4–12.9	Polyhydric phenol	–
Turbidity (NTU)	–	Oil and Grease	–
EC (μS/cm)	310–2535	CN	0.01
TS	–	SCN	–
TFS	–	Salinity	–
TVS	–	Fe	0.02–6.7
TSS	12–140	Ni	0.03–2.1
TDS	142–1743	Mn	0.01–36.4
TH	105–985	Cd	0.001–0.8
Ca	32–17,250	Cr	0.004–3.3
Mg	0.4–352	Co	0.05–1.1
Na	17–601	Cu	0.01–1.7
K	35–892	Sr	0.56–16.5
S^{2-}	–	Pb	0.01–2.8
SO_3	–	Zn	0.01–37
SO_4	49–916	Al	1.4–373
Cl	3–95	As	0.003–0.28
CO_3	–	Ba	0.1–1.32
HCO_3	–	Be	0.01–0.2
Alkalinity	41–681	Hg	0.0002–0.038
TKN	–	Se	0.002–2.4
NO_3	–	Si	3.7–215
NH_3-N	0.04–0.64	B	0.3–109
BOD5	–	Cs	0.01–0.03
COD	–	Sb	0.01–3.4
TOC	–	Mo	0.06–3.92
Filtered TOC	–	Li	0.68–26.3
P	0.01–0.08	V	0.3–1

EC electrical conductivity, *TS* total solids, *TFS* total fixed solids, *TVS* total volatile solids, *TSS* total suspended solids, *TDS* total dissolved solids, *TH* total hardness, *Ca* calcium, *Mg* magnesium, *Na* sodium, *K* potassium, S_2 sulphide, SO_3 sulphite, SO_4 sulphate, *Cl* chloride, CO_3 carbonate, HCO_3 bicarbonate, *TKN* total kjeldahl nitrogen, NO_3 nitrate, NH_3*-N* ammonium nitrogen, *BOD* biological oxygen demand, *COD* chemical oxygen demand, *TOC* total organic carbon, *CN* cyanide, *SCN* thiocyanate, *Fe* Iron, *Ni* Nickel, *Mn* manganese, *Cd* cadmium, *Si* silicon, *Se* selenium, *Hg* mercury, *Be* berrelium, *Ba* barium, *As* arsenic, *Al* aluminium, *Zn* zinc, *Pb* lead, *Sr* strontium, *Cu* copper, *Co* cobalt, *Cr* chromium, *Li* lithium, *Mo* molybdenum, *Sb* antimony, *Cs* cesium, *B* boron, *P* phosphorus, *Li* lithium, *Sb* antimony, *V* vanadium

Leachability of FA particles depends on the degree of attachment of the metals to the particles, pH of the extractant (acidic conditions shows more leaching) as well as leaching time. Prior to ash disposal it should be ensured that any contaminant present in it would not become a future source of pollution through the leachates generated from the dump. Leaching studies are the most suitable tests which help to know the metal mobility in the material. The different types of leaching tests mentioned in the literature are stated here [32]. Leaching methods are differentiated on the basis of one time added leaching fluid (static extraction tests) or replaced several times (dynamic tests). Another criterion for classifying leaching tests are batch leaching tests in which a weighed amount of material is placed in a measured volume of an extracting leaching fluid or column leaching tests wherein the fluid is passed through a column of the material. These methods have been mainly developed by EPA and ASTM [67]. The laboratory leaching tests are further classified into types which: (1) simulate metal release in a particular condition like acidic rain, (2) demonstrate sequential chemical extraction, (3) and find out fundamental leaching properties. Tests which denote metal release in specific conditions do not provide real information of metal release in actual environmental scenarios. Therefore, simulation-based tests can solve the purpose to some extent. Usually, the factors which affect metal release are leaching time, liquid-solid ratio (L/S) and the leachate pH [1, 34, 68]. Literature has also reported many single and sequential extraction techniques for estimation of metal leaching from any FA. These include extraction with mineral acids like 1N HCl or HNO_3, salt solutions (for example, calcium chloride), buffer solutions (for example, ammonium acetate) and chelating agents (for example, DTPA) [69]. According to USEPA, the toxicity characteristic leaching procedure (TCLP) should be preferred, for determining harmful metal release from FA [70]; done by extracting 5 g sample in 100 mL extractant @30 rpm (rotation per minute) for 18 h and analyzing the filtered leachates for metals [71]. Some tests are also known as bio-availability tests which find the amount of metal which can get released easily and may affect living organisms. For example, Zhou et al. [71] prepared soil extracts by extracting a soil in magnesium chloride solution in 1:10 ratio w/v by shaking @150 rpm at room temperature for 2 h and @4000 rpm for 10 min. DTPA, Mehlich I and III solutions are also used for the same purpose [72] to extract soil micronutrients in 1:4 and 1:8 ratio of w/v respectively. Mehlich I is composed of sulphuric and hydrochloric acid while Mehlich III is composed of acetic acid, ammonium nitrate, nitric acid, ammonium fluoride and EDTA [73]. Leaching of metals from FA depends on the degree of solubility of the oxides, hydroxides and carbonate salts. Lower pH of the aqueous medium increases the solubility of the oxides but some studies also show leaching of metals from alkaline pH. Dutta et al. [74] observed higher mobility of metals at a low pH from FA samples obtained from a TPP in West Bengal, India. They observed a consistent release of metals like Mn, Cu, K, Cr, Pb and as for 180 days. Fernandez-Turiel et al. [75] reported that 2–36% of the total metals in FA can leach under natural conditions. Metals like As, Mo, and Se show more mobility at higher pH [41]. For example, Jegadeesan et al. [76] showed moderate leaching of Fe, Pb, Cr from Class F FA in low pH and a considerable as release at $pH < 4$ and $pH > 9$ due to dissolution and desorption respectively. Ugurlu [77] reported that alkaline FA with pH 11–12 showed maximum leachability of Na, K, Ca, Fe, Mn, S and Pb. Under humid conditions, weathering on ash dumps is more rapid which solubilise the metals and ions in it to leach and contaminate the groundwater [78]. Table 5

Table 5 Concentrations of metals in leachates of fly ash extracted by different extractants* [42, 70–81]

Metals (mg kg^{-1})	Extractant used					STLC
	DW	DTPA	HNO_3	Mehlich I	TCLP	
Ratio (w/v)*	1:10	1:2	1:10	1:4	1:20	1:20
pH	5.8–6.3	7.3	3.5–4	2.5	4.93	4.93
Time period	6 h	2 h	2 h	4 h	18 h	18 h
Fe	0.01–1.7	8.2–21.9	22.8	161	5.4–8.2	NA
B	0.1–1.1	25	4	3	2.9	NA
Mn	0.03–0.13	0.41–3.5	23.1	19	2.1	NA
Co	0.02–0.06	<0.05	1.9	1.07	0.11	80
Cu	0.01–0.02	0.85–6.2	4.5	11	0.48	25
Zn	0.02–4.8	1.2	7.4	4.6	2.4	250
Ni	0.01	0.09–0.56	1.4	2.3	0.25	20
Se	0.1	1.1	0.84	0.35	0.12	0.3–1
As	0.1	0.6–4.7	0.91	16	0.29	0.3–5
Cd	0.01–0.71	0.14–0.34	0.14	0.12	0.02	0.3–1
Cr	0.03–0.86	0.94–2.1	1.6	1.3	0.05	5
Hg	BDL	0.04	0.07	0.07	0.08	0.2
Pb	0.07–1.10	0.38–1.8	2.2	<0.1	0.004	0.3–5
Ba	0.04	0.2–15	0.2–15	0.63	1.7	100
Ag	BDL	NA	NA	<0.25	BDL	5

NA not available, *BDL* below detection limit, *STLC* soluble threshold limit concentration

shows the extent of metals released from FA under the influence of various extracts and compared to the values of regulatory limits given by different countries, known as soluble threshold limit concentration (STLC). If the amount of metals released from the FA exceeds the limits, the FA material is marked hazardous. However, all the metals, except Se and as are often observed within limits [77].

Bhattacharyya et al. [80] reported a simple, low-cost, effective in situ chemical treatment technique which can stabilize hazardous metals in FA. Their experiment used ferrous sulfate solutions to immobilize the metals in FA which can then be safely disposed. In another study Duchesne and Reardon [81] used separate addition of lime, Portland cement and cement kiln dust to Class C and F FA, to see their remediation potential for removal of toxic elements from leachates. They found better results for Class F FA than Class C FA, because the elemental composition of FA impacts its reaction with the additive. They also reported the formation of anionic clay (hydrocalumite) in Class F FA treated with lime; and can accommodate anions into its interlayer, while treatment with Class C FA produced hydrogarnet, which had fewer sites for anion accommodation. Overall lime was considered the best additive for improvement in fly ash leachate water quality. Additionally, another premier method in the context FA leachate management is the prevention of leachate generation, which can be done in the following ways:

- Incorporating topsoil during the restoration of FA dumps, in virtue of soil's pH and buffering capacity, which also attenuates trace metals in organic matter.
- Transport of FA as highly concentrated slurry.
- Controlled surface water flow

Following the above techniques, the next step in the management of leachates is preventing their migration, which can be done as follows [43]:

- An adequate liner system. Astrup et al. [82] proposed that attenuating barriers can be an alternative to traditional liners regarding leachate collection systems at FA disposal sites. They investigated the effectiveness of a barrier composed of sand, Fe(0), and bentonite in laboratory experiments to remove Cr(VI) from leachates of higher pH typical to FA and found that Cr concentrations decreased from 25 mg/L to 0.0025 mg/L.
- A leachate collection system.
- Reuse of ash pond effluents

The collected leachates can then be treated by various methods as given below. Treatment relies on the degree of pollution in it. Minor parameters like hardness, TDS, Fe, and Mn can be eradicated from water by using simple treatment techniques.

- Acidic leachates can be made to undergo precipitation using various reagents.
- Bio-electrochemical technology is a recent technique which can carry out efficient metal recovery by both oxidation and reduction processes, depending on the metal to be used as electron acceptor in the cathode chamber or electron donor in the anode chamber [83]. Metal extraction by this technique is carried out by various modifications like abiotic cathodes and biocathodes.
- Jayaranjan et al. [84] used an innovative method to utilize flue gas desulfurization (FGD) gypsum for the treatment of fly ash leachate from dumpsites. The mechanisms of the process included H_2S production (in a sulfide-rich effluent with concentration up to 150 mg/L) by SRB using SO_4 from solubilized FGD gypsum, followed by leaching of heavy metals from FA and subsequent precipitation of metals as sulfides up to 40 to 100%.
- Biosorption is yet another promising technology which uses cheap, non-pollutant materials, to sequester metal and metalloids from leachates. Extensive research is being carried out at this end to establish its efficiency for multi-metal systems to be used in a larger scale [85].
- Biological treatments don't indulge extra input of chemicals in the treated water. Moreover, improved water treatment methods that give purity up to drinking water level can also effectively incorporated in best practices. Thus, successful implementation of technologies would produce water of desired quality.

5 Conclusion

In developing countries like India, water is crucial for industrial development. Water treatment strategies are the most studies topics across the globe. Various sophisticated techniques have been found till date which can remove impurities in every drop of water. However, prevention of water pollution should be the primary concern which is more useful for the economic development of a country and it further reduces treatment costs. In this context, manipulation of the fly ash leachate quality before its origination should be the primary concern of the thermal power authorities. Research on demineralization of coal can be a suitable alternative in this context. Additionally, the idea of switching over to alternative sources of energy yielding fuels (viz., biomass to biofuel, algae to biofuel, biohydrogen etc) is of tremendous importance. It is a highly demanding area of research in the era of global warming and climate change caused by fossil fuels. However, to conclude in the present context leaching tests are an utmost parameter which should be followed to indicate possible future groundwater contamination. The leaching test method should be wisely chosen keeping in view the climate, geological strata, groundwater table, surface water bodies in the area where FA has to be disposed in future. Incorporation of attenuation barriers underneath fly ash ponds and establishment of a proper leachate collection system are of utmost requirement before FA disposal. These strategies can be followed up with a suitable treatment technique as per leachate quality, so as to further discharge or recycle the water as per industry requirements and minimize freshwater consumption.

6 Recommendations

Following recommendations have been listed for preventing water pollution from fly ash waste disposal sites:

- FA disposal by the Thermal Power Authorities should include prior involvement of a series of leaching tests which will indicate future possibilities of groundwater contamination. The leaching test method should be chosen on the basis of climate and geological strata. Also before disposal of the fly ash Thermal Power Authorities should note the groundwater table and position of surface water bodies from the disposal site. This will determine the strategy of pollution prevention to be used like attenuation barriers underneath fly ash ponds and establishment of a proper leachate collection system.
- After disposal of fly ash, regular manipulation of the originating FA leachates should be done before discharging the leachates. This can be done through a suitable treatment technique as per the leachate quality, so as to further discharge or recycle the water as per industry requirements and minimize freshwater consumption.

- Regular monitoring of water quality near the disposal site and planning of remediation measures accordingly.

Acknowledgments Authors are grateful to Dr. Pradeep Kumar Singh, Director, CSIR-CIMFR, Dhanbad, for continuous support and motivation.

References

1. Belviso C, Cavalcante F, Di Gennaro S, Palma A, Ragone P, Fiore S (2015) Mobility of trace elements in fly ash and in zeolitised coal fly ash. Fuel 144:369–379
2. Mittal ML, Sharma C, Singh R (2012) Estimates of emissions from coal fired thermal power plants in India. In: International emission inventory conference
3. Yang Q, Ma S, Zhang R, Zheng S (2012) Research progress of extracting alumina from high-aluminum fly ash. Multipurpose Utilization of Mineral Resources:3–7
4. Ram LC, Masto RE (2014) Fly ash for soil amelioration: a review on the influence of ash blending with inorganic and organic amendments. Earth Sci Rev 128:52–74
5. CEA (Central Electricity Authority) (2017) Annual report on fly-ash utilization, report on fly ash generation at coal/lignite based thermal power stations and its utilization in the country for the year 2016–17, New Delhi
6. Singh A, Sharma RK, Agrawal SB (2008) Effects of fly ash incorporation on heavy metal accumulation, growth and yield responses of Beta vulgaris plants. Bioresour Technol 99:7200–7207
7. Jayasinghe GY, Tokashiki Y (2012) Influence of coal fly ash pellet aggregates on the growth and nutrient composition of Brassica campestris and physicochemical properties of greysoils in Okinawa. J Plant Nut 35:47–453
8. Mishra DP, Das SK (2010) A study of physico-chemical and mineralogical properties of Talcher coal fly ash for stowing in underground coal mines. Mater Charact 61:1252–1259
9. Tiwari MK, Bajpai S, Dewangan UK (2016) Fly ash utilization: a brief review in indian context. Int Res J Eng Technol 3(4):949–956
10. Wang P, Wang J, Qin Q, Wang H (2017) Life cycle assessment of magnetized fly-ash compound fertilizer production: a case study in China. Renew Sust Ener Rev 73:706–713
11. Wong MH, Wong JWC (1989) Germination and seedling growth of vegetable crops in fly ash-amended soils. Agric Ecosys Environ 26:23–35
12. Kim BJ, Back JH, Kim YS (1997) Effect of fly ash on the yield of Chinese cabbage and chemical properties of soil. KJSSF 30:161–167
13. Alonso JL, Wesche K (1991) Characterization of fly ash. In: Wesche K (ed) Fly ash in concrete properties and performance, E & FN SPON. An Imprint of Chapman & Hall
14. Nyambura MG, Mugera WG, Felicia PL, Gathura NP (2011) Carbonation of brine impacted fractionated coal fly ash: implications for CO_2 sequestration. J Environ Manag 92:655–664
15. El-Mogazi D, Lisk DJ, Weinstein LH (1988) A review of physical, chemical, and biological properties of fly ash and effects on agricultural ecosystems. Sci Total Environ 74:1–37
16. Pandey VC, Abhilash PC, Singh N (2009) The Indian perspective of utilizing fly ash In phytoremediation, phytomanagement and biomass production. J Environ Manag 90:2943–2958
17. Ram LC, Masto RE (2010) An appraisal of the potential use of fly ash for reclaiming coal mine spoil. J Environ Manag 91:603–617
18. Pandian NS (2004) Fly ash characterization with reference to geotechnical applications. IISc J 84(6):18–33
19. Gupta DK, Rai UN, Tripathi RD, Inouhe M (2002) Impacts of fly ash on soil and plant responses. J Plant Res 115:401–409

20. Mittra BN, Karmakar S, Swain DK, Ghosh BC (2005) Fly ash—a potential source of soil amendment and a component of integrated plant nutrient supply system. Fuel 84(11): 1447–1451
21. Kumar M, Philip L (2006) Adsorption and desorption characteristics of hydrophobic pesticide endosulfan in four Indian soils. Chemosphere 62(7):1064–1077
22. Chahal N, Siddique R, Rajor A (2012) Influence of bacteria on the compressive strength, water absorption and rapid chloride permeability of fly ash concrete. Constr Build Mater 28(1): 351–356
23. Kaur R, Goyal D (2014) Soil application of fly ash based biofertilizers for increased crop production. Vegetos Int J Plant Res 27(2):291–300
24. Stutzxna PE, Centeno L (1995) Compositional analysis of beneficiated fly ashes. NISTIR 5598. National Institute of Standards and Technology, Gaithersburg
25. Maiti SK, Maiti D (2015) Ecological restoration of waste dumps by topsoil blanketing, coir-matting and seeding with grass–legume mixture. Ecol Eng 77:74–84
26. Tripathi RD, Vajpayee P, Singh N, Rai UN, Kumar A, Ali MB, Kumar B, Yunus M (2004) Efficacy of various amendments for amelioration of fly-ash toxicity: growth performance and metal composition of Cassia siamea Lamk. Chemosphere 54(11):1581–1588
27. Page AL, Elseewi AA, Straughan IR (1979) Physical and chemical properties of fly-ash from coal-fired power plants with reference to environmental impacts. Residue Rev 71:83–120
28. Adriano DC, Page AL, Elseewi AA, Chang A, Straughan IA (1980) Utilization and disposal of fly ash and other coal residues in terrestrial ecosystem: a review. J Environ Qual 9:333–344
29. Jambhulkar HP, Juwarkar AA (2009) Assessment of bioaccumulation of heavy metals by different plant species grown on fly ash dump. Ecotox Environ Safe 72:1122–1128
30. Kumar KV, Patra DD (2013) Effect of metal tolerant plant growth promoting bacteria on growth and metal accumulation in Zea mays plants grown in fly ash amended soil. Int J Phytoremediation 15(8):743–755
31. Pandey SK, Bhattacharya T, Chakraborty S (2016) Metal phytoremediation potential of naturally growing plants on fly ash dumpsite of Patratu thermal power station, Jharkhand, India. Int J Phytoremediation 18:87–93
32. Tiwari MK, Bajpai S, Dewangan UK, Tamrakar RK (2015) Suitability of leaching test methods for fly ash and slag: a review. J Radiat Res Appl Sci 8(4):523–537
33. Lee H, Ho SH, Lee CH, Lee YB, Kim PJ (2006) Fly ash effect on improving soil properties and rice productivity in Korean paddy soils. Bioresour Technol 97:1490–1497
34. Hartuti S, Fadhillah Hanum F, Takeyama A, Kambara S (2017) Effect of additives on arsenic, boron and selenium leaching from coal fly ash. Fortschr Mineral 7(6):99
35. Shi J, Li Q, Li H, Li S, Zhang J, Shi Y (2017) Eco-design for recycled products: Rejuvenating mullite from coal fly ash. Resour Conserv Recy 124:67–73
36. Patra KC, Rautray TR, Nayak P (2012) Elemental analysis of coal and coal ash by PIXE technique. Appl Radiat Isotopes 70:612–616
37. Prasad B, Maiti D, Singh KKK (2019) Impact of fly ash placement in an abandoned opencast mine on surface and ground water quality: a case study. Mine Water Environ 1–9
38. Sandeep P, Sahu SK, Kothai P, Pandit GG (2016) Leaching behavior of selected trace and toxic metals in coal fly ash samples collected from two thermal power plants. India Bull Environ Contam Toxicol 97(3):425–431
39. Rao RD, Nanda SK, Kareemulla K (2012) Technological forecasting and assessment of future fly ash use in agriculture in India. DST Project Report, National Academy of Agricultural Research Management, Hyderabad, India, 140
40. Praharaj T, Swain SP, Powell MA, Hart BR, Tripathi S (2002) Delineation of groundwater contamination around an ash pond geochemical and GIS approach. Environ Int 27:631–638
41. Jankowski J, Ward CR, French D, Groves S (2006) Mobility of trace elements from selected Australian fly ashes and its potential impact on aquatic ecosystems. Fuel 85:243–256
42. Das M, Agarwal P, Singh R, Adholeya A (2013) A study of abandoned ash ponds reclaimed through green cover development. Int J Phytoremediation 15(4):320–329

43. Maiti D, Prasad B (2016) Revegetation of fly ash–a review with emphasis on grass-legume plantation and bioaccumulation of metals. AEER 14(2):185–212
44. Maiti D, Prasad B (2017) Studies on colonisation of fly ash disposal sites using invasive species and aromatic grasses. J Environ Eng Landsc 25(3):251–263
45. Choi SK, Lee S, Song YK Moon HS (2002) Leaching characteristics of selected Korean fly ashes and its implications for the groundwater composition near the ash disposal mound. Fuel 81(8):1083–1090
46. U.S. Senate Committee on Environment and Public Works (2009) Testimony of Stephan A. Smith, DVM, Executive Director, Southern Alliance for Clean Energy, accessed online from < https://www.epw.senate.gov/public/index.cfm/2009/1/post-b6f29af3-802a-23ad-43cd-d132952f070b > (on 4.08.2018)
47. Baba A, Kaya A (2004) Leaching characteristics of fly ash from thermal power plants of Soma and Tuncbilek. Turkey Environ Monit Assess 91(1–3):171–181
48. Sarode DB, Jadhav RN, Khatik VA, Ingle ST, Attarde SB (2010) Extraction and leaching of heavy metals from thermal power plant fly ash and its admixtures. Polish J Environ Stud 6: 1325–1330
49. Chu TYJ, Ruane RJ, Krenkel PA (1978) Characterization and reuse of ash pond effluents in coal-fired power plants. JWPCF 2494–2508
50. Silva LF, Ward CR, Hower JC, Izquierdo M, Waanders F, Oliveira MLS, Li Z, Hatch RS, Querol X (2010) Mineralogy and leaching characteristics of coal ash from a major Brazilian power plant. CCGP 2:51–65
51. Eisenberg SH, Tittlebaum ME, Eaton HC, Soroczak MM (1986) Chemical characteristics of selected fly ash leachates. J Environ Sci Health A 21(4):383–402
52. Nathan Y, Dvorachek M, Pelly I, Mimran U (1999) Characterization of coal fly ash from Israel. Fuel 78(2):205–213
53. Nnaji PC, Okolo BI, Menkiti MC (2014) Nephelometric performance evaluation of oxidized starch in the treatment of coal washery effluent. Nat Resour 5(3):79
54. Menkiti MC, Onukwuli OD (2011) Impact of pH variation on Coag-flocculation behaviour of chitin derived Coag-flocculant in Coal washery effluent medium. JMMCE 10(15):1391
55. Ghose MK (2001) Design of cost-effective coal washery effluent treatment plant for clean environment. JSIR 60:40–47
56. Das B, Prakash S, Biswal SK, Reddy PSR (2006) Settling characteristics of coal washery tailings using synthetic polyelectrolytes with fine magnetite. J South Afr Inst Min Metall 106(10):707
57. Zhao D, Lun W, Wei J (2017) Discussion on wastewater treatment process of coal chemical industry. In: IOP conference series: earth and environmental science, IOP Publishing 100(1): 012067
58. Yu X, Wei C, Wu H, Jiang Z, Xu R (2015) Improvement of biodegradability for coking wastewater by selective adsorption of hydrophobic organic pollutants. Sep Purif Rev 151:23–30
59. Yu X, Xu R, Wei C, Wu H (2016) Removal of cyanide compounds from coking wastewater by ferrous sulfate: Improvement of biodegradability. J Hazard Mater 302:468–474
60. Sharma NK, Philip L (2016) Combined biological and photocatalytic treatment of real coke oven wastewater. Chem Eng J 295:20–28
61. Ozyonar F, Karagozoglu B (2015) Treatment of pretreated coke wastewater by electrocoagulation and electrochemical peroxidation processes. Sep Purif Technol 150:268–277
62. Wei XX, Zhang ZY, Fan QL, Yuan XY, Guo DS (2012) The effect of treatment stages on the coking wastewater hazardous compounds and their toxicity. J Hazard Mater 239:135–141
63. Pal P, Kumar R (2014) Treatment of coke wastewater: a critical review for developing sustainable management strategies. Sep Purif Rev 43(2):89–123
64. Zhang N, Xing-Dong HE, Yu-Bao G, Yong-Hong LI, Hai-Tao W, Di MA, Zhang R, Yang S (2010) Pedogenic carbonate and soil dehydrogenase activity in response to soil organic matter in Artemisia ordosica community. Pedosphere 20(2):229–235

65. Zhang T, Ding L, Ren H, Xiong X (2009) Ammonium nitrogen removal from coking wastewater by chemical precipitation recycle technology. Water Res 43(20):5209–5215
66. BIS-10500 (2012) Drinking water specifications IS 10500. Bureau of Indian Standards, India
67. Wang QR, Li YC, Klassen W (2003) Effects of soil amendments at a heavy loading rate associated with cover crops as green manures on the leaching of nutrients and heavy metals from a calcareous soil. J Environ Sci Heal B 38(6):865–881
68. Neupane G, Donahoe RJ (2013) Leachability of elements in alkaline and acidic coal fly ash samples during batch and column leaching tests. Fuel 104:758–770
69. Van der Watt HVH, Sumner ME, Cabrera ML (1994) Bioavailability of copper, manganese and zinc in poultry manure. J Environ Qual 23:43–49
70. USEPA (1992) Method 1311 Toxicity characteristic leaching procedure. Test methods for evaluating solid waste physical/chemical methods. Publ. SW-846, USEPA, Washington, DC
71. Zhou H, Zhou X, Zeng M, Liao BH, Liu L, Yang WT, Wu YM, Qiu QY, Wang YJ (2014) Effects of combined amendments on heavy metal accumulation in rice (*Oryza sativa* L.) planted on contaminated paddy soil. Ecotoxicol Environ Saf 101:226–232
72. Lindsay WL, Norvell WA (1978) Development of DTPA soil test for Zn, Fe, Mn and Cu. SSSA 42:421–428
73. Mylavarapu RS, Sanchez JF, Nguyen JH, Bartos JM (2002) Evaluation of Mehlich-1 and Mehlich-3 extraction procedures for plant nutrients in acid mineral soils of Florida. Commun Soil Sci Plan 33:807–820
74. Dutta BK, Khanra S, Mallick D (2009) Leaching of elements from coal fly ash: assessment of its potential for use in filling abandoned coal mines. Fuel 88(7):1314–1323
75. Fernandez-Turiel JL, De Carvalho W, Cabanas M, Querol X, Lopez-Soler A (1994) Mobility of heavy metals from coal fly ash. Environ Geol 23:264–270
76. Jegadeesan G, Al-Abed SR, Pinto P (2008) Influence of trace metal distribution on its leachability from coal fly ash. Fuel 87:1887–1893
77. Ward CR, French D, Jankowski J (2003) Comparative evaluation of leachability test methods and element mobility for selected Australian fly ash samples. CCSD 1–22
78. Shim YS, Rhee SW, Lee WK (2005) Comparison of leaching characteristics of heavy metals from bottom and fly ashes in Korea and Japan. Waste Manag 25:473–480
79. Punshon T, Adriano DC, Weber JT (2002) Restoration of drastically eroded land using coal fly ash and poultry biosolid. Sci Total Environ 296:209–225
80. Bhattacharyya S, Donahoe RJ, Patel D (2009) Experimental study of chemical treatment of coal fly ash to reduce the mobility of priority trace elements. Fuel 88(7):1173–1184
81. Duchesne J, Reardon EJ (1999) Lime treatment of fly ash: characterization of leachate composition and solid/water reactions. Waste Manage 19(3):221–231
82. Astrup T, Stipp SLS, Christensen TH (2000) Immobilization of chromate from coal fly ash leachate using an attenuating barrier containing zero-valent iron. Environ Sci Technol 34(19): 4163–4168
83. Wang H, Ren JZ (2014) Bioelectrochemical metal recovery from wastewater: a review. Water Res 66:219–232
84. Jayaranjan MLD, Annachhatre AP (2012) Precipitation of heavy metals from coal ash leachate using biogenic hydrogen sulfide generated from FGD gypsum. Water Sci Technol 67(2): 311–318
85. Gautam RK, Mudhoo A, Lofrano G, Chattopadhyaya MC (2014) Biomass-derived biosorbents for metal ions sequestration: adsorbent modification and activation methods and adsorbent regeneration. JECE 2(1):239–259

Assessment of Water Quality by Evaluating Water Quality Index Over Kolkata Using Statistical Approach: A Step Towards Water Pollution Management

Sayantika Mukherjee

Abstract It is well known fact that water is being polluted enormously. Though it is impossible for us to sustain without water still we are responsible for addition up of pollutants in water as well. To start with proper management of water pollution, it is very much necessary to analyze the physical, chemical and biological parameters revealing the water quality, level of degradation and remedial measures for sustenance as well. The present study aims at identifying the physical as well as chemical characterization using statistical methods over Kolkata (22°32′N; 88°20′E) from 2011 to 2017. Total eight sites have been selected for the study. Different sites show significant enrichment with Zinc, Iron, Chromium, Magnesium indicating input from industrial sources. The observed values of different physicochemical characteristics like pH, temperature, turbidity, total hardness, iron, chloride, total dissolved solids, Calcium, Sulphate, total alkalinity, Chemical Oxygen Demand (COD), Biological Oxygen Demand (BOD), phosphate values were compared with standard values recommended by Bureau of Indian standard (BIS). In addition to that, water quality index (WQI) has been evaluated for dry and wet season.

Keywords Water Quality · Chemical Oxygen Demand · Biological Oxygen Demand · Water quality index

1 Introduction

Water quality is a significant principles in evaluating the demand and supply of water. While taking necessary steps towards Water pollution management, it is essential to access the WQI of that particular site. Kolkata 'the city of Joy' is blessed with the Ganga beside its western end and wide wetland in its eastern periphery which is known to naturally treats its wastewater. The huge amount of the waste water is the result of adding up of the number of individuals day by day along with

S. Mukherjee (✉)
Department of Environmental Science, Amity University Kolkata, Kolkata, India
e-mail: sgmukherjee@kol.amity.edu

S. Yadav et al. (eds.), *Water Quality, Assessment and Management in India*, Earth and Environmental Sciences Library, https://doi.org/10.1007/978-3-030-95687-5_12

rapid industrialization and due to the movement of rural mass to the urban areas [1]. It is causing a stress on the aquatic ecosystems. It is observed from different studies that Indian rivers are polluted mainly due to the discharge of untreated sewage and industrial effluents. In Kolkata Groundwater reserve is under stress [2]. Poor quality of groundwater can be a menace to human health especially when it is combined with hazardous pollutants like heavy metals [3]. The emerging delinquent of degradation of the river ecosystem has initiated the insight for analyzing and implementing environmental auditing and environmental management system to achieve sustainable management of water resource. Along with that it is necessary to monitor water quality of various rivers all over the country to appraise their production capacity, utility potential and to plan healing trials as well [4–6]. In this regards, a study was aimed to assess the water quality of Rangit river basin for human consumption and agricultural purposes [7]. Similarly, a study has been conducted using quite a few common physico-chemical parameters to measure the water quality for Brahmaputra and the Ganges river. It was observed from the study that due to the contamination, the water is not fit for drinking purposes but might be utilized for agricultural activities [8]. Authors have developed artificial intellenge in predicting WQI prediction and to evaluate the water quality over India [9]. The overview of the integrated two-model system demonstrats a significant advance in estimating the water quality over Kaoping River Basin [10].

The main objective of the present study is to find out the variability of physico-chemical parameters over Kolkata. In addition to that the water quality index has also been assessed.

2 Site Location

Kolkata city is spread over around 200 square kilometres and has a population of 4.5 million people. It is one of the most water-rich cities in India.In this present study Kolkata had been selected as the site. The site is comprising of 21 different locations all over Kolkata over the Ganga river. They are Adi Ganga at Bansdroni, Adi Ganga at Jirat Bridge, Adi Ganga at Kalighat, Adi Ganga at Karunamoyee, Adi Ganga at Kudghat, Adi Ganga at SahidKhudiram, Behala, Central Kolkata, Cossipore-North Kolkata, Dhapa Calcutta, Ganga at Dakshineswar, Ganga at Garden Reach, Garia Calcutta, Inside Kolkata Leather Complex, Victoria (Pond No 1,2,3,4), RabindraSarovar National Lake, Tangra, Topsia.

3 Materials and Methods

The data for the present study has been collected from the West Bengal Pollution Control board from the year 2011 to 2017. The variables that are being analyzed are BOD, Calcium, Chloride, Chromium, COD, Conductivity, Fluoride, Iron,

Magnesium, Mercury, Nitrate-N, pH, Phosphate P, Potassium, Sodium, Sulphate, Temperature, Total Alkalinity,Total Coliform, Total Dissolved Solids (TDS), Total Fixed Solids (TFS), Total Hardness as $CaCO_3$,Total Suspended Solids (TSS), Turbidity,Zinc. For the analysis, statistical methods are being used for evaluation of data. Box whisker plots are drawn to find out the variability of the parameters. The scheming of the Water Quality Index (WQI) was performed on the basis of WQI estimation projected by Horton (1965) and developed by Brown et al. (1972) [11]. The WQI is calculated by assimilating the complex water quality data into a numerical score that defines the overall water quality standing. In this regards; pH, Turbidity, Total Dissolved Solids, Total Hardness, Chloride and Iron concentration of the selected sites are considered for calculating the Water Quality Index (WQI) [12]. To achieve this the weighted arithmetic water quality index has been formulated and calculated as well [13–15].

Box Whisker plot has drawn for the above mentioned parameters to evaluate the variability. The standards of the water quality parameter as per IS-10500 (1991) [11] and their respective weight used in the present study are highlighted in this study.

The weighted arithmetic water quality index (WQI) is in the following form:

Calculation of WQI;

$$WQI = \sum_{i=0}^{n} w_i q_i / \sum_{i=1}^{n} w_i \tag{1}$$

where n is the number of variables or parameters, w_i is the relative weight of the ith parameter, and q_i is the water quality rating of the ith parameter. The unit weight (wi) of the various water quality parameters are inversely proportional to the referred standards for the corresponding parameters. According to Brown et al. (1972), the value of q_i is calculated using the following equation:

$$q_i = 100\ [(V_i - V_{id})/(S_i - V_{id})]$$

where, V_i is the observed value of the ith parameter, Si is the standard allowable value of the ith parameter, and V_{id} is the ideal value of the ith parameter in pure water. All the ideal values (V_{id}) are taken as zero for drinking water except pH [16]. The classification of water quality has been incorporated in this study based on its quality index [17, 18].

4 Implementation Procedure

For this purpose data has been tabulatedfrom 2011 to 2017. Then the data set has been categorized in to Dry season and Wet Season. Mean, maximum and minimum values are estimated using statistical analysis followed by the selection of parameters for calculating WQI. Box-whisker plot has been done to observe the variability of each of the parameters during the dry and wet season. Ultimately WQI has been evaluated for dry season and wet season as well.

5 Results and Discussion

The values of all the parameters for 21 sites are being tabulated for calculation. The Data has been categorized into Dry season and Wet Season to find out the Hydrochemistry of the water quality of two different seasons. The data of different variables that ensure the water quality of Kolkata are evaluated for the present study. A brief analysis along with the description of all the physico-chemical attributes related to water quality is tabulated and discussed as follows (Tables 1 and 2). Along

Table 1 Parameters during dry season

	Mean	std	max	min
BOD	11.93	11.89	54.50	1.05
COD	33.12	28.36	150.92	2.00
DO	7.95	2.87	17.90	0.20
Fecal Coliform	1693890.78	2506291.60	9000000.00	200.00
pH	7.99	0.61	9.58	7.02
Temperature	27.56	4.48	36.00	16.00
Total Coliform	3090359.47	4596793.94	17000000.00	400.00
TSS	76.67	103.33	948.00	4.00
Conductivity	334.25	53.58	545.00	124.60
Nitrate-N	0.35	0.29	1.77	0.004
Ammonia-N	0.30	0.24	1.85	0.10
Boron	0.56	0.66	3.20	0.10
Calcium	36.77	11.99	68.00	9.60
Chloride	30.27	14.09	78.27	6.00
Fluoride	0.24	0.08	0.54	0.10
Magnesium	12.18	7.55	46.17	2.43
Phenolpthalein Alkanity	0.30	0.72	6.00	0.00
Phosphate-P	0.08	0.07	0.43	0.01
Potassium	5.86	2.63	20.00	1.00
Sodium	33.41	60.61	700.00	2.35
Sulphate	17.69	13.60	161.32	3.15
Total Alkalinity	139.93	24.36	200.00	55.00
Total Dissolved Solids (TDS)	218.58	81.94	682.00	8.00
Total Fixed Solids (TFS)	184.25	92.76	722.00	12.00
Total Hardness as $CaCo_3$	142.04	46.48	290.00	60.00
Total Kjeldahl Nitrogen (TKN)	2.49	6.72	55.00	0.10
Turbidity	57.66	58.93	454.00	1.47
Iron	0.73	0.83	3.67	0.07

Table 2 Parameters during wet season

	Mean	std	max	min
BOD	7.607959	6.531774	35.75	1.05
COD	28.85214	21.04503	104.56	2.06
DO	5.704673	2.962959	13.9	0.3
Fecal Coliform	1,494,884	2,173,250	9,000,000	700
pH	7.864595	0.528478	9.83	7.06
Temperature	29.79054	2.039586	36	25
Total Coliform	3,107,496	4,486,143	16,000,000	1700
TSS	130.1343	171.9933	1134	2
Conductivity	286.871	195.8523	1941	173
Nitrate-N	0.391347	0.284105	1.44	0.01
Ammonia-N	0.201556	0.106472	0.36	0.10
Boron	0.524921	0.447975	2.92	0.10
Calcium	29.2577	8.498931	52.00	11.43
Chloride	29.67809	16.25878	64.06	6.00
Fluoride	0.273734	0.113055	0.68	0.10
Magnesium	11.60	7.28	29.16	2.43
Phenolpthalein Alkanity	0.00	0.00	0.00	0.00
Phosphate-P	0.08	0.07	0.39	0.01
Potassium	6.30	3.62	20.00	2.78
Sodium	26.94	16.67	80.00	4.43
Sulphate	14.44	6.16	42.28	3.28
Total Alkalinity	128.03	31.47	240.00	76.00
Total Dissolved Solids (TDS)	185.99	49.21	356.00	84.00
Total Fixed Solids (TFS)	202.64	130.97	952.00	32.00
Total Hardness as $CaCo_3$	120.86	38.39	200.00	57.14
Total KjeldahlNitrogen (TKN)	1.42	1.85	11.50	0.06
Turbidity	128.96	133.77	656.00	1.92
Iron	0.64	0.83	2.01	1.56

with that a comparison with the standard guideline values as recommended by the WHO for drinking and public health purposes has been made to find out the percentage of data is within/exceeding the limit (Table 3).

5.1 *Temperature*

The values of temperature varyfrom 16 to 36 having an average value of 27.56 in dry season whereas in wet season the average temperature 29.79, and the range exists between 25 to 36. It reveals that 99% and 98% data are falling within the permission range during dry and wet season respectively.

Table 3 Comparison of the observed data with desirable limit

SL No	Parameter	Desirable limit	Dry season Within limit	Dry season exceeding limit	Wet season Within limit	Wet season exceeding limit
1	Temperature	≤ 35 °C	99.25	0.75	98.65	1.35
2	pH	6.5–8.5	75.8	24.2	75.3	24.7
3	EC	300 µs/cm	24.1	75.9	73.0	27.0
4	TDS	500 mg/L	99.2	0.8	100.0	0.0
5	TSS	< 10 mg/l	9.02	90.98	8.11	91.89
6	DO	≥ 5 mg/l	96.24	3.76	66.22	33.78
7	BOD	≤ 3 mg/l	24.00	81.95	23.29	76.71
8	COD	< 10 mg/l	40.60	59.40	18.92	81.08
9	Total Hardness	200 mg/l	87.22	12.78	0.00	0.00
10	Total alkalinity	200–600 mg/l	100.00	0.00	100.00	0.00
11	Ca	75 (mg/L)	100.00	0.00	100.00	0.00
12	Mg	30 (mg/L)	94.74	5.26	100	0
13	Phosphate-P	5 mg/l	100.00	0.00	100.00	0.00
14	Potassium	<20 mg/l	100.00	0.00	100.00	0.00
15	Na	200 (mg/L)	98.50	1.50	100.00	0.00
16	No3	45 mg/L	100.00	0.00	100.00	0.00
17	Ammonia	≤ 1.2 mg/l	100.00	0.00	100.00	0.00
18	Sulphate	200 mg/l	100.00	0.00	100.0	0.00
19	Cl	250 (mg/L)	100.00	0.00	100.00	0.00
20	F	1.0 (mg/L)	100.00	0.00	100.00	0.00
21	Turbidity	≤ 30 NTU	33.86	66.14	33.78	66.22
22	Iron	03–1.0				

5.2 *pH*

In dry season the values of pH remain within 7.02 to 9.58 having the average value of 7.99 in dry season. In wet season the values of pH varies from 7.06 to 9.83 having the mean value of 7.86. The present results reveal that though the average value remains within the desirable limit, some values are exceeding the higher range indicating that Water with a pH > 8.5 could indicate that the water is hard. Hard water does not pose a health risk, but can also cause aesthetic problems. The result reveals that during dry season the values of pH of 75.8% of sample is within the permissible limit and 24.2% of sample is observed to exceed the limit. In addition to that, in wet season 75.3% of sample is within the permissible range, and 24.7% is exceeding the limit.

5.3 *Conductivity*

The mean values of conductivity are 334.25 µs/cm and 286 µs/cm for dry and wet season respectively. In the dry season, the values remain within the range of 124.6 to 545 µs/cm whereas in wet season the same varies from 173 to 1941 µs/cm. It reveals that less variability is observed for dry season. 24.1% and 73% of data set are falling in desirable limit during dry and wet season respectively.

5.4 *Total Dissolved Solids (TDS)*

TDS plays a very important role in determining the quality of water. The value is observed to be within 8–682 having the average value of 218.58 and 84–356 obtaining an average value of 185.99 in dry and wet season, respectively. In dry season 99.2% of the values are within limit wherein wet season whole the data obtained are within the permissible limit of TDS.

5.5 *TSS*

In the dry season the values of TSS remains within the range of 4–948 having an average value of 76.67. On the other hand, the TSS values remain within the range of 2–1134, having anaverage value of 130. For both of the seasons, more than 90% data are exceeding the permissible limit.

5.6 *Dissolved Oxygen*

The variability is observed to be more in the dry season in comparison to wet. The value of dissolved oxygen varies from 0.20 to 17.90 with the mean value of 7.95 in dry season and wet season it remains within the range of 0.3 to 13.9 having an average value of 5.70. Though the seasonal mean remains within the permissible limit for both of seasons, more than 33% data is observed to be exceeding the limit.

5.7 *Biological Oxygen Demand*

The values of BOD vary from 1.05 to 54.50 having the average value of 11.93 in the dry season. In wet season it remains within 1.05 to 37.35. It reveals from the study that, only 24% of the data is falling within the permissible limit for dry season as well as wet season. This is observed from the present analysis that the BOD level is in bad condition.

5.8 Chemical Oxygen Demand

The values of COD vary within the range of 2 to 150.92, obtaining the average value of 33.12 in the dry season where it lies within 2.06 to 104.56 in the wet season. It depicts that 59% data are exceeding the limit in dry season whereas 81% data are within the exceeding limit category.

5.9 Total Hardness as $CaCO_3$

In the dry season the average value of Total Hardness as $CaCO_3$ is142.04 and the range is 60–290. The same is obtained for wet season as well, and the values remain within 57.14 to 200 having an average value of 120.86. It depicts that all the data set both wet seasons remain within the permissible range during the study period whereas in dry season 87.22% of data are well to be obtained within the range.

5.10 Total Alkalinity

In dry season the value of Total Alkalinity remains within the range of 55 to 200 mg/l having the average value of 139.93 mg/l whereas it remains within 76–240 mg/l having an average value of 128.03 mg/l. Total alkalinity is the measure of the capacity of water to neutralize a strong acid. The major source for alkalinity is due to landfills and pipelines. For both of the seasons, the values are within the desirable range.

5.11 Calcium

In the dry season, the values of Calcium remains within 9.60–68.00 with a mean value 36.77 whereas in wet season the range is 11.43 to 52 having an average value of 29.25. All the values are observed to be within permissible range.

5.12 *Magnesium*

In the dry season, the value of Magnesium remains within 2.43–46.17 having an average value of 12.18. In wet season it remains within 2.43 to 29.16 with an average value of 11.60. 94.74% of the data are well versed within the threshold value for the dry season.

5.13 *Phosphate-P*

The average value of 0.08 is observed to be for both dry and wet season as well. It varies from 0.01 to 0.43 and 0.01 to 0.39 in dry and wet season respectively. All the data are within the range.

5.14 *Potassium*

In the dry season, the value is observed to be within 1.00 to 20.00 with an average value of 5.86. In addition to that, in wet season the values remain within 2.78 to 20.00 having an average value of 6.30. Irrespective of season all the data are within permissible range.

5.15 *Sodium*

The value of Na ranges from 2.35 to 700 having a mean value of 33.41 in the dry season whereas it remains within the range of 4.43 to 80.00 having an average of 26.94. It is observed that more than 98% of the data set are within the permissible range and only 1.50% are exceeding the limit in dry season. In wet season all level of Sodium is within the permissible range.

5.16 *Nitrate-N*

The values of Nitrate-N remains within the range of 0.0042 to 1.77 having the average value of 0.35 whereas in the wet season the same varies from 0.01 to 1.44. It reveals that all the values are within the specified range. Nitrate in the study area is found to be comparatively very low in concentration. This may be due to the fact that nitrate is going more in runoff (Rao et al. 2004; Kumar et al. 2016).

5.17 Ammonia-N

In the dry season, the range varies from 0.1–1.85 having the average value of 0.30. In wet season it remains within the range of 0.1–0.36 securing the average value of 0.20. It is observed from the study present that100% of the collected data are within the permissible range.

5.18 Sulphate

The value of Sulphate ranges from 3.15–161.32 having a mean value of 17.69 in the dry season whereas it remains within the range of 3.28–42.28 having an average of 14.44. All the values are observed to be within the permissible range for both of the seasons.

5.19 Chloride

In the dry season, the values of Chloride remains within 6.00 to 78.27 with a mean value 30.27 whereas in wet season the range is 6–64.06 having an average value of 29.67. Both seasons are showing the value of chlorine within the permissible limit.

5.20 Fluoride

The value of Fluoride ranges from 0.10–0.54 having a mean value of 0.24 in the dry season whereas it remains within the range of 0.1–0.68 having an average of 0.273. The values depict that the Fluoride level is permissible in dry as well as wet season.

5.21 Turbidity

In the dry season the values of 1.47 to 454 having an average value 57.66 whereas it remains within the range of 1.92 to 656 having mean value of 128.96 in wet season.

5.22 Iron

In the dry season, the average value is 0.73 whereas in wet season the same is 0.64.

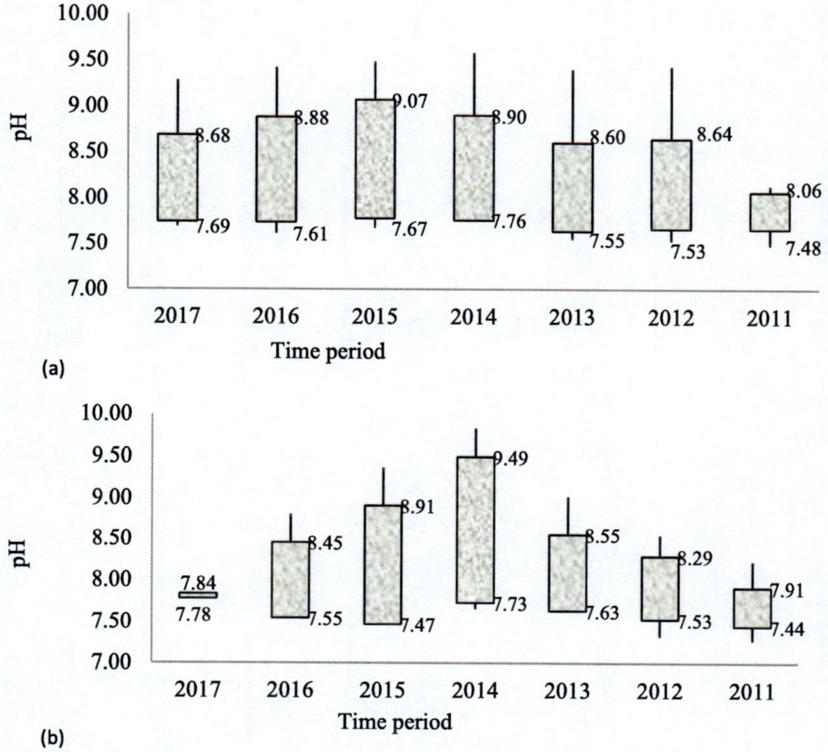

Fig. 1 Box plot showing the variability of pH during dry season means yearly period of low rainfall (**a**) and wet season means rainly (**b**)

Statistical Box whisker plots are drawn for the six selected parameters required for the computation of WQI for the dry season and wet season as well.

Figure 1 shows the variability of the pH values for dry season (a) and wet season (b). For dry season pH varies from 7.69 to 8.68, 7.61 to 8.88, 7.67 to 9.07, 7.76 to 8.90, 7.55 to 8.60, 7.53 to 8.64 and 7.58 to 8.06 for 2017, 2016, 2015,2014,2013,2012 and 2011 respectively. For Wet season pH varies from 7.78 to 7.84, 7.55 to 8.45, 7.47 to 8.91, 7.73 to 9.49, 7.63 to 8.55, 7.53 to 8.29 and 7.44 to 7.91 for 2017, 2016, 2015,2014,2013,2012 and 2011 respectively.

Figure 2 shows the variability of the Turbidity values for dry season (a) and wet season (b). For dry season the values of turbidity varies within the range of 0.28–53.49, 0.94–205.08, 19.41–144.34,
16.54–414.81,
28.77–255.62, 0.51–256.57,
2.12–269.87 during 2017, 2016, 2015, 2014, 2013, 2012, 2011 respectively. Maximum variability is observed during 2014 for dry as well as wet season.

The variability of TDS values for dry season (a) and wet season (b) are shown in Fig. 3. The value is observed to be within 8–682 having the average value of 218.58

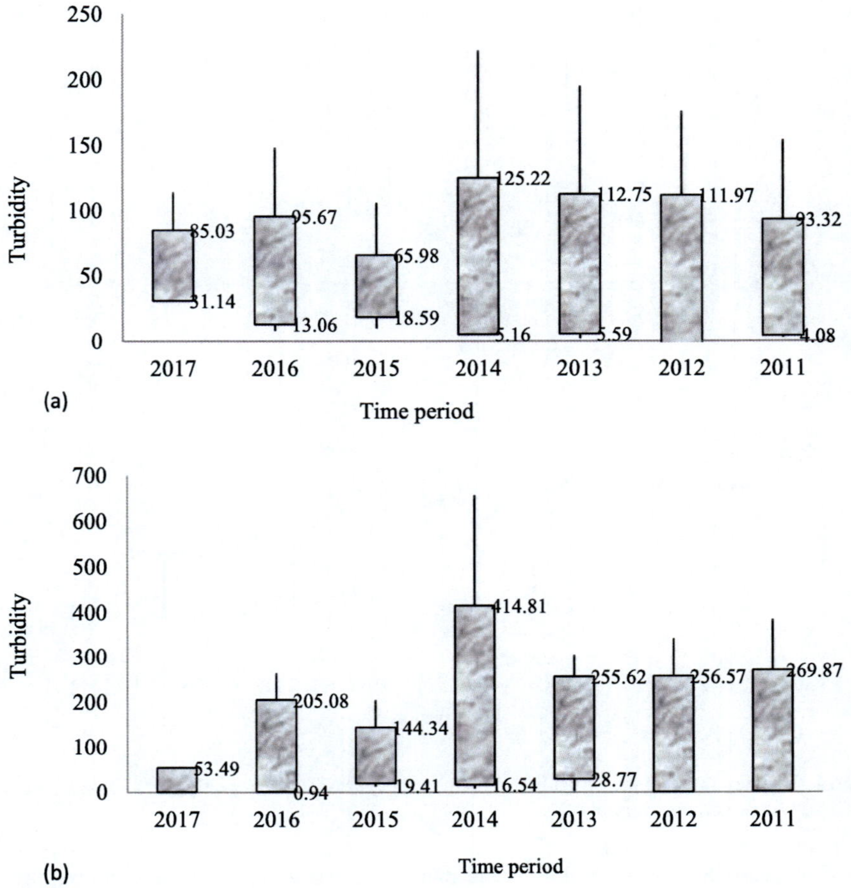

Fig. 2 Box plot showing the variability of turbidity during dry season means yearly period of low rainfall (**a**) and wet season means rainly (**b**)

and 84–356 obtaining an average value of 185.99 in dry and wet season, respectively. During the period of 2014 it is reflecting to have the maximum variability.

Figure 4 shows the variability of the Total Hardness values for dry season (a) and wet season (b). In the dry season the average value of Total Hardness as $CaCO_3$ is142.04 and the range is 60–290. The same is obtained for wet season as well, and the values remain within 57.14 to 200 having an average value of 120.86. The variability to its maximum value is observed to be during the year 2013 for dry season and the same is for wet season is noticed during 2012.

In addition to that, the variability of the Chlorine values for dry season (a) and wet season (b) are shown in Fig. 5. In the dry season, the values of Chloride remains within 6.00 to 78.27 with a mean value 30.27 whereas in wet season the range is

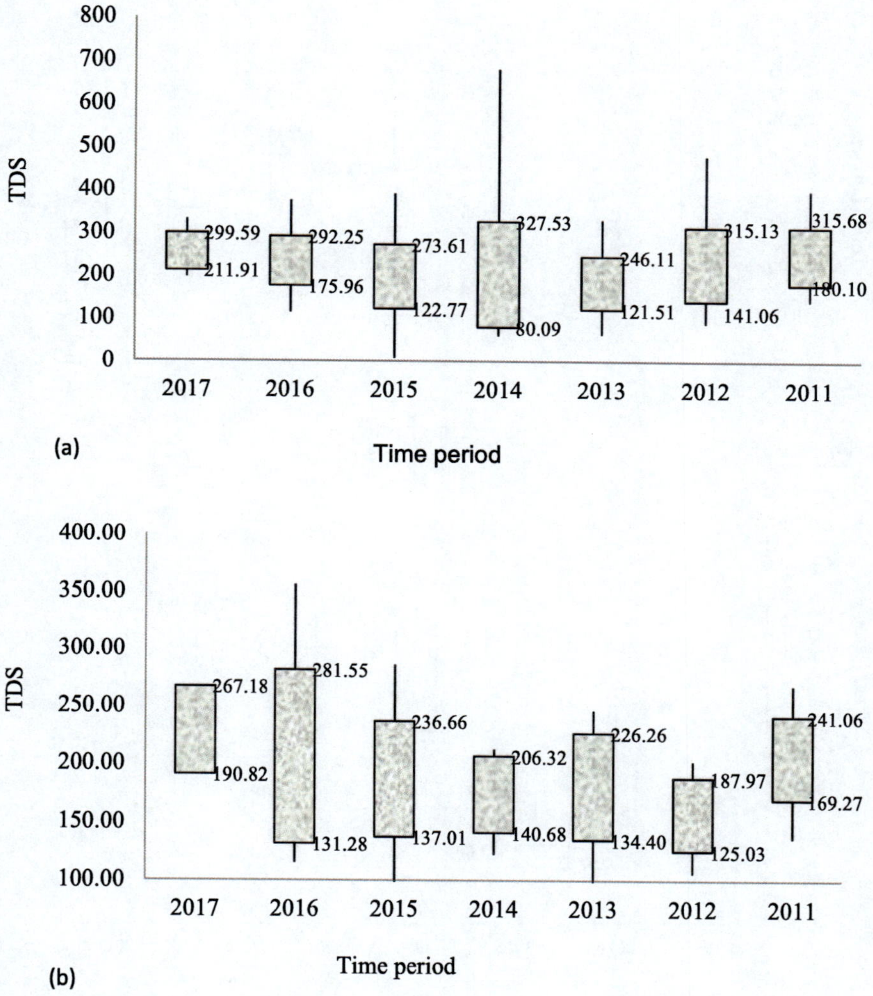

Fig. 3 Box plot showing the variability of TDS during dry season means yearly period of low rainfall (**a**) and wet season means rainly (**b**)

6–64.06 having an average value of 29.67. Both seasons are showing the value of chlorine within the permissible limit.

Similarly, Fig. 6 shows the variability of the iron values for dry season (a) and wet season (b) as well. In the dry season, the average value is 0.73 whereas in wet season the same is 0.64. For dry season the maximum variability is observed during the year 2015, followed by 2012.

The WQI has been calculated using the formula given. The respective weight of the selected parameters used in the present study are highlighted in Table 4. In addition to that the classification of WQI has been tabulated (Table 5). Tables 6 and 7

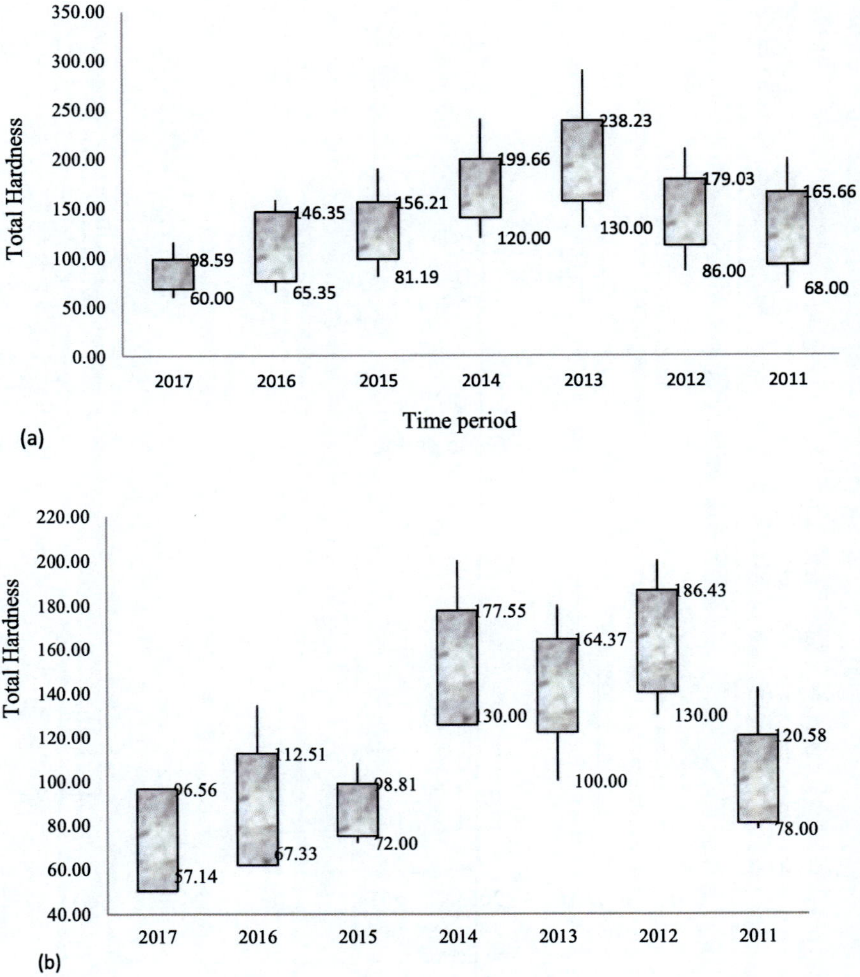

Fig. 4 Shows the variability of the Total Hardness values for dry season means yearly period of low rainfall (**a**) and wet season means rainly (**b**)

show the values of WQI for dry season and wet season respectively. Table 6 reveals that the values of WQI varyfrom 72 to 101 in dry season where it fluctuates within 39.33 to 77.33 in wet season (Table 7). Figure 7 depicts that comparison of the values of WQI during dry and wet season from 2011 to 2017. In dry season the Water Quality Index values varywithin the category of Good, Excellent and Medium Category whereas in Wet season the WQI values varywithin Bad, Medium and Good Category. Thus it is observed that in Wet season the Water Quality is in deteriorating condition within the study area during the study period.

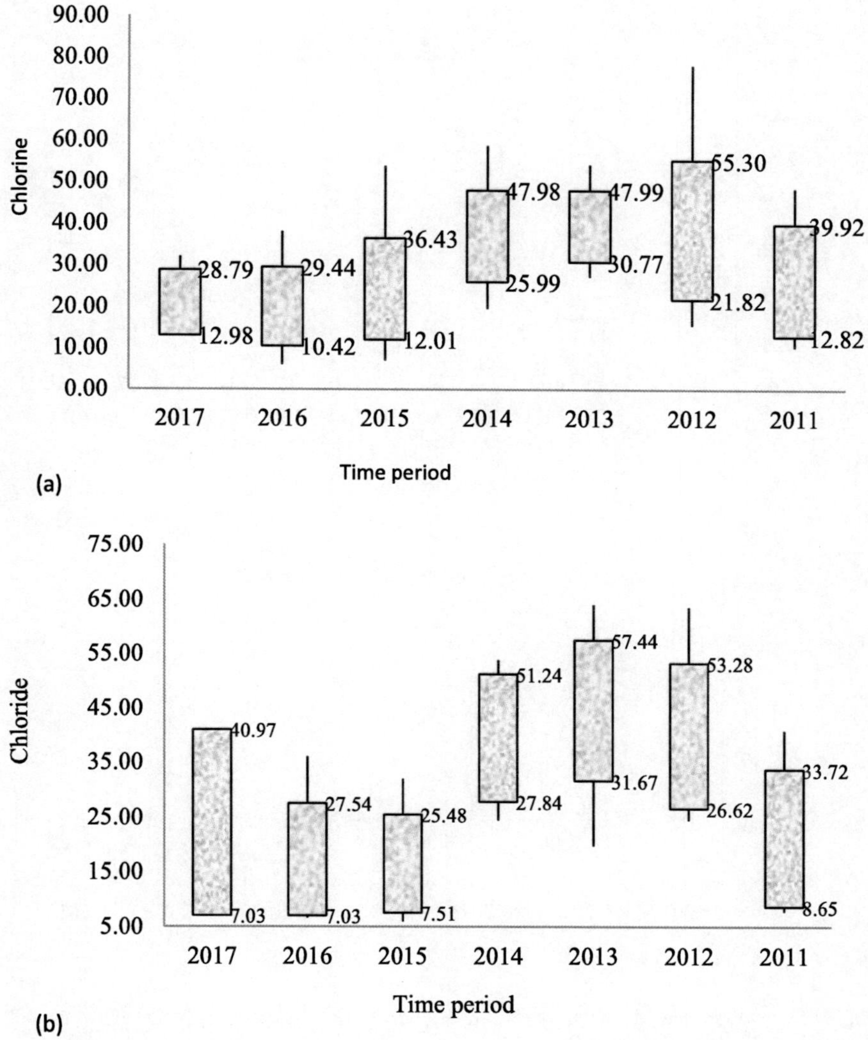

Fig. 5 Shows the variability of the chlorine values for dry season means yearly period of low rainfall (**a**) and wet season means rainly (**b**)

6 Conclusions

The study aims the assessment of Water Quality by evaluating water quality index over Kolkata. For this purpose, the analysis of different physicochemical parameters are considered. From the analysis of the parameters a distinct seasonal variability has been observed. In addition to that water quality has been monitored through calculation of Water Quality index. It is observed that during wet season the water quality

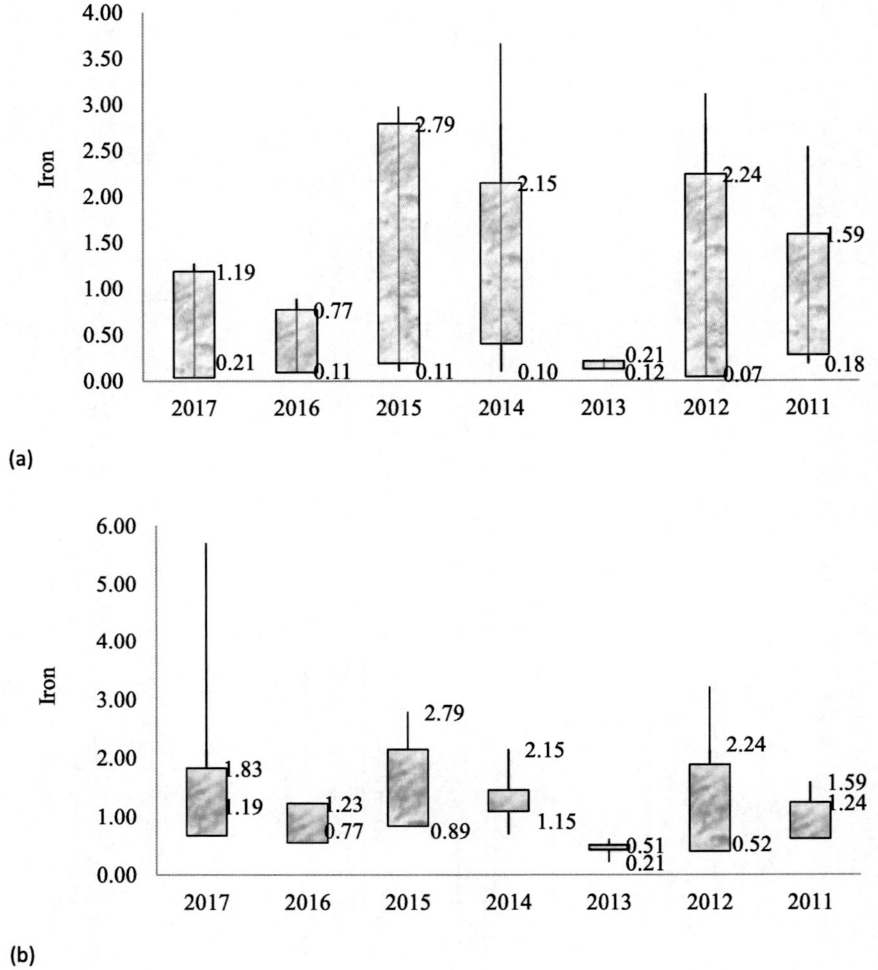

Fig. 6 Shows the variability of the values of Iron for dry season means yearly period of low rainfall (**a**) and wet season means rainly (**b**)

is found to be rather bad condition in terms of WQI. Thus it can be stated thatthe endeavour of the present study is toidentify the changes in water quality so that the stakeholders can take necessary actions to deal with.

Table 4 Water quality parameter, standards and weight used in the present study

SL No.	Parameters	Standards (IS-10500) (1991)	Weight (Wt)	Unit weight (Wi)
	pH	6.5–8.5	4	0.266667
	Turbidity (in NTU)	5–10	2	0.133333
	Total dissolved solids (TDS) (in mg/L)	500–2000	4	0.266667
	Hardness as $CaCo_3$ (in mg/L)	300–600	2	0.133333
	Chloride as Cl- (in mg/L)	250–1000	2	0.133333
	Iron as Fe (in mg/L)	0.3–1.0	1	0.066667
			$\Sigma Wt = 15$	$\Sigma Wi = 1$

Table 5 Classification of water quality index

Range	Quality
90–100	Excellent
70–90	Good
50–70	Medium
25–50	Bad
0–25	Very bad

Table 6 Average values of physico-chemical parameters and WQI value during dry season

	pH	Turbidity	TDS	Hardness	Chloride	Iron	WQI
2017	8.21	58.09	255.75	83.50	20.89	0.61	72.00
2016	8.30	54.36	234.11	111.31	19.93	0.43	68.00
2015	8.42	42.28	198.19	127.45	24.22	1.30	79.99
2014	8.32	65.19	203.81	170.00	36.99	0.87	94.67
2013	8.12	59.17	183.81	197.62	39.38	0.17	92.00
2012	8.15	54.70	228.10	145.52	38.56	1.10	101.00
2011	7.86	48.70	247.89	129.00	26.37	0.65	69.30

Table 7 Average values of physico-chemical parameters and WQI value during Wet Season

	pH	Turbidity	TDS	Hardness	Chloride	Iron	WQI
2017	7.81	26.89	229.00	73.47	24.00	0.61	42.00
2016	8.00	103.01	206.42	87.13	17.28	0.43	66.67
2015	8.19	81.88	186.83	86.92	16.50	1.30	72.00
2014	8.61	199.14	173.50	151.67	39.54	0.87	39.33
2013	8.09	142.19	180.33	143.33	44.55	0.17	77.33
2012	7.91	128.54	156.50	163.33	39.95	1.10	73.33
2011	7.68	136.00	205.17	100.67	21.19	0.65	50.04

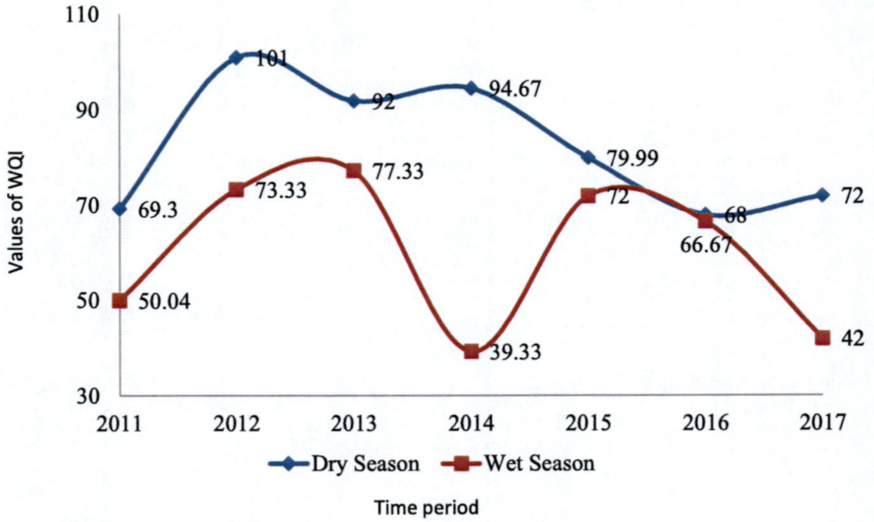

Fig. 7 Variablity of the values of WQI during yearly period of low rainfall [dry season] (**a**) and wet season means rainly (**b**)

7 Recommendations

Establishment of Water Quality Monitoring and Assessment Collaboration Across Jurisdictions is necessary in this regard. To determine the socio economic impact setting up of Research and Monitoring program can be recommended. Availability of data on daily basis would give more acuuracy to evaluate the water quality parameters over a specific location.

8 Acknowledgments

I would like to offer my sincere gratitude to West Bengal pollution control board for sharing the data used in his study.

References

1. Mohanta T, Goel S (2014) Assessment of water quality of three different aquatic environments over three seasons. IPCBEE 64
2. Banerji S, Mitra D (2018) Geographical information system based groundwater quality index assessment of Northern part of Kolkata. India for Drinking Purpose Geocarto International 34: 1–27. https://doi.org/10.1080/10106049.2018.1451922

3. Ehsan Kamali Maskooni, Mehran Naseri-Rad, Ronny Berndtsson and Kei Nakagawa; Use of Heavy Metal Content and Modified Water Quality Index to Assess Groundwater Quality in a Semiarid Area, 2020, 12, 1115
4. Datar MO, Vashistha RP (1992) Indian J Environ Protect 12:577–580
5. Das NK, Sinha RK (1993) Environ Ecol 11(4):829–832
6. Marschner H (1995) Mineral nutrition of higher plants. London Academic
7. Gupta S, Nayek S, Chakraborty D (2016) Hydrochemical evaluation of Rangit River, Sikkim, India: using water quality index and multivariate statistics. Environ Earth Sci 75(7):567
8. Dutta S, Nayek S (2020) Water Quality of the Ganges and Brahmaputra Rivers: An Impact Assessment on Socioeconomic Lives at Ganga–Brahmaputra River Basin. Sust Environ Eng Sci.237–241
9. Aldhyani THH, Al-Yaari M, Alkahtani H and Maashi M. 2020. Water quality prediction using artificial intelligence algorithms. Adv Comput Eng Bionics Med Appl https://doi.org/10.1155/2020/6659314
10. Lai YC, Yang CP, Hsieh CY, Wu CY, Kao CM (2011) Evaluation of non-point source pollution and river water quality using a multimedia two-model system. J Hydrol 409(3–4):583–595
11. IS-10500 (1991) Indian standard for drinking water. Bureau of Indian standards, New Delhi, India
12. Das S, Roy PK, Mazumdar A (2013) Development of water quality index for ground water in Kolkata City, West Bengal, India. ARPN J Eng Appl Sci 8(12):1054–1058
13. Tiwari TN, Mishra MA (1985) A preliminary assignment of Water Quality Index of major Indian rivers. Indian J Environ Protect 5(4):276–279
14. Singh DF (1992) Studies on the water quality index of some major river of pune, Maharastra. Proc Acad Environ Biol 1(1):61–66
15. Das KK, Panigrahi T, Panda RB (2012) Evaluation of water quality index (WQI) of drinking water of Balasore district, Odisha, India. Discovery Life 1(3):48–52
16. Tripathy JK, Sahu KC (2005) Seasonal hydrochemistry of groundwater in the barrier-spit system of Chilika lagoon. J Environ Hydrol 12(7):1–9
17. Brown RM, McClelland NI, Deininger RA, O'Connor MF (1972) Water quality index-crashing, the psychological barrier, Proc. 6th annual conference, advances in water pollution research: 787–794
18. Chatterjee C, Raziuddin M (2002) Determination of water quality index of a degraded river in Asanol Industrial area, Raniganj, Burdwan, West Bengal. Nat Environ Poll Technol 1(2): 181–189

Assessment of Water Quality of Ramganga River in Moradabad, India

Vineet Tirth, Ram Karan Singh, Amit Tirth, and Saiful Islam

Abstract The study includes the assessment of the water quality of the Ramganga River, which originates from Dudhatoli-Pauri Garhwal (Uttrakhand) and merges in the Ganges in Farukhabad (Uttar Pradesh) in North India. The length of the river is about 540 km and it has a catchment area of about 30,800 km^2. While flowing through the Moradabad region, the water quality of the river deteriorates severely due to the discharge of domestic wastewater and effluents from metal, food, paper, and sugar industries in Moradabad. The water quality of the Ramganga River has been estimated at three points in 2019–20; before the entrance to the Moradabad city while flowing through the city and at the exit from the city. The water quality has been assessed in terms of the dissolved oxygen, fecal contamination, BOD, pH, and TDS, on the samples taken during the summer, pre-monsoon, post-monsoon, and winter season. The water quality has been assessed and major determinants responsible for water pollution have been identified. A few suggestions to curb water pollution have been suggested in the study.

Keywords Water pollution · Water quality · Ramganga River · Water treatment · Deterioration of water quality · Sewage discharge

---

V. Tirth (✉)
Mechanical Engineering Department, College of Engineering, King Khalid University, Abha, Kingdom of Saudi Arabia
e-mail: vtirth@kku.edu.sa

R. K. Singh
Vice-Chancellor, The ICFAI University Dehradun, Dehradun, Uttrakhand, India

A. Tirth
Department of Public Health Dentistry, Kothiwal Dental College and Research Center, Moradabad, UP, India

S. Islam
Department of Civil Engineering, Universiti Tecknologi Malaysia, Johar Bahru, Malaysia
e-mail: isaiful2@graduate.utm.my

S. Yadav et al. (eds.), *Water Quality, Assessment and Management in India*, Earth and Environmental Sciences Library, https://doi.org/10.1007/978-3-030-95687-5_13

1 Introduction

Rivers are the primary source of water for mankind. Due to urbanization, policy deficit and unsustainable living habits, Rivers are getting polluted day by day. The water quality is represented in terms of water quality index (WQI) and is represented by the eight primary biological and physio-chemical factors namely; Dissolved Oxygen (DO), pH, Biological Oxygen Demand (B.O.D.), Total Dissolved Solids (TDS), Alkalinity, Fluoride and Chloride content, Hardness, Fecal Coliform (FC) and Total Coliform (TC), and heavy metals [1]. The major causes of pollution of River water are sewage discharge, run-off from agriculture, wastewater from households and cattle bath, industrial effluents, etc. [2–4]. The polluted river water also contaminates the underground water belts making it hazardous for the consumption of humans, animals, industries, and irrigation [5]. As per the Indian Standard IS:2296, the water is divided into five classes from A to E. Class A water is appropriate for consumption (drinking) without any treatment, class B water is appropriate for cleaning, class C is appropriate for consumption and domestic use only after treatment and disinfection, class D is appropriate for aquatic life and class E is recommended for cooling, waste disposal and dilution, industrial and irrigation applications [6].

Moradabad is an important city at the border of western Uttar Pradesh (UP), India. As one travels from the national capital New Delhi towards Lucknow, the Capital of UP, Moradabad is one of the largest cities on the way. It may be termed as the gateway of eastern UP. The city was established in 1625 AD and named after Murad Baksh, the son of Emperor Shah Jahan. Due to its brass industries, it is also called as Peetal Nagri alias Brass City [7]. About 40% of the handicraft export from India originates from Moradabad [8]. Ramganga River is one of the main tributaries of the Ganges, which originates from Dudhatoli-Pauri Garhwal (Uttrakhand) and merges in the Ganges in Farukhabad (UP) in North India [9]. The pollution of Ramganga water adds to the pollution of Ganges to a large extent [10]. The length of the river is about 540 km and it has a catchment area of about 30,800 km^2. Moradabad, Bareilly, and Farukhabad are the most polluted stretches of Ramganga [11]. In Moradabad city, barring a few proper industries, about 80% of the brass and steel articles are made in foundries and makeshift factories within the congested residential areas or within the households, resulting in unhygienic living conditions and pollution. In the Moradabad city, main industries, which discharge polluted water are metal surface treatment, slaughterhouses, meat processing, sugar, distilleries, dyeing and textile, dairy and milk processing, bottling plants etc. [6]. The registered industries have wastewater treatment in place but the wastewater and byproducts from the unorganized and unregistered brass and other metal industries, the sewage discharge from the households and the wastes from the slaughterhouses are discharged in drains, which fall in the river Ramganga. There are twenty fours identified drains in Moradabad city, which fall in Ramganga River. Hence, the River Ramganga is polluted to a large extent while it passes from Moradabad. A schematic diagram of the major pollutants and their sources are shown in Fig. 1.

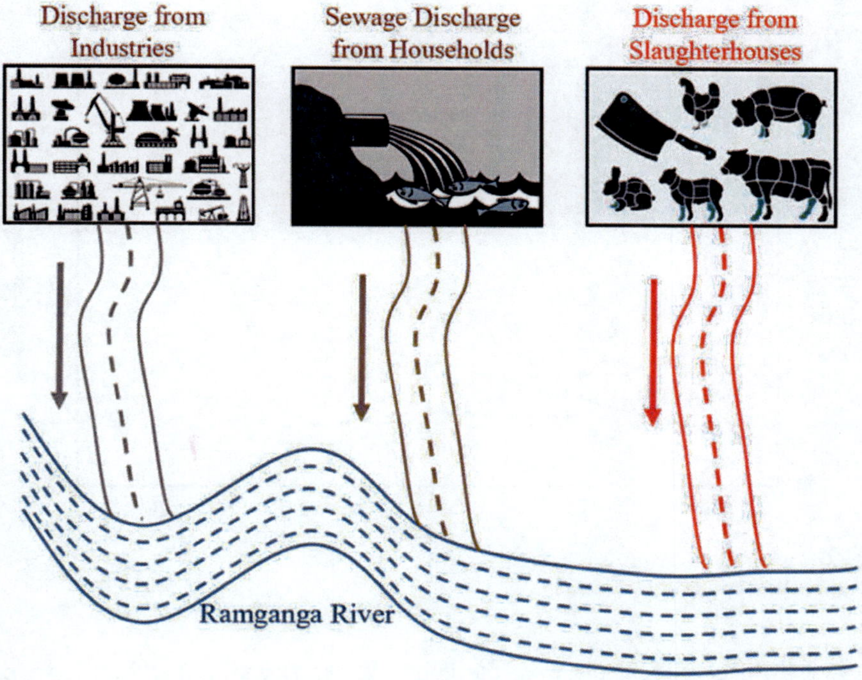

Fig. 1 Schematic diagram representing the discharge from industries, households, and slaughterhouses in Ramganga River in Moradabad (U.P.), India

The Ranganga water is not potable by any standards for aquatic life or for human consumption [12]. The color of the water of Ramganga River turns blackish towards the downstream due to sewage and industrial contamination [9].

1.1 Demography of Moradabad Region

Moradabad city is developing at a brisk rate. The city is expanding quickly under the planning of the Moradabad Development Authority (MDA) and new residential colonies, commercial complexes and special economic zones (SEZ) are coming up. Due to these developments, the population of the city is rising at a faster rate. Figure 2 shows the demography of Moradabad city in five years from 2014 to 2018. The numbers in Fig. 2 represent the population in lakhs in the corresponding year. The percentage increase in population from the preceding year is also shown in the figure. There is a sharp increase of about 12% in the population of Moradabad city from 2017 to 2018.

The landmass of the Moradabad city is insufficient to cater to such an increase in population and hence the natural resources are under extreme pressure. The domestic

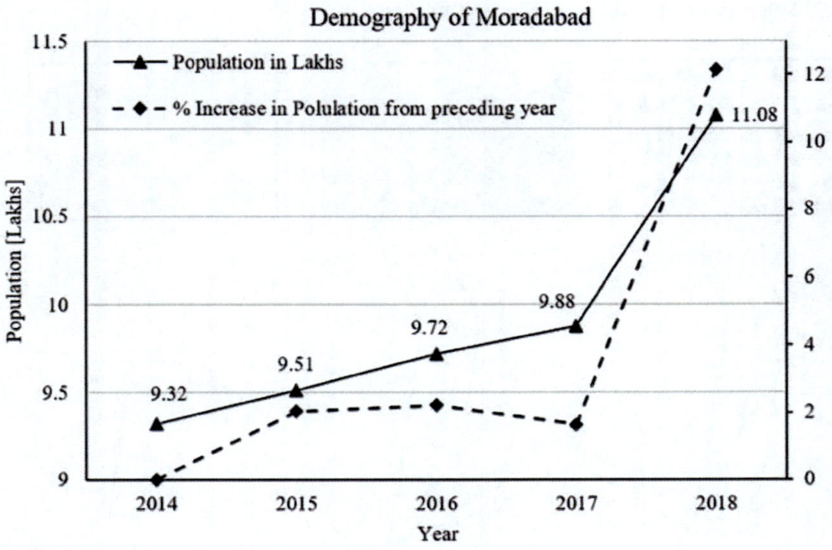

Fig. 2 Demography of Moradabad [8]

and industrial water supply from the Public Works Department (PWD) is from groundwater pumping. The irrigation water is also largely pumped from the tube wells. Ramganga River is the primary source of groundwater recharge in the city.

1.2 Determinants of Water Pollution

In India, Rivers are the major source of industrial or domestic water supply, either directly or indirectly through groundwater recharge. Unfortunately, the cities are being expanded along the banks of the rivers leading to encroachment of the River basin. Furthermore, the effluents from the industry and the domestic wastewater, including sewage water are directly discharged into the rivers without any treatment. Hence, the water quality of the Rivers is deteriorating at a faster pace. The efforts of the government to control water pollution are inadequate and the schemes for purification of water are least effective. The River water starts getting contaminated by chemicals, heavy metals like Cr, Cu, Ni, Cd, Fe, Pb, Zn, Mn [13–15]. The contamination of fecal coliform renders river water unsuitable to sustain aquatic life and making it extremely harmful for human beings. To determine the average water quality of the River round the year, the sampling is done across different seasons. The summer, pre-monsoon, post-monsoon, and winter seasons are generally considered for sampling. The sampling is also done at different locations across the river channel [16].

2 Material and Methods

The water quality data for Ramganga River was obtained from the Uttar Pradesh Pollution Control Board (UPPCB) and, the Central Pollution Control Board (CPCB), India for ten years from 2007 to 2016. The data of 2017 onwards was not published yet by CPCB. For the most recent real-time analysis, three water samples were collected from three locations in the Moradabad region and their average was reported as the result. The scope and test methods are described in the following subsection.

2.1 Study Area and Scope

The flow of the Ramganga River through Moradabad city is shown in Fig. 3 [17]. Three locations from which, the water samples were collected for the analysis are earmarked in circles in Fig. 3 as Entry 1, City 2, and Exit 3.

The water samples were collected at four-time intervals, during; the peak of the Summer (31.05.2019), Pre-monsoon (30.06.2019), Post-monsoon (30.09.2019), and

Fig. 3 Map of Moradabad, the flow of Ramganga with earmarked sample sites [17]

peak Winter (15.01.2020). The water quality data is the latest and has not been published yet. The scope of the study is limited to Moradabad city only.

2.2 *Procedure of Sampling and Analysis*

The water quality data for DO, pH, B.O.D., FC, and TC was obtained from the website of CPCB for ten years from 2007 to 2016. Three samples of Ramganga water were analyzed for DO, pH, B.O.D., FC, TDS, and Fluoride concentration for 2019–20 in the Research Testing & Calibration Laboratory at Metal Handicrafts Service Centre, Moradabad. The results are compared and discussed. The DO was measured by AquaPlus Optical DO, having measurement range 0–50 mg/l and based on Dynamic Luminescence Quenching. The pH was measured by a pH pen, make Bluelab, model PENPH, make bluelab, New Zealand. The TDS was measured by TDS&EC meter, make Health Metric, Malaysia, having a range of 0–9999 ppm and an accuracy of $\pm 2\%$. The B.O.D. was measured on electrochemical YSI Pro20 instrument make YSI, USA. The FC was measured on the Hach MEL/MPN E. Coli laboratory (test kit) model 2,569,800, make Thomas Scientific, USA. The Fluoride concentration was measured by Extech ExStik FL 700 Fluoride meter, make FLIR, USA. The Tisab Reagent tablets were used for preparing the sample solutions.

The average of the three samples at three locations over four seasons is discussed in the following section.

3 Results and Discussion

The water quality data for the ten years obtained from CPCB has been compared with the real-time water quality analysis from the most recently collected samples (2019–20).

3.1 *DO in Ramganga Water*

The free oxygen present in water due to aeration from the surface air and release from the aquatic photosynthesis is termed as DO, represented as mg/l. It is particularly crucial for aquatic life and considered one of the most important indicators of water quality. The DO is Ramganga water has shown a fluctuation over ten years, shown in Fig. 4. The net DO have shown an increasing trend from 2007 to 2016. Compared with the data up to the year 2016, the mean DO (across all the seasons) has decreased from 7.5 mg/l to about 5 mg/l (see Fig. 5), which is an alarming indicator of deteriorating water quality. The DO levels in winter increases, hence the

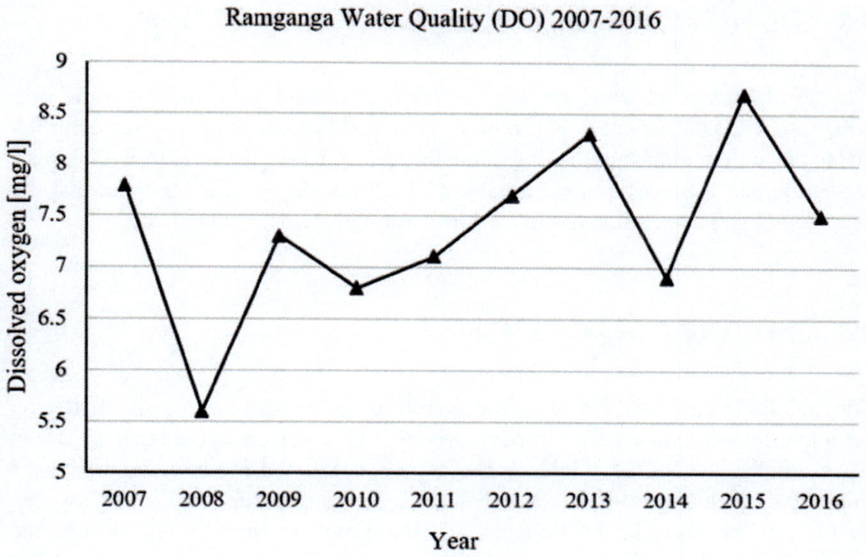

Fig. 4 DO in Ramganga water from 2007–16 [18–27]

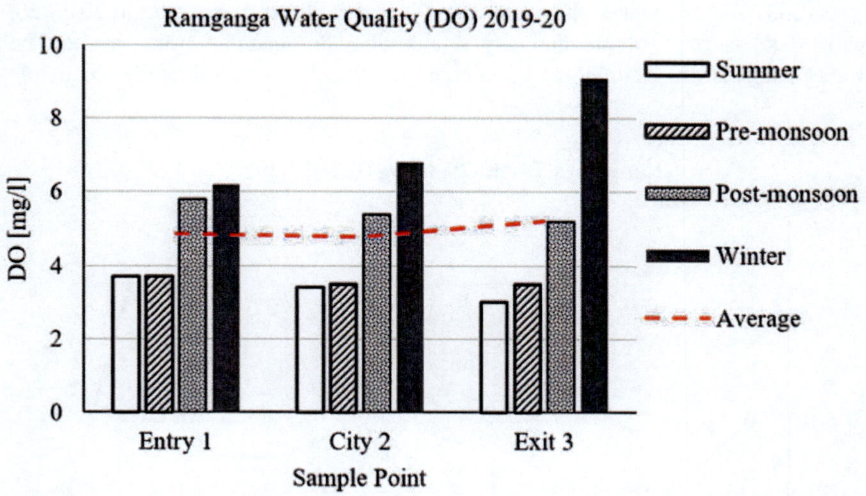

Fig. 5 DO in Ramganga water in Moradabad city in 2019–20 at entry, in the city, and at the exit

values in winters are higher. There is not much difference in the DO levels at the entry, within the city and exit from the city.

3.2 The pH of Ramganga Water

The pH of Ramganga water has been consistent between 7.7 to 8.1 from the year 2007 to 2014, as observed in Fig. 6. In the year 2015, the mean pH increases to 8.5 due to the high rainfall and flooding of the Ramganga River, but in 2016 it observes a correction and achieves its lowest value (7.1). From Fig. 7, it is observed that the average pH increases when the River flows through the Moradabad city.

3.3 B.O.D. of Ramganga Water

The B.O.D. is the dissolved oxygen required to decompose the organic matter by aerobic micro-organisms. The lower levels of B.O.D. are desirable for high DO in water. Higher the B.O.D., higher is the oxygen consumed by the microorganisms (bacteria) and lower is the oxygen available for aquatic life. Figure 8 gives the B.O.D. for ten years in the Ramganga River. From the year 2008 onwards, the B.O.D. decreases, which complements the increase in DO, observed from Fig. 4 due to attention UPPCB and CPCB towards the severity of water pollution. The values of B.O.D. in Ramganga water, published by CPCB ranges from 4.1 to 7.1 mg/l in ten years (2007–16). However, the actual sampling of the Ramganga water in 2019–20, while it flows from Moradabad city is alarming, as observed from Fig. 9. The B.O.D. ranges from 18 to 99 mg/l, which is significantly greater than the acceptable

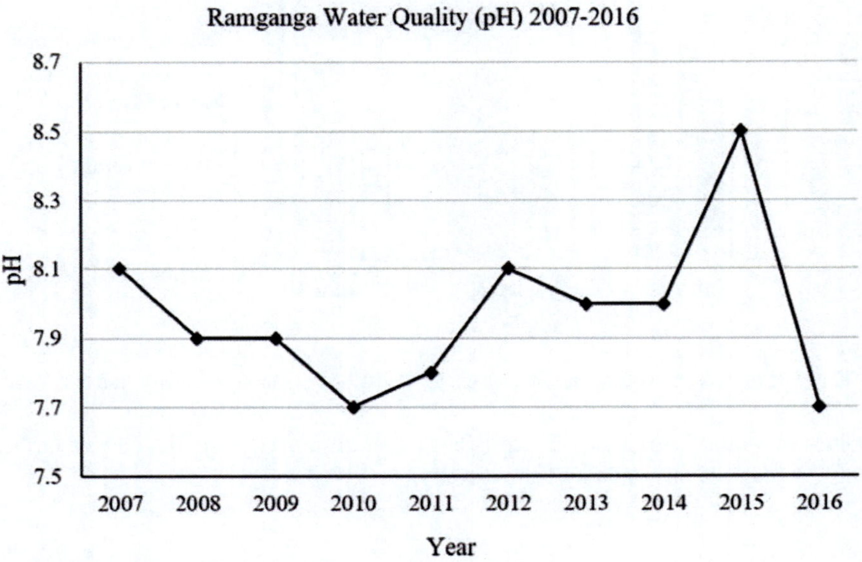

Fig. 6 pH in Ramganga water from 2007–16 [18–27]

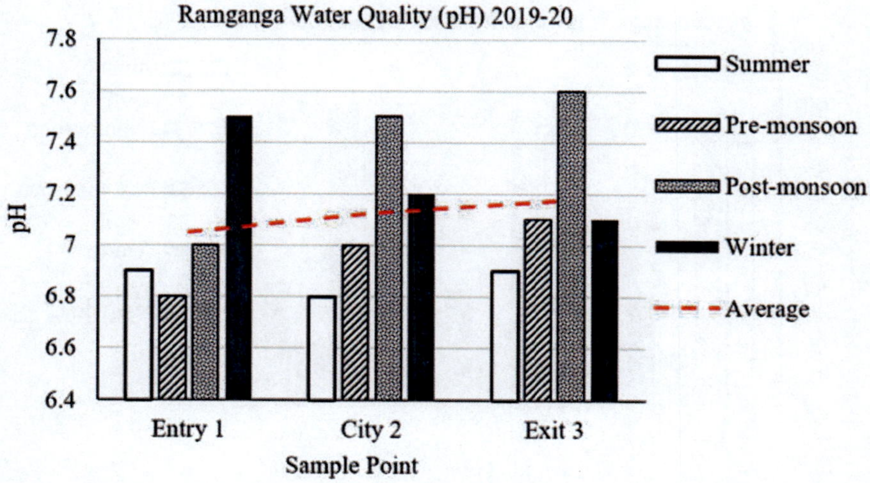

Fig. 7 pH in Ramganga water in Moradabad city in 2019–20 at entry, in the city and at the exit

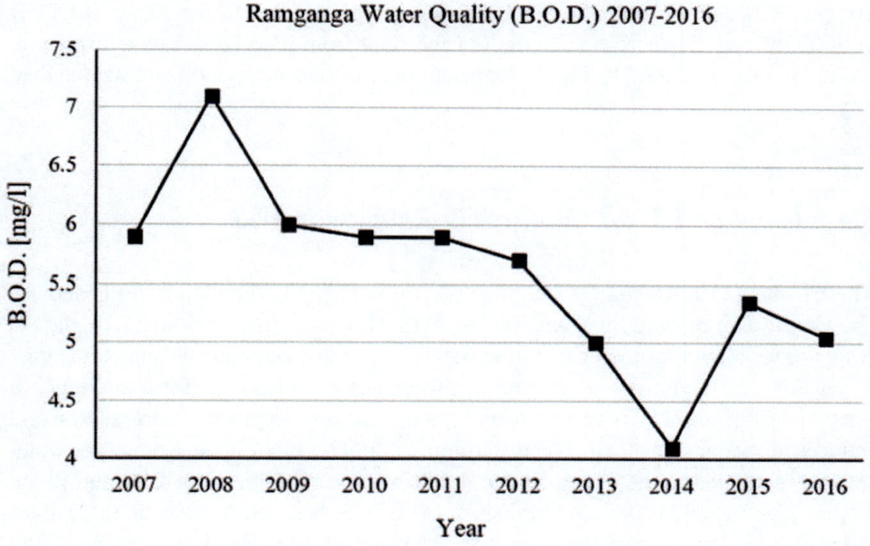

Fig. 8 B.O.D. in Ramganga water from 2007–16 [18–25]

limits (3–5 mg/l) [6]. The higher values of B.O.D. indicate deterioration of water quality, higher growth of bacteria, and difficult conditions for aquatic life. Higher B.O.D. may be due to the release of untreated sewage and wastewater from the slaughterhouses. The B.O.D. is quite high during summer and pre-monsoon, and subsequently subside as the monsoon arrives, due to dilution of the impurities and increase in the quantity of freshwater. The variation in the CPCB data and the results

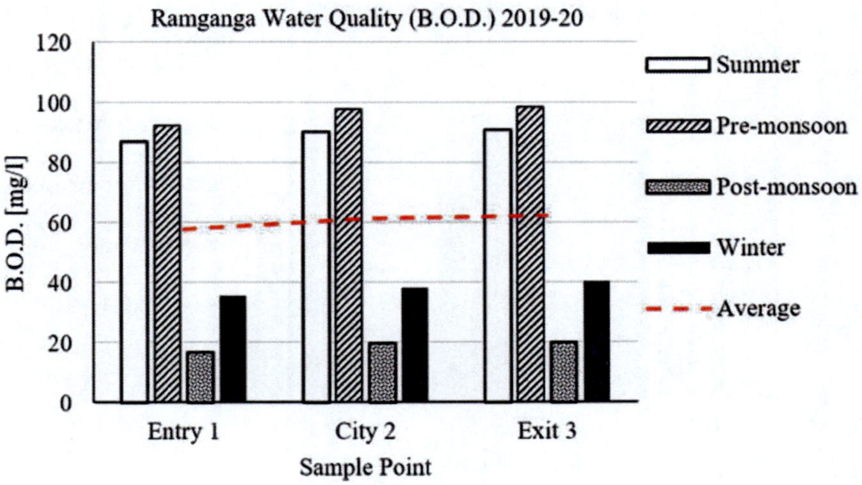

Fig. 9 B.O.D. in Ramganga water in Moradabad in 2019–20 at entry, in the city, and at the exit

obtained from the examination of the water samples is due to the fact that CPCB collects the samples across the length of the river from its origin to its termination, while the results plotted in Fig. 9, represent the samples collected from Moradabad city.

3.4 Fecal and Total Coliform in Ramganga Water

The FC and TC in Ramganga water are plotted in Fig. 10. The values of FC and TC show quite a large variation from 2007 to 2013. However, from 2014 to 2016, the FC levels have shown an increase, while the TC is almost stagnant. When Ramganga River flows through the Moradabad city, there is a rise of about 100% in the FC as observed from Fig. 11. Its value is very high in summer and pre-monsoon season and reduces post-monsoon season. The alarming high FC content in the Ramganga water may be attributed to the direct discharge of sewage and domestic wastewater in the River. The mean FC values published by CPCB are about 50% of the values determined by actual measurement in the Moradabad city. This is a clear indication that the water quality of the Ramganga River is far from the acceptable limits and requires immediate attention.

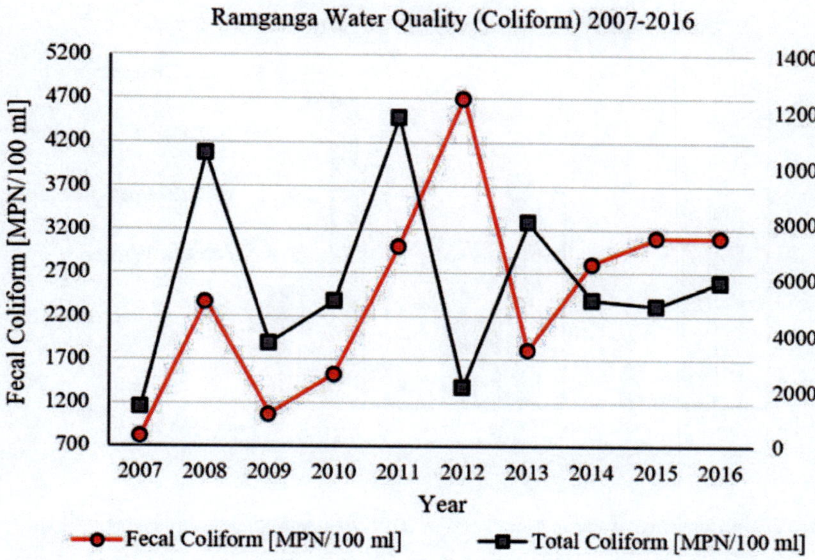

Fig. 10 Coliform in Ramganga water from 2007–16 [18–27]

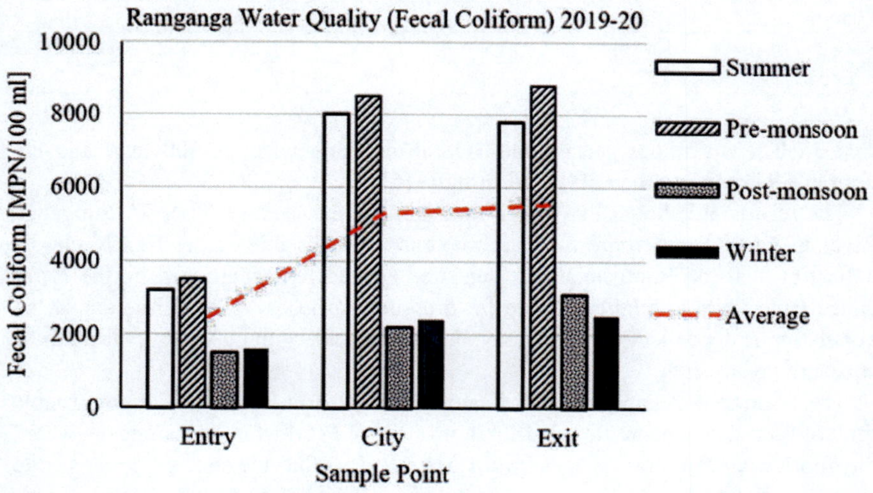

Fig. 11 Fecal coliform in Ramganga in Moradabad (2019–20) at entry, in the city and at the exit

3.5 TDS and Fluoride Concentration in Ramganga Water

TDS is a direct indicator of drinking water quality. Figure 12 shows that the TDS of Ramganga water varies from 96 (post-monsoon, Entry 1) to 2560 (pre-monsoon,

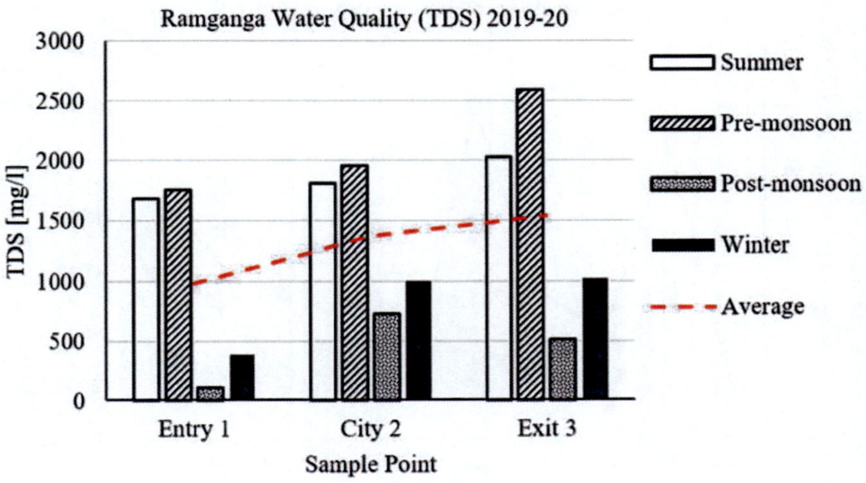

Fig. 12 TDS of Ramganga water in Moradabad in 2019–20 at entry, in the city and at the exit

Table 1 Fluoride in Ramganga water in Moradabad in 2019–20 at entry, in the city and at the exit

Sample Site	Summer	Pre-monsoon	Post-monsoon	Winter
Entry 1	1.7	1.7	1.6	1.6
City 2	1.8	1.9	1.7	1.6
Exit 3	1.8	1.9	1.7	1.7

Exit 3). The maximum permissible TDS in drinking water is 500 mg/l and the desirable TDS is less than 300 mg/l in India [6].

The results are shown in Fig. 12 indicate that the average TDS of Ramganga River is 2 to 3 times more than the maximum permissible value. Hence, clearly indicating that the Ramganga water and the groundwater recharged by the Ramganga River fails to qualify as safe for drinking purposes. A large fraction of the population is dependent on the bore wells and the underground water, putting them at severe health risks.

The fluoride concentration in the samples is given in Table 1. The permissible limit of fluoride concentration in drinking water is 1 PPM while in Ramganga water, the fluoride concentration ranges from 1.6 to 1.9. This fluoride concentration is quite high even before the Ramganga enters the city. Higher fluoride concentration leads to severe to mild dental fluorosis and so, such water is unsafe for drinking purposes. Besides other contaminants, high fluoride content adds another dimension to the water pollution in the Ramganga River.

In similar studies on water quality of rivers, the primary source of pollution was found to be aerobic pollution, chemical oxygen demand and ammonia-nitrogen [28, 29].

3.6 Remedial Measures to Curb the Water Pollution of the Ramganga River

The UPPCB under the coordination of CPCB is making efforts to restore the water quality of the Ramganga River thorough an action plan. The plan comprises the study of Ramganga River upstream and downstream, its banks, water quality assessment, and monitoring the treatment methods. Still the results are unsatisfactory. The real-time sample analysis gave even more severe water pollution levels. The main reason for unsatisfactory results is due to the contribution of the unorganized sector in water pollution. In a recent study, the most effective measure to control water pollution in the Rivers was suggested as wastewater collection and treatment [30]. A few most significant measures required to restore the water quality of the Ramganga River are proposed in Fig. 13. The most important aspect is monitoring and assessment of water quality from the origin of the River to the point of its termination. The previous studies have highlighted that the Ramganga River is a large contributor to the pollution of Ganges and its contamination becomes severe from Moradabad onwards. The assessment of River water quality on a real-time basis using physical testing by a third body (outsourced) and its monitoring through a satellite may enable better control on encroachment and silt removal for construction activities. The people need to be made aware of the consequences of water pollution and the levels of pollution in the River through advertisement by print, electronic, and social media platforms periodically. Training on new methods of water quality analysis and pollution control may help to increase the effectiveness of the restoration process by knowledge sharing, the use of state-of-the-art methodology and replacement of obsolete and outdated information as well as water quality analysis procedures with the most recent ones. The unorganized industrial units are required to be identified and dealt with accordingly. Such sectors should be at first identified and earmarked. Thereafter the promotors, managers, and workers should be upraised about the environmental and health consequences of their practices. The help of the society and common people may be taken for feedback and surveillance. In case, the change in the procedures require equipment and machinery, the provision for soft loans shall be made in consultation with the banks and district administrative authorities. The treatment plants of registered food processing, metal and, pulp and paper industries need to be inspected by third parties or by the teams of CPCB. Sewage treatment is required to be ensured. Dumping of solid wastes near the Riverbanks should be stopped completely. The rainwater harvesting and development of local water bodies will reduce the pressure on the River water.

4 Conclusions

The following major conclusions are derived from the study.

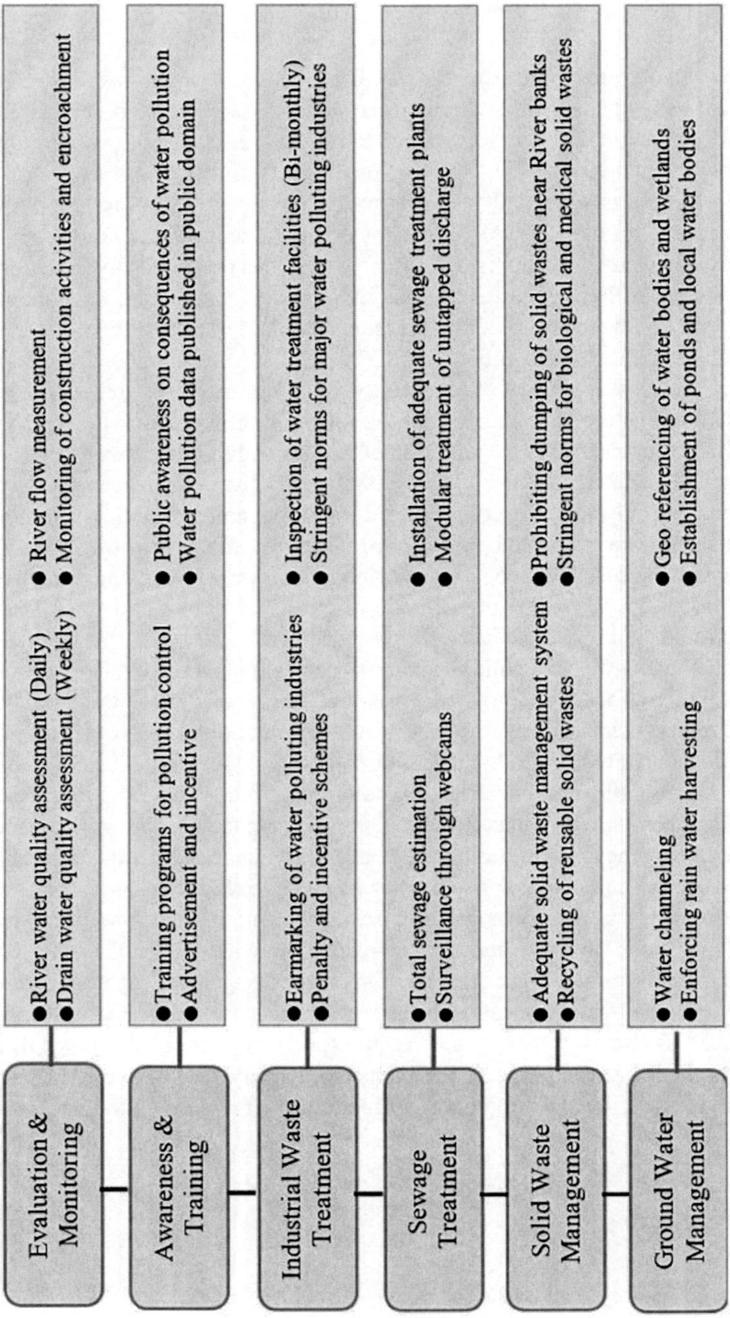

Fig. 13 Remedial measures for restoration of class A water quality in Ramganga River

1. Ramganga River gets polluted severely while it flows from the Moradabad city to Farukhabad. It adds significant pollution to the River Ganga.
2. The industries such as metal surface, food processing, slaughterhouses, meat processing, pulp and paper, and milk processing are major contributors to the water pollution of Ramganga. The sewage, agriculture runoff, solid wastes, domestic wastewater, and cattle bath discharge are other main contributors.
3. The efforts of UPPCB and CPCB are insufficient due to a large number of unorganized industry units or subsidiaries, which are not registered. Such small units are operated in the households in Moradabad city and have no treatment facilities.
4. A comprehensive remedial method is proposed with a large emphasis on the monitoring of the River flow through satellite imagery and assessment of its water quality on a real-time basis. The assessment results should be put in the public domain to make people aware and responsible.
5. Awareness and training of the public are required to a large extent.
6. To control the population of Ramganga River and to meet the further pressure on it due to the increase in population, rainwater harvesting and development of local water bodies is of utmost importance. Pollution control measures need further improvement.

Acknowledgments The authors gratefully acknowledge resources and facilities used in the present work at College of Engineering, King Khalid University, Abha-Asir, Kingdom of Saudi Arabia, such as the chemical analysis laboratory, reprographic and computing, the Saudi Digital Library, etc. to complete the study.

References

1. Alam M, Pathak JK (2010) Rapid assessment of water quality index of Ramganga River, Western Uttar Pradesh (India) using a computer programme. Nat Sci 8(11):1–8
2. Singh P (2018) Water quality assessment of River Ramganga at Bareilly. J Ecol Environ Sci 6(4):33–40
3. Singh P (2018) Hydrobiological study on summer season of river Ramganga at Bareilly. Environ Pollut Clim Change 2(3):1–4
4. Gangwar RK, Singh J, Singh AP, Singh DP (2013) Assessment of water quality index: a case study of river Ramganga at Bareilly U.P. India. Int J Sci Eng Res 4(9):2325–2329
5. Murty MN, Kumar S (2011) India infrastructure report, Chapter 19: water pollution in India an economic appraisal, pp 285–298
6. UPPCB (2018) Action plan for restoration of polluted stretch of river Ramganga from Moradabad to Kannauj. Uttar Pradesh Pollution Control Board, Vibhuti Khand, Gomtinagar, Lucknow, pp 1–131
7. Srivastava A, Gupta V, Agarwal G, Srivastava S, Singh I (2011) Water quality assessment of Ramganga River at Moradabad by physico-chemical parameters analysis. VSRD Int J Tech Non-Tech Res 2(3):119–127
8. India Population (2019) Population of Moradabad. https://indiapopulation2019.com/population-of-moradabad-2019.html

9. Chandra R, Gupta M, Pandey A (2011) Monitoring of river Ram Ganga: physico-chemical characteristic at Bareilly. Recent Res Sci Technol 3(6):16–18
10. Pathak D, Whitehead PG, Futter MN, Sinha R (2018) Water quality assessment and catchment-scale nutrient flux modeling in the Ramganga River Basin in north India: an application of INCA model. Sci Total Environ 631–632:201–215
11. Khan MYA, Gani KM, Chakrapani GJ (2016) Assessment of surface water quality and its spatial variation. A case study of Ramganga River, Ganga Basin, India. Arab J Geosci 9(28): 1–9
12. Channdra R, Kashyap K, Pandey A (2010) Pollution status of river Ramganga: physico-chemical characteristics at Bareilly. J Exp Sci 1(5):20–21
13. Nizami G, Rehman S (2018) Assessment of heavy metals and their effects on quality of water of rivers of Uttar Pradesh, India: a review. J Environ Chem Toxicol 2(2):65–71
14. Research Matters (2017) Tracing pollution along the length of Ramganga river. https://researchmatters.in/shots/tracing-pollution-along-length-ramganga-river
15. https://www.downtoearth.org.in/news/water/huge-amounts-of-toxic-heavy-metals-swim-in-indian-rivers-60545
16. Samantray P, Mishra BK, Panda CR, Rout SP (2009) Assessment of water quality index in Mahanadi and Atharabanki Rivers and Taldanda Canal in Paradip Area. India J Hum Ecol 26(3):153–161
17. https://www.google.com/maps/d/viewer?mid=1J2vCZzJIFHjU-VJ8Es-G9nAyZY62azq7&ll=28.788937465925674%2C78.85379020873575&z=13
18. Water Quality Database (2014) National Water Quality Monitoring Programme (NWMP) water quality data. http://www.cpcbenvis.nic.in/waterpollution/2014/RIVERWATER%20DATA%202014_7.htm
19. Water Quality Database (2013) National Water Quality Monitoring Programme (NWMP) water quality data. http://www.cpcbenvis.nic.in/waterpollution/2013/RIVERWATER%20DATA%202013_7.htm/8
20. Water Quality Data (2012) CPCB ENVIS. http://www.cpcbenvis.nic.in/waterpollution/2012/TRIBUTARY%20STREAM%20%20SUSWA,GOLA.htm
21. Water Quality Data (2011) CPCB ENVIS. http://cpcbenvis.nic.in/waterpollution/2011/TRIBUTARY%20STREAMS%20SUSWA.htm
22. Water Quality Data (2010) CPCB ENVIS. http://cpcbenvis.nic.in/waterpollution/2010/TRIBUTARY%20STREAMS%20RAMGANGA.htm
23. Water Quality Data (2009) ENVIS CPCB. http://cpcbenvis.nic.in/waterpollution/2009/TRIBUTARY%20STREAMS%20RAMGANGA%20GOMTI.htm
24. Water Quality Data (2008) CPCB ENVIS. http://cpcbenvis.nic.in/waterpollution/2008/RIVER%20KALINADI.htm
25. Water Quality Data (2007) CPCB ENVIS. http://cpcbenvis.nic.in/waterpollution/2007/kalinidi.htm
26. Water Quality Data (2015) CPCB ENVIS. http://cpcbenvis.nic.in/waterpollution/2015/kalinidi.htm
27. Water Quality Data (2016) CPCB ENVIS. http://cpcbenvis.nic.in/waterpollution/2016/kalinidi.htm
28. Zhao SY, Wang DD, Wang YL, Chen L (2019) Current status and integrated pollution control of black and odorous water bodies in Dongxinkai river basin of China. Appl Ecol Env Res 17(6):14475–14486
29. Chen L, Han L, Ling H, Wu J, Tan J, Chen B, Zhang F, Liu Z, Fan Y, Zhou M, Lin Y (2019) Allocating water environmental capacity to meet water quality control by considering both point and non-point source pollution using a mathematical model: Tidal river network case study. Water 11(900):1–13
30. Zhang W, Fang S, Li Y, Dong F, Zhang C, Wang C, Wang P, Xiong W, Hou X (2019) Optimizing the integration of pollution control and water transfer for contaminated river remediation considering life-cycle concept. J Clean Prod 236:117651

Assessment of Surface Water Quality of Indian Rivers in Terms of Water Quality Index (WQI)

Vaishali Sahu and Ram Karan Singh

Abstract Assessment of water quality is imperative because clean water is essential for human use and for balanced ecosystem. In view of continuous pollution of water bodies due to point and non-point sources, it is required to protect and manage freshwater resources. Various monitoring program and modelling approaches have been developed and suggested in past for effective management. In modelling the surface water body is studied for an extended period of duration and is experienced as tedious and complex task with enormous amount of quality parameter data. Therefore, water quality index (WQI) has been used extensively as a tool to access the river water quality. It can help decision makers and stake holders to analyze the water quality easily. Water quality assessment has been performed on many Indian rivers in terms of WQI and the suitability of water resource for human consumption has been addressed. The proposed chapter will provide a compressive review of water quality of Indian rivers and can provide an effective approach for assessment and management of water quality.

Keywords Water quality index (WQI) · Monitoring · Health risk · Assessment · Management

Introduction

What Is Potable Water?

Fresh water sources are the key for life and sustainable development on earth. These open sources are susceptible to human pollution and is increasing with increased population. Its quality is continuously degrading due to significant increase in the pollution caused through point and non-point sources. This has led to more frequent

V. Sahu (✉) · R. K. Singh
Department of Civil & Environmental Engineering, The NorthCap University, Gurugram, Haryana, India
e-mail: vaishalisahu@ncuindia.edu; vc@iudehradun.edu.in

S. Yadav et al. (eds.), *Water Quality, Assessment and Management in India*, Earth and Environmental Sciences Library, https://doi.org/10.1007/978-3-030-95687-5_14

monitoring and assessment of the river quality. Protection of water bodies have become major responsibilities for researchers and scientists.

It has been reported that 71% of the earth's total surface is covered with water and only 2.5% of this amount can be considered as freshwater [1]. It has estimated that nearly1.6 billion population is fronting economic water shortage and two-thirds of the world's population is suffering from water shortage at least one month in a year [2, 3]. Water stress and scarcity is increasing due to excessive usage of aquifers and it has been found that majority of the world's largest aquifers have exhaust their sustainable points [4]. It has been estimated that nearly 1.8 billion world's population may face absolute water scarcity by 2025 [5].

To sustain life on earth, support agriculture and industrial processes, recreational purposes, surface water is one of the most valuable resource. Many parameters, like, lithology of basin, atmospheric conditions, climate change and human activities affect the quality of the river water from time to time [6–10]. Rivers not only supports the life but also plays a key role in dispersing and diluting huge amount of pollutants from the domestic and industrial discharges and agricultural runoff [11–13].

Augmented population growth and inefficient environmental management plan have further increased the problem of extreme levels of pollution in developing countries. Surface water in urban areas is more susceptible to pollution owing to relatively less self-purification capacities and potentially larger pollution loads. The discharge of untreated sewage and industrial waste water into water bodies has manifold the problem. Consequently, water pollution in urban rivers is becoming a major environmental problem in many countries around the world. Water stressed countries, like, India, observes water pollution problems owing to eslacating urban population, increased per capita demand and changed lifestyle land use pattern [14, 15].

Monitoring Water Quality

The need to monitor water quality was recognized long back in Germany (1850) where it was attempted to draw some connection between the level of purity and presence of contaminants. In 1912, the Royal Commission on Sewage Disposal classifies rivers on the basis of visible condition of a watercourse; it includes parameters, such as, odour, turbidity, suspended matter and algal growth etc.

The deteriorated water quality has reduced the biodiversity and stream functions of the streams and poses a huge threat on the river ecosystem health. Therefore, it is necessary to safe guard the fresh water resources and to research on identifying the factors influencing water quality in streams. It has become important to assess and monitor the river water quality. It has been reported that the assessment of water quality is normally based on a comparison of physical, chemical, and biological parameters with prescribed water quality guidelines. Research has been performed and various methods has been proposed for assessment of surface water and the present chapter provides a comprehensive review on such methods in perspective of Indian situation.

In today's era, water scarcity is one of the most important issues wherein countries are becoming water stressed. India is the second largest populated country with population of nearly 1.34 billion in year 2017 (World bank) and is also severely experiencing water pollution and scarcity. A report [16] by Niti Aayog on Composition Water Management Index (CWMI) mentioned that about 600 million people face high-to extreme water stress in India and 75% of households do not have drinking water in premise. The conditions are more worsen in rural areas where about 84% rural households do not have a tap connection.

Conventional methods to assess river water quality emphasize to compare the experimental parameter with the prescribed guidelines. In many situations, this practice allows identifying the contamination sources precisely and helps to check its legal compliance; conversely, the overall water quality of source involves spatial and temporal trends and thus conventional methods is not considered to provide a comprehensive approach for it. Use of a Water Quality Index (WQI) has been proposed by many researchers as a means to develop a mathematical expression to monitor the surface water quality.

Additionally, to the deteriorating quality of river water, it is imperative to evaluate the human health risk due to poor water quality. WQI of rivers can be used to determine the hazard quotient (HI) as a measure of noncarcinogenic risk to the consumers.

The advantages, applicability, limitations of WQI method and health hazard due to poor water quality has been dealt in various sections in this chapter.

Methodology and Methods to Determine Water Quality Index

Numerous authors have suggested the use of a Water Quality Index (WQI) as a means to develop a numerical expression for the general quality of surface water [17–22]. A single WQI value makes information to be understood rapidly with ease as compared a long list of numerical values for a large variety of parameters. Additionally, WQI's also simplify comparison between different sampling sites and/or events. Subsequently, they are considered better for conveying information to general audiences. When their specific characteristics and limitations are taken into consideration [18, 21, 22], WQI's can be very beneficial for the purpose of management and decision-making. The usage of WQI for assessing water quality was firstly proposed by Horton [23] and Brown [24]. Subsequently there has been development of various other methods to compute WQI. The physical and chemical input parameters for WQI have been observed to be same, however, different approached for statistical integration and interpretation have been used.

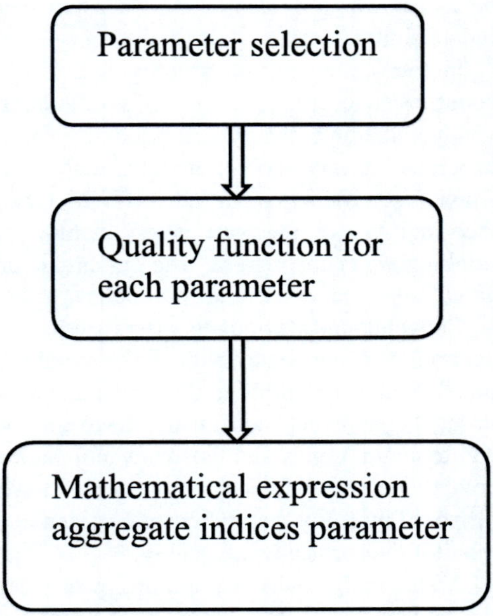

Fig. 1 Approach for WQI

Related studies have been performed on water quality index (WQI) and its modeling. It has been proved that WQI is a valuable and unique rating system to portray the overall water quality status in a single term. This is supportive for the selection of appropriate treatment technique to meet the concerned issues [25]. Water quality indices are tools to determine conditions of water quality.

The general approach to compute WQI is shown in Fig. 1. To carry out assessment, the significant parameters are required to be selected based on the need and requirement of the study. Selection of parameters can be performed by a team of professional experts, government institution or any legislative agency. It has been reported that parameters, like dissolved oxygen (D.O.) content, eutrophication level, physical characteristics of water body, dissolved substance have significant impact on water quality and hence should be considered for WQI computation. Each considered parameter is termed as sub-index and is measured in different units (ppm, percentage, count, volume) therefore; it is required to express them in similar units or in terms of unit less parameter. Quality function is to be determined through curve or graph for each sub index. All such sub-indices are aggregated together by mathematical expression using arithmetic or geometric average.

Various national and International organizations have developed and proposed index to measure water quality. The widely used indices are shown in Fig. 2.

The above proposed indices are area specific and are comparable to their respective standards. WQI have been effectively used to establish seasonal, temporal and spatial variation in water quality even at lower concentrations. The detailed procedure for these methods has been discussed in following sections:

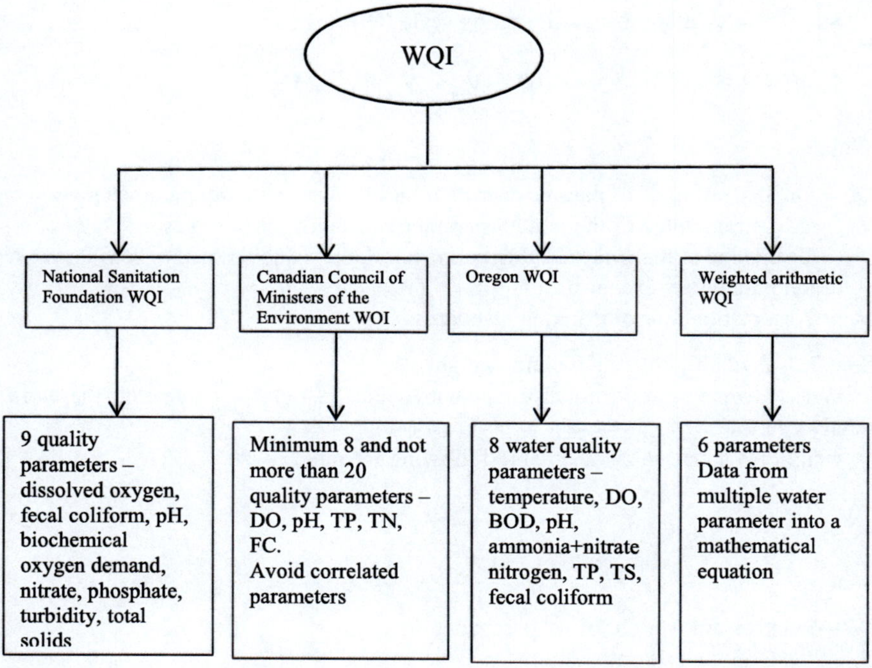

Fig. 2 Water quality indices [21–24]

Determination of WQI Using National Sanitation Foundation Water Quality Index (NSFWQI)

A regressive WQI model is the one which is based on sufficient number of important parameters with common scale and effective way to assign weights. Basic arithmetic weighting was proposed without the multiplicative variables [24]. This effort was well supported by the National Sanitation Foundation (NSF) and is also referred as NSFWQI. The statistical tool, Delphi method was used to select the water quality variables [26]. The NSF WQI helps in converting the water quality parameter results into sub-index values.

It was used for critically polluted water bodies and major parameters used were, physical quality parameters like, pH, temperature and turbidity. Chemical parameters, like, dissolved oxygen biochemical oxygen demand, total phosphates, nitrates and total solids, biological paramters - fecal coliform. It is one of the most comprehensive water quality indexes and has been discussed in various papers. The step by step procedure is described here:

Step 1: Calculating the quality rating scale (Qi)

$$Qi = 100 * (V_m - V_i)/(V_s - V_i) \tag{1}$$

Where,

Qi = Quality rating of ith parameter for a total of n water quality parameters
V_m = Measured value of the quality parameter of the water samples
V_i = Ideal value of that water quality parameter, Ideal value is equal to zero for most of the parameters except for pH (pH = 7)
V_s = Acceptable limit of drinking standards

Step 2: Calculating the relative unit weight (W_i).

Wi is inversely proportional to standard value (S_i) of the parameter, therefore relative unit.

weight (W_i) was calculated using following formula

$$W_i = K/\ S_i \tag{2}$$

Where,

W_i—Relative unit weight of nth parameter
S_i—Standard value for nth parameter
K = Proportionality constant = 1

Step 3: Calculating water quality index (WQI)

$$WQI = \Sigma\ Q_i * W_i\ /\Sigma W_i \tag{3}$$

Where,

Q_i = Quality rating
W_i = Relative weight

Determination of WQI Using Canadian Council of Ministers of the Environment Water Quality Index (CCME WQI)

Canadian jurisdictions framed a reliable way to determine the water quality information. WQI was developed by a committee which was established under the Canadian Council of Ministers of the Environment [27]. The proposed WQI can be used by any water agency of different countries with slight modification. The CCMEWQI compares observations to a benchmark, where the benchmark can be a standard of water quality or filed concentration [28, 29].

This method has been established to appraise water quality and aquatic life protection as per the guidelines of different places. The measurement parameters may differ from place to place and it is proposed to have minimum four parameters sampled at least four times to determine WQI. The following formula can be used to estimate WQI [29]:

$$WQI = 100 - \frac{\sqrt{F_1^2 + F_2^2 + F_3^2}}{1.732} \tag{4}$$

Where,

F_1 = Number of variables, whose objectives are not met
F_1 = [No. of failed variables/Total no. of variables]*100
F_2 = Number of times by which the objectives are not met
F_2 = [No. of failed tests/Total no. of tests]*100
F_3 = Amount by which the objectives are not met

Determination of WQI Using Oregon Water Quality Index (OWQI)

To assess the general water quality of Oregon's stream, a score has been created (OWQI) and its applicability to other regions was evaluated. This method suggests using eight water quality variables, like, temperature, dissolved oxygen (DO), biochemical oxygen demand (BOD), pH, ammonia and nitrate nitrogen, total phosphorus, total solids and fecal coliform to determine WQI. Following expression can be used to compute WQI [29].

$$WQI = \sqrt{\frac{n}{\sum_{i=1}^{n} \frac{1}{SI_i^2}}} \tag{5}$$

Where,

n = number of subindices
SI = subindex of ith parameter

Determination of WQI Using Weighted Arithmetic Water Quality Index Method

This method categorized the water quality in terms of degree of purification for most frequently measured water quality parameters. The step by step procedure is detailed here:

Step 1: Unit weight of each quality parameter, W_i, can be calculated as:

$$W_i = \frac{K}{S_i} \tag{6}$$

S_i is recommended standard value of ith parameter

Where, K is the proportionality constant and can be estimated as

$$K = \frac{1}{\sum \frac{1}{S_i}} \tag{7}$$

Step 2: Quality rating scale (Qi) for each parameter is calculated by using this expression the calculation of WQI was made by using the following equation:

$$Q_i = 100\left(\frac{V_i - V_o}{S_i - V_o}\right) \tag{8}$$

Where,

V_i is estimated concentration of ith parameter in the analyzed water
V_o is the ideal value of this parameter in pure water
$V_o = 0$ (except pH =7.0 and DO = 14.6 mg/l)

Step 3: Calculation of WQI,

$$WQI = \frac{\sum Q_i W_i}{\sum W_i} \tag{9}$$

Each of the above defined methods categorizes water bodies based on WQI values. Figure 3 shows the scale and range for different methods as reported by many researchers [21–24]. It can be observed that OWQI is more stringent in defining quality and WQI less than 79 falls into poor category whereas NSFWQI seems to categorize in a generalize manner.

From the study of various water quality indices, it can be noted that the purpose of WQI is to provide a unique value to the quality of water from any source. It provides a simple expression using the various parameters and help in easy understanding of water quality monitoring data. These quality indices uses different physical, chemical, biological parameters and provide a quality check.

No matter all of the efforts and distinct discussed indices being used globally, no index has been recognized to be been universally accepted. Reasearch for greater useful and typical water high-quality index continues to be occurring, so that water agencies, customers and water managers in distinctive international locations may use and followed it with little change [29].

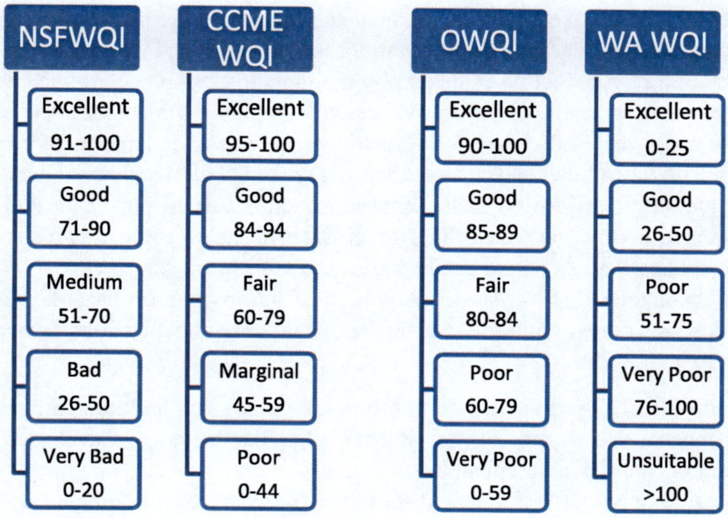

Fig. 3 Water Quality Index and Rating [21–24]

WQI and Health Risk Assessment

Health risk assessment studies are important to evaluate the probability of adverse health effects in humans due to exposure of any contaminant.

Health risk assessment (HRA) emerged after the 1980s. It uses risk as an evaluation index to link environmental pollution with human health and quantitatively describing the risk of pollution to human health [25]. Various studies related to HRA has been conducted using the method recommended by the USEPA around the world [26]. By evaluating the level of harm to human health from water pollution and acceptable risk level of the human body, health risk assessment (HRA) can use as a guidance by administrative sector in water environment protection, pollution remediation and water environment risk management [27].

The risk assessment studies provide a theoretical evaluation of risk to human health from any contaminant and are divided into two major categories: carcinogenic and non-carcinogenic health risks. The carcinogenic risk, which is associated with cancer causing chemical exposure, is expressed in terms of an increased probability of developing cancer during a person's lifetime, whereas the health effects other than the cancer are categorized as a non-carcinogenic health risk. A risk assessment study is a 4-step process as detailed below:

- Hazard Identification—It is a process of gathering data, identifying the toxicity and its adverse effects to the populations that might be affected. It is to determine if a certain contaminant present in water can cause adverse health effects on human body.

- Dose–Response Assessment—It is the determination of the relationship between the magnitude of an applied dose and a specific biological response. The probability of adverse reaction of the exposed population can be determined here.
- Exposure Assessment—It is the process of estimating the contact between the contaminant and individual. It calculates frequency and duration of exposure along with the number and characteristics of the population exposed. Quantitative and qualitative estimation of the exposure factors carried out. Here the risk is considered to drinking water, thus, direct water ingestion is the only pathways in these studies. The exposure can be expressed with formula:
- Risk characterization—It is the integration of information on hazard, exposure, and dose-response to provide an estimate that adverse effects will occur in people who are exposed.

The health effects are divided into two types, carcinogenic and non-carcinogenic. Health impacts due to most of the water quality parameters are evaluated using non-carcinogenic risk assessment.

Non-carcinogenic risk assessment studies are based on the hazard quotient (HQ). It is the ratio of the estimated dose of a contaminant to the reference dose level (RfD). The United States Environmental Protection Agency (US EPA) has defined RfD as the acceptable safety level for chronic non-carcinogenic and developmental effects. In simpler words, it "is the amount of a chemical to which a person can be exposed each day for a long time (usually lifetime) without suffering harmful effects" [28].

The probability of non-carcinogenic health risks resulting from exposure to a chemical is generally assessed in terms of Hazard Quotient (HQ). It is mostly assessed by comparing an exposure estimate to a reference dose (RfD) for oral exposure and comparing an estimated chemical specific concentration to the reference concentration (RfC) for direct inhalation. Here drinking water is an oral exposure to the contaminant, therefore RfD is to be used. It is a daily oral intake rate that is estimated to pose no appreciable risk of adverse health effects, even to sensitive populations, over specific exposure duration. Further, HQ is calculated by estimating the average daily intake (ADI) and the reference dose (RfD) for a contaminant using the Eq. (10).

$$HQ = ADI/RfD \tag{10}$$

ADI is the mean concentration of contaminant that is absorbed by a person's body depending upon the intake of contaminant, body weight, exposure duration and frequency, etc. It is calculated using following Eq. (11).

$$\mathrm{ADI} = \mathrm{C\ x\ I\ x\ EF\ x\ ED/BW\ x\ AT} \tag{11}$$

Where,

C = Concentration of chemical in drinking water (mg/L)
I = Daily water intake (L/day)
EF = Exposure frequency (days/year)

ED = Exposure duration (years)
BW = Average body weight (kg)
AT = Averaging time (days)

The hazard quotient has been defined so that if it is less than 1.0, there should be no significant risk of contaminant toxicity. Ratio above 1 could represent a potential risk to human health. Further risk from any non-carcinogen per year can be defined in terms of R^n and it can be calculated as per the method recommended by USEPA [29] as described below in Eq. (12).

$$R^n = \left(HQ^* \, 10^{-6}\right)/70 \tag{12}$$

The greater the value of R^n, the greater the health risk caused by the contaminant, namely the higher the probability of adverse effects from it. This health hazard risk model can be used to understand the health risks of drinking water quality in various urban and rural areas.

Results and Discussion on WQI of Indian Rivers

Study has been undertaken [30] to investigate suitability of the Kolong River water of Assam based on WQI values. Samples have been collected from seven sites along the river channel to analyze seasonal water quality in terms of WQI values using most important physico-chemical parameters. Weighted arithmetic index have been used to compute WQI and water quality status was then determined. It was reported that majority of the samples fall under very poor ($75 < WQI < 100$) and unsuitable category ($WQI > 100$) and a high WQI was observed during monsoon season. Using WQI, sites with contaminated unsuitable water was also identified and it was suggested to use immediate effective treatment measures to enhance the river water quality. It was recommended that water quality of such rivers can be reestablished by implementing measures like limiting waste water flow, restricting direct storm water discharge and averting solid waste dumping along the river.

Attempt has been made [31] to study the historical changes in water quality of River Ganga at Haridwar. Variation in the water quality at monitored location was assessed over an 11 year period using three water quality indices, namely, weighted multiplication form, NSF-WQI and arithmetic weighted index. Total sixteen parameters were used to estimate WQI from 2000 to 2010. It has been reported that the River Ganga water quality has ranged from poor to good over the 11-year study period. It is inline to the several studies reported on the WQI of the river. With the progressing urbanization and escalated pollution loads on rivers, essential actions should be taken to limit any futher pollution entering in the river. The research confirms that solid and liquid waste contaminants are the prime sources of pollution [31].

The WQI for Sabarmati River in western India is studied [32] through weighted arithmetic method using six quality parameters (pH, dissolved oxygen, biochemical oxygen demand, electrical conductivity, nitrate nitrogen and total coliform). Anthropogenic activity, illegal sewage and industrial effluent disposal impacted most of the parameters under study. Worst water quality has been observed for the station located in highly urban area then for moderately urban area and lastly for moderately rural area. Systematic and thorough water quality monitoring of the Sabarmati River is required and also it is essential to the identify changes or trends in water quality over time and space.

Study [33] has been performed to access water quality of Gomti River which has known to be polluted heavily through anthropogenic activities. WQI has been estimated using Arithmetic weighted method with 13 physico-chemical parameters. Multivariate statistical technique was used to find effectiveness of WQI for nutrient polluted river like, Gomti. It has been observed that regular monitoring through WQI is not enough for water quality assessment. The water quality was found to be deteriorated from upstream to downstream to middle stretch. It has been suggested that spatial and temporal variation can be analyzed through box plots.

Loktak lake is situated in North East India and the water is used for drinking purposes by the nearby residents. Scientific study [34] has been carried out through WQI estimation (weighted arithmetic index) for five sampling sites of the lake. Based on importance in quality assessment, total eleven parameters have been selected. Based on the analysis the lake was found to be in poor category owing to discharge from two major rivers Nambul and Nambol and the agricultural wastes. The water was thus found to be unfit for drinking purposes. It was recommended to identify the pollutants in the lake and for continuous monitoring.

Study [35] has been carried out in upper and middle regions of Damodar river basin (DRB). Damodar river is a source of water to millions of people in and around it. Authors have developed simplified indices, WQI_m, WQI_{DO} and WQIpca to access spatial and temporal variation and suitability of river water on the basis of six quality parameters over four years. WQIm is proposed in case of minimum quality parameters available for analysis. Dissolved oxygen (DO) is the most important parameter for water bodies and thus WQIDO can be used in many situation where other parameters are not available. Two methodologies, namely, mix aggregation function (arithmetic and multiplicative) and PCA (principal component analysis) have been combined together to form WQI_{pca}. It has been suggested that this index can provide a better index by minimizing problems during parameter categorization in the conventional methodologies. Results from WQI revealed that the quality was river was better in post monsoon season as compared to pre monsoon and monsoon season owing to industrial and domestic discharge. It has been suggested to use these simplified indices to study the suitability of water for domestic use and for aquatic life.

Water quality of Mahananda River was evaluated [36] to check its appropriateness for domestic, agricultural and industrial uses. Sampling has been performed at fourteen locations during pre and post monsoon periods. Quality indices in terms of WQI, agricultural and industry related have been computed. Twelve parameters were considered to estimate WQI and sampling stations were ranked from very bad (downstream) to excellent (upstream) category. Hydrogeological analysis has been performed using Piper's diagram and it shows the dominance of calcium, magnesium and carbonate ions and depicts it to be typical shallow waters. Sodium absorption ratio (SAR) has been used to classify the water for irrigation purposes. The SAR ranged from 0.22 to 1.58 during pre-monsoon season and from 0.46 to 1.32 during post monsoon season thus are considered as excellent for irrigation purposes. Langelier saturation index (LI), aggressive index (AI) and Ryznar stability index (RSI) was used to check its suitability for industrial purposes and indicated the water to be moderately aggressive to non-aggressive in nature. For industrial purposes the water should be non-aggressive. The study revealed that the overall quality of river was good with exception at few places. It shows requirement of protection of river and precautionary management plans.

Yamuna river has undergone cleaning under Yamuna action plan phase I and II since 1993. The aftereffects of cleaning have been studied by [37] using WQI of the river in the National Capital Region (NCT) for 10 years (2000–2009). Six quality parameters have been analyzed to compute WQI using CCME method. The results showed that even after the implementation of YAP I and II, the river is very polluted and is not suitable to intended use. WQI was observed to be in poor categories at most of the locations and good at only one location. It was also noted that presence of free ammonia was affecting the water quality. Recommendations have been made to minimize point and non-point pollution sources, improving the sewerage system and ways to augment the assimilative capacity of the river.

The tremendous increase in water demand for various purposes has not only depleted the water resources but also deteriorated the water quality. In Himalayan regions the water demand in urban and rural area is fulfilled by fresh water springs. Climate change and global warming has affected the water resources of Himalayas and consequently springs are widely. Authors [38] have studied the spring water quality of Basantar watershed in Jammu and Kashmir. Total 60 water samples were collected from three kinds of terrains and 13 parameters were analyzed for hydrochemistry through Piper Trilinear diagram. Suitability of springs for drinking purposes has been determined by computing WQI. It has been found that the ion concentration was dominated by calcium, magnesium, bicarbonate and sulphate ions and the water was mainly governed by rock dominance. It was concluded that nearly 27% of the spring samples fell under excellent category, 30% in good and rest 30% were found to be in poor category. This shows that most of the springs are source of fresh drinking water and also requires a regular assessment due to increased water dependency.

Water Quality Index is a tool that provides an overall analysis of water quality and provides information if a water body can support life or it poses a potential threat to various uses of water [39]. If it indicates to be a poor water quality then parameters need to be identified having higher than permissible concentration. Health risk due to consumption of this increased level concentration is required to be assessed for such parameters. Limited literature has reported use of hazard quotient (HQ) and hazard index (HI) to access non-carcinogenic risk posed to human health. Few researchers

[40–46] have reported the use of HQ and HI as a reliable risk assessment tool. However, more detailed work is required to be done to relate water quality with health risk rather than only commenting on the water quality.

Conclusion

This chapter shows the simplicity and ease of water quality index to access any water bodies. WQI is a statistical tool that provides an single number for several water quality parameters. Most of the reviewed Indian Rivers shows a wide range of quality varying from bad to excellent category. Downstream ends of most of the rivers have been fallen under bad category owing to anthropogenic pollution and illegal sewage and effluent discharges. Such bad quality water is not suitable to domestic, agricultural and industrial purposes. In these scenarios it is necessary to estimate the health risk of people living in the polluted areas of the riverside. Health-hazards assessment cannot merely describe whether the pollution is serious, but should quantify the severity of the pollution.

Recommendation

1. Monitoring: Regular and stringent monitoring of water bodies is required to safeguard them from pollution.
2. Improving the sewerage system: Cities located near rivers should be covered with sewerage system and the waste water produced from the catchment areas should be sent to the sewage treatment plants. If required the capacity of the existing sewage treatment plants can be increased to cater the augmented sewage influent.
3. Training on waste water management: Efficient waste water treatment and disposal techniques have been proved worthwhile in improving the river quality. Training related to sustainable disposal ways of waste water may help in improving the river water quality.
4. Flow augmentation to increase capacity: To increase the flow capacity of the rivers, water stored during monsoon period should be used and during dry seasons allow water to be released in rivers. Efficient aeration should be done for streams and various open drains carrying the waste water.
5. Risk: Health risk assessment is required to be evaluated for all parameters having concentration higher than the permissible range to limit the water borne diseases. This can provide a basis for local governments to effectively manage water quality and provide a reference for water quality management in urban and rural areas.

References

1. Gleick PH (1993) Water in crisis: a guide to the world's fresh water resources. Oxford University Press, New York
2. FAO (2007) Coping with water scarcity: challenge of the twenty-first century UN water
3. WHO (2012) Guidelines for drinking-water quality (4th ed). World Health Organization. ISBN 978-92-4-154995-0
4. Richey AS, Thomas BF, Lo MH, Reager JT, Famiglietti JS, Voss K (2015) Quantifying renewable groundwater stress with GRACE. Water Resour Res 51:5217–5238
5. WWAP (2011) The United Nations world water development report 4: managing water under uncertainty and risk
6. Biglin A, Konanc MU (2016) Evaluation of surface water quality and heavy metal pollution of Coruh river basin (Turkey) by multivariate statistical methods. Environ Earth Sci 75:1029
7. Bricker OP, Jones BF (1995) Main factors affecting the composition of natural waters. In: Salbu B, Steinnes E (eds) Trace elements in natural waters. CRC, Boca Raton, pp 1–5
8. Reza R, Singh G (2010) Assessment of heavy metal contamination and its indexing approach for river water. Int J Environ Sci Technol 7(4):785–792
9. Herojeet RK, Rishi MS, Kishore N (2015) Integrated approach of heavy metal pollution indices and complexity quantification using chemometric models in the Sirsa Basin, Nalagarh valley. Chin J Geochem 34(4):620–633
10. Jung KY, Lee KL, Im TH, Lee IJ, Kim S, Han KY, Ahn JM (2016) Evaluation of water quality for the Nakdong River watershed using multivariate analysis. Environ Technol Innov 5:67–82
11. Qadir A, Malik RN, Husain SZ (2008) Spatio-temporal variations in water quality of Nullah Aik-tributary of the river Chenab, Pakistan. Environ Monit Assess 140:43–59
12. Wang Y, Wang P, Bai Y, Tian Z, Li J, Shao X, Mustavich LF, Li BL (2013) Assessment of surface water via multivariate statistical techniques: a case study of the Songhua river Harbin region, China. J Hydro-Environ Res 7:30–40
13. Malik RN, Hashmi MZ (2017) Multivariate statistical techniques for the evaluation of surface water quality of the Himalayan foothills streams, Pakistan. Appl Water Sci. https://doi.org/10.1007/s13201-017-0532-6
14. Vipan KV, Raj KS, Mohinder PSK (2013) Ugly face of urbanization and industrialization: a study of water pollution in Buddha Nala of Ludhiana city, India. J Environ Conserv Res 1(1): 6–11
15. Herojeet RK, Rishi Madhuri S, Sharma R, Lata R (2015) Hydrochemical characterization, ionic composition and seasonal variation in groundwater regime of an alluvial aquifer in parts of Nalagarh valley, Himachal Pradesh, India. Int J Environ Sci 6(1):68–81
16. Composition water management index – A tool for water management, NITI Aayog, 2018
17. Brown RM, McClelland NI, Deininger RA, Tozer RG (1970) A water quality index: do we dare? Water Sewage Works 117:339–343
18. Otto M (1998) Multivariate methods. In: Kellner R, Mermet JM, Otto M, Widmer HM (eds) Analytical chemistry. Wiley-VCH, Weinheim, p 916
19. Bordalo AA, Nilsumranchit W, Chalermwat K (2001) Water quality and uses of the Bangpakonk River (Eastern Thailand). Water Res 35(15):3635–3642
20. Cude C (2001) Oregon water quality index: a tool for evaluating water quality management effectiveness. J Am Water Resour Assoc 37(1):125–137
21. Hallock D (2002) A water quality index for ecology's stream monitoring program, technical report, P No. 02-03-52, Washington Department of Ecology, Environmental Assessment Program, Olympia, WA
22. Hubler S, Miller S, Merrick L, Leferink R, Borisenko A (2009) High level indicators of Oregon's forested streams. Lab. Environ. Assess. Div, Hillsboro, Oregon
23. Horton RK (1965) An index number system for rating water quality. J Water Pollut Control Fed 37(3):300–305

24. Brown RM, McClelland NI, Deininger RA, Tozer RG (1970) Water quality index-do we dare? Water Sewage Works 117(10):339–343
25. Han X (2011) Urban water quality risk assessment and emergency treatment methods [D]. Xi'an University of Architecture and Technology
26. Kavcar P, Sofuoglu A, Sofuoglu SC (2009) A health risk assessment for exposure to trace metals via drinking water ingestion pathway. Environ Health 212:216–227
27. Khadam IK, Kaluarachchi JJ (2003) Multi-criteria decision analysis with probabilistic risk assessment for the management of contaminated groundwater. Environ Impact Assess Rev 23:683–721
28. Human health risk assessment, United States Environmental Protection Agency, http://www.epa.gov/risk/human-health-risk-assessment
29. Tyagi S, Sharma B, Singh P, Dobhal R (2013) Water quality assessment in terms of water quality index. Am J Water Resour 1(3):34–38
30. Bora M, Goswami DC (2016) Water quality assessment in terms of water quality index (WQI): case study of Kolong river, Assam, India. J Appl Water Sci 7:3125–3135
31. Bhutani R, Khanna DR, Kuljarni DB (2016) Assessment of Ganga river ecosystem at Haridwar, Uttarakhand, India with reference to water quality indices. Appl Water Sci 6:107–113
32. Shah KA, Joshi GS (2017) Evaluation of water quality index for river Sabarmati, Gujrat, India. Appl Water Sci 7:1349–1358
33. Iqbal K, Ahmad S, Dutta V (2019) Pollution mapping in an urban segment of a tropical river: is water quality. Appl Water Sci 9:197
34. Kangabam RD, Bhoominathan SD, Kanagaraj S, Govindaraju M (2017) Development of water quality index (WQI) for Loktak lake in India. Appl Water Sci 7:2907–2918
35. Verma RK, Murthy S, Tiwary RK, Verma S (2019) Development of simplified WQIs for assessment of spatial and temporal variations of surface water quality in upper Damodar river basin, eastern India. Appl Water Sci 9:21
36. Shil S, Singh UK, Mehta P (2019) Water quality assessment of a tropical river using water quality index (WQI), multivariate statistical techniques and GIS. Appl Water Sci 9:168
37. Sharma D, Kansal A (2011) Water quality analysis of River Yamuna using water quality index in the national capital territory, India (2000–2009). Appl Water Sci 1:147–157
38. Taloor AK, Pir RA, Adimalla N, Ali S, Manhas DS, Roy S, Singh AK (2020) Spring water quality and discharge assessment in the Basantar watershed of Jammu Himalaya using geographic information system (GIS) and water quality Index (WQI). Groundw Sustain Dev 10: 100364
39. Akkaraboyina M, Raju B (2012) A comparative study of water quality indices of river Godavari. Int J Eng Res Dev 2(3):29–34
40. Njuguna SM, Onyango JA, Githaiga KB, Gituru RW, Yan X (2020) Application of multivariate statistical analysis and water quality index in health risk assessment by domestic use of river water. Case study of Tana River in Kenya. Proc Safety Environ Protecton 133:149–158
41. Mamun AA, Haoladar F, Sohail A (2019) Assessment of surface water quality Fuzzy Analytic hierarchy process (FAHP): a case study of Piyain Rivers sand and gravel quarry mining area in Jaflong, Sylhet. Groundwater Sust Dev 9:100208
42. Ustaoğlu F, Tepe Y, Tas B (2020) Assessment of stream quality and health risk in a subtropical Turkey river system: a combined approach using statistical analysis and water quality index. Ecol Indic 113:105815

43. Wu H, Xu C, Wang J, Xiang Y, Hantong MR, Zhang Y, Yao R, Lin A (2021) Health risk assessment based on source identification of heavy metals: a case study of Beiyun River, China. Ecotoxicol Environ Saf 213:112046
44. Chai N, Yi X, Xiao J, Liu T, Liu Y, Deng L, Jin Z (2021) Spatiotemporal variations, sources, water quality and health risk assessment of trace elements in the Fen River. Sci Total Environ 757:143882
45. Ahmed ASS, Hossain B, Omar SM, Babu F, Rahman M, Sarker SI (2021) Human health risk assessment of heavy metals in water from the subtropical river, Gomti, Bangladesh. Environ Nanotechnol Monit Manag 15:100416
46. Setia R, Dhaliwal SS, Kumar V, Singh R, Kukal SS, Pateriya B (2020) Impact assessment of metal contamination in surface water of Sutlej River (India) on human health risks. Environ Pollut 265(Part B):114907

Part V
Groundwater Quality Assessment

Groundwater Quality Assessment in the Semi-Arid Blocks of Rajasthan, India: A Combined Approach of Fuzzy Aggregation Technique with GIS

Ajit Pratap Singh and Kunal Dhadse

Abstract In developing countries, there has been an increase in dependencies of various water uses on groundwater (GW) due to ease in access, low cost, and steadily increasing domestic and industrial water demands. The excessive consumption of sub-surface water resources has resulted in the degradation of groundwater quality and substantial stress on its quantity, especially in arid regions of India. GW degradation's main reasons are rapid population growth, urbanization development, increased pollution discharge, change in irrigation and land use patterns and inefficient water supply systems, and high water demand. In addition, unpredictable and scanty GW recharge have resulted in uncertain GW hydrological cycles with unreliable water availability and demand. This study deals with the overall assessment of groundwater quantity, quality, and sustainability at the regional level. The study develops a fuzzy-based GW status assessment framework that can be used to quantify severity levels and identify the spatial regions of critical importance in terms of sustainability. Using the GIS-integrated fuzzy index model, the groundwater quality of Rajasthan, India, has been assessed by deriving three important GW indices viz., GW scarcity index, GW quality index, and GW sustainability index. Classified maps of the study area in terms of priority zoning have also been developed at the end of the study that the policymakers and decision-makers can use for sustainable development of groundwater resources in Rajasthan, India.

Keywords Groundwater · Sustainability · Scarcity · Water Quality · Fuzzy set theory · GIS

A. P. Singh (✉) · K. Dhadse
Civil Engineering Department, Birla Institute of Technology and Science, Pilani, Rajasthan, India
e-mail: aps@pilani.bits-pilani.ac.in

S. Yadav et al. (eds.), *Water Quality, Assessment and Management in India*, Earth and Environmental Sciences Library, https://doi.org/10.1007/978-3-030-95687-5_15

1 Introduction

Groundwater has been a vital resource for developing a society; it supports many consumptive water demands and provides a direct potable source for freshwater. The quality of groundwater has a considerable effect on human health, irrigation, and industrial water usability. It is crucial to assess groundwater conditions qualitatively and quantitatively for long-term use and maintain its environmental sustainability [1, 2]. Groundwater is highly vulnerable to contaminant intrusions caused by anthropogenic pollution sources or geological actions, mainly because water is a good solvent [3, 4]. Therefore, the groundwater quality assessment is essential with respect to pollutants like Nitrates, Fluorides, and Iron, etc. Vasanthavigar et al. (2010) used these parameters to estimate groundwater suitability for various purposes [5]. It is also evident that the usability of hydrological resources like groundwater at regional levels is highly dependent on human activity-driven environmental changes [6].

Groundwater contamination occurs through various processes like groundwater interaction with weathering of mineral rocks and dissolution of salt concentrations in the groundwater. Other mechanisms include sub-surface intrusion of pollutants resulting in the accumulation of dissolved salts in groundwater. The quality of groundwater in aquifers is also affected by non-point source pollution caused due to agricultural activities. Significant contributors to non-point source pollution are fertilizers, excessive pesticides, and nitrogen and phosphorus-based enhancers in agricultural fields [7, 8].

In arid regions, significant causes of groundwater resource degradation are the over-exploitation and subsurface interaction of groundwater in aquifers with mineral deposits. A comprehensive literature on groundwater resource quality and quantity assessment in arid regions of India can be found in work by [9–13]. In some arid blocks of Rajasthan, India, the groundwater levels have dropped up to 30 m along with constant water quality degradation [14–16]. The dropping of groundwater levels increases the water-mineral interaction and has increased salinity and concentrations of nitrates, iron, and fluorides [2, 9, 10, 17]. Groundwater movement in aquifers is very slow, resulting in contaminants accumulating over a long period [18–20]. There are currently various practices to assess groundwater quality, and researchers have studied groundwater quality using geochemical interactions, numerical and correlation analysis [14, 15, 20, 21]. Apart from groundwater chemistry, the researchers have developed a groundwater quality index using fuzzy logic to assess subsurface water resources' overall quantity and quality status in aquifers [22].

In the western arid region of India, Rajasthan receives a deficient annual average rainfall of 200–400 mm and is scarce in surface water resources. Due to natural hydrological cycles and arid geography, most of the water requirements of Rajasthan are highly dependent on groundwater resources. Poor water resource management (inefficient water pumps, irrigation systems, etc.) has led to a critical condition of groundwater resources in Rajasthan. There is also excess groundwater extraction because of its low price and a belief in abundant supply from aquifers [23].

In recent years, groundwater quality and quantity in some of the blocks of Rajasthan have reached a point of criticality in terms of sustainability. Some of the blocks have experienced a severe shortage of GW resources with a decline in GW levels. The risk of GW contamination has also been increasing day by day. Most Rajasthan blocks comprise alluvial aquifer systems and are surrounded by highly permeable well-drained coarse sandy soil. The alluvial aquifers systems have high risks of contaminating GW because of their shallow characteristics and high permeability.

Since the groundwater resources in Rajasthan are vulnerable to quality, quantity, and overall sustainability, there is a need to quantify the current status of the groundwater quality and quantity so that sustainability measures can be adopted in the specified regions/blocks of Rajasthan. The present study develops various indices to assess GW sustainability regarding quality, quality, and long-term usability as a valuable resource. The study develops a fuzzy and GIS-based groundwater sustainability assessment framework that helps to study the current GW status of Rajasthan, its distribution, and areas of critical importance in terms of sustainable groundwater management.

2 Study Area

Rajasthan is located in India's western arid regions and comprises 32 administrative districts and 236 Blocks. The study area of Rajasthan with Digital Elevation Map (DEM) is shown in Fig. 1. With a 342,239 km^2 area, the state receives a deficient average rainfall of about 300 mm, with some regions receiving as low as 100 mm annual average precipitation. The state has diverse aquifer systems, and the overall groundwater hydrology of the state is very complex to study due to numerous mineral deposits throughout the state. Broadly, the state is divided into three geological strata consisting of consolidated rocks, semi-consolidated sediments, and unconsolidated sediments.

Under unconsolidated sediments, the state has abundant alluvial and aeolian deposits, which exist mainly in the blocks of Barmer, Jalore, and Jodhpur districts that consist mainly of clay, sand, cobbles, and gravel. The formations like Valley fills are present in the different blocks of Ajmer, Jhunjhunu, Udaipur, and Bhilwara districts. The consolidated rocks and aquifers systems of phyllites, gneiss, marble granites, schist, and sandstone, limestone, quartzite, and basaltic flows are present in eastern Rajasthan.

The data for the GW levels for the month of May were collected from Central Pollution Control Board, New Delhi for the Rajasthan state, and the GW fluctuations have been calculated in terms of rising and fall using the past time-series data on average monthly GW levels. The GW levels in western and lower northern regions of Rajasthan have been found out to be critically low (below 40 m), as shown in

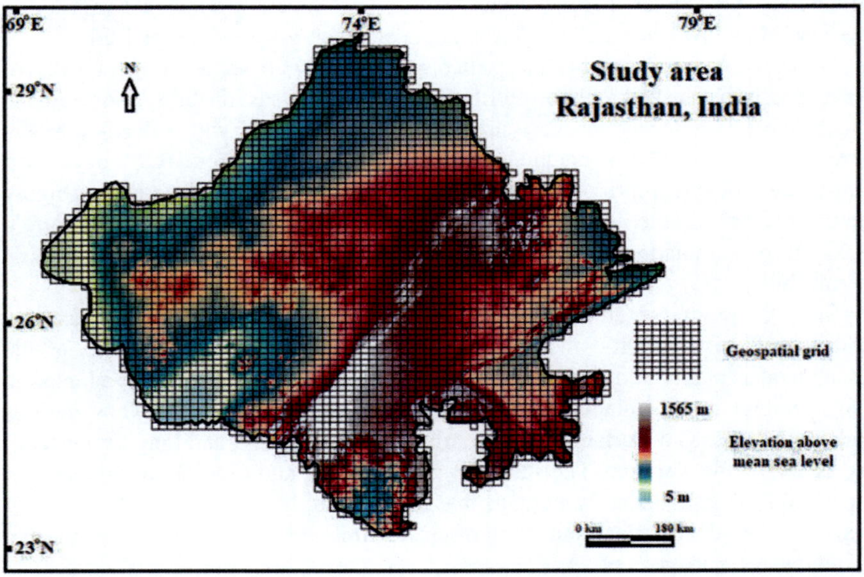

Fig. 1 Digital Elevation Map (DEM) of Rajasthan state of India

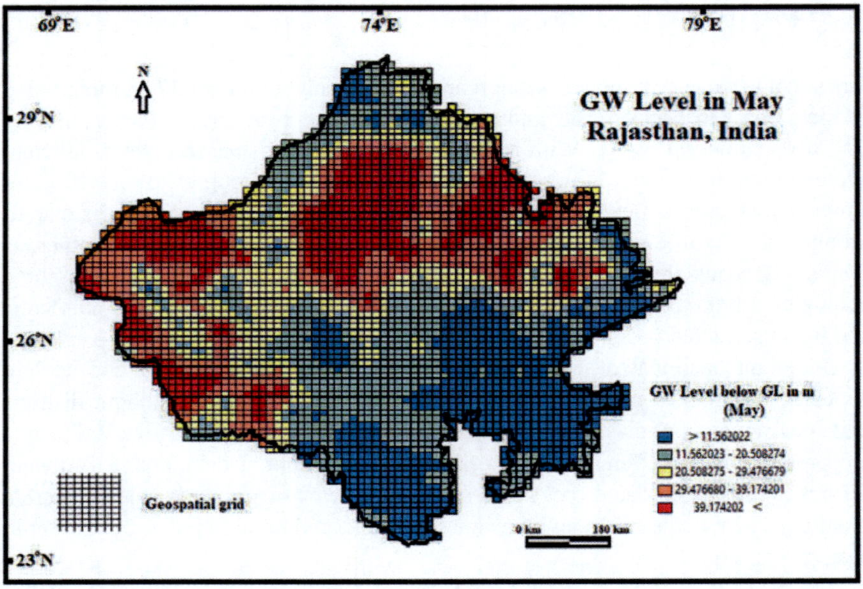

Fig. 2 Average GW levels of Rajasthan in the month of May

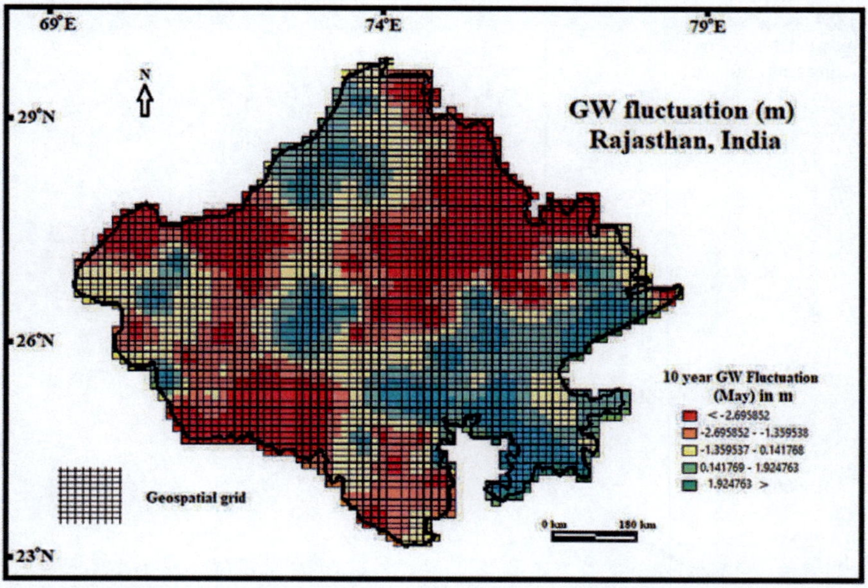

Fig. 3 Average GW levels fluctuations (past ten years) of Rajasthan in the month of May

Fig. 2. Figure 3 shows that the north-eastern regions of Rajasthan are in critical endanger of GW resources as the GW levels have gone down more than 3 m in the past ten years.

3 Data and Methodology

There are different ways of assessing the GW quality, and one of them is by defining the water quality index for groundwater. Since the traditional approaches do not consider the errors and uncertainties in data collection and reporting, the current study uses fuzzy-based mathematical programming to develop a series of indexes for groundwater sustainability and quality and quantity status indicators.

The fuzzy-based techniques for the various application have been used [24–28]. The above studies introduced fuzzy logic while dealing with qualitative terms using "Fuzzy sets". The fuzzy approach helps to devise modeling tools for scenarios with inherent uncertainties and a high degree of natural variability. A flexible and robust expert support system can be developed for solving real-life problems using fuzzy set tools with the capabilities of providing a dynamic range of input and output attributes. Linguistic water quality grades are among the many examples in which fuzzy set theory is useful to model realistic scenarios.

The attributes of the fuzzy set theory-based model can be expressed using selected membership functions. Examples include triangular, Gaussian, sigmoid,

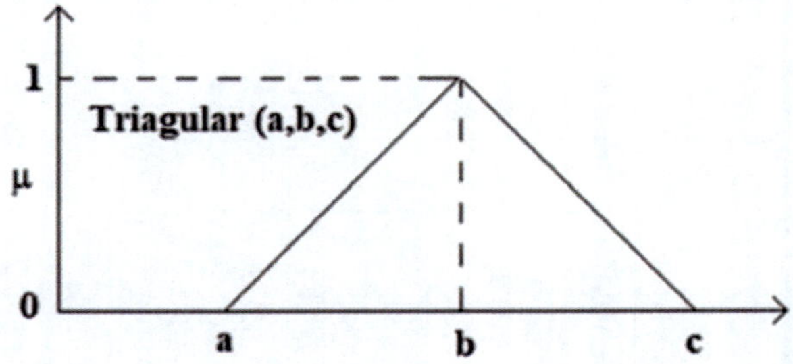

Fig. 4 A typical representation of triangular membership function

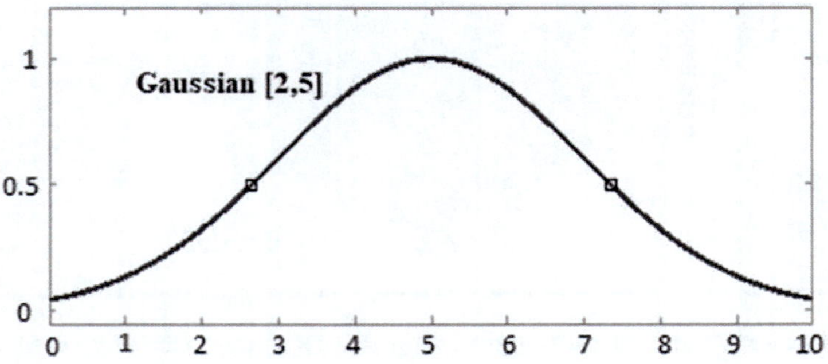

Fig. 5 A typical representation of Gaussian membership function

and trapezoidal functions. In the present work, triangular and Gaussian membership functions define the attribute sets. The triplets with three variables, a, b, and c in Fig. 4, define triangular membership functions as given in Eq. (1):

$$\mu_A(x, a, b, c) = \begin{cases} 0 \; if \; x \leq a \\ \frac{x-a}{b-a} \; if \; a \leq x \leq b \\ \frac{c-x}{c-b} \; if \; b \leq x \leq c \\ 0 \; if \; x \leq c \end{cases} \tag{1}$$

Similarly, the Gaussian membership function (Fig. 5) is defined by using two parameters, i.e., c and σ, and can be expressed mathematically as given in Eq. (2):

$$f(x, \sigma, c) = e^{\left[\frac{-(x-c)^2}{2\sigma^2}\right]} \tag{2}$$

where c corresponds to a point on abscissa where the Gaussian curve has maximum membership value on its ordinate, and σ is the width/thickness of the curve.For devising various GW sustainability indicators, a combination of attributes has been devised to develop an index-based fuzzy inference relationship model. The study mainly focuses on defining three indices, GW scarcity index, GW quality index, and GW sustainability index. For the GW scarcity index, two fuzzy attributes, namely

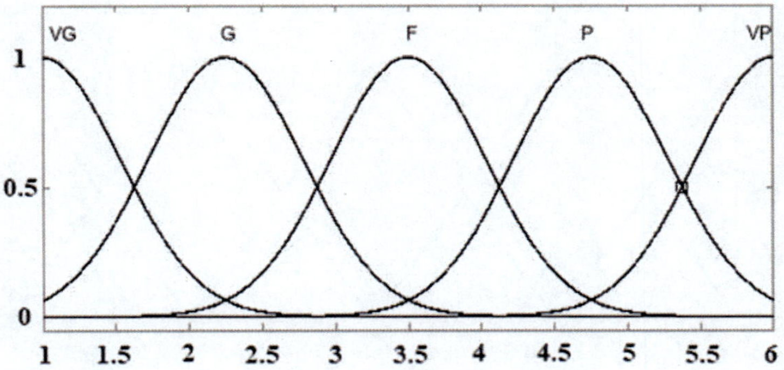

Fig. 6 Gaussian membership function used for describing attributes

the GW levels and past decade GW fluctuation, have been taken to obtain a GW scarcity index (0–100). For the GW quality index, five fuzzy water quality attributes, namely Chloride, EC, Fluoride, Iron, and Nitrate, have been considered to obtain a GW quality index (0–100). For the GW sustainability index, all the seven fuzzy attributes, namely the GW levels, past decade GW fluctuation, concentrations of Chloride, EC, Fluoride, Iron, and Nitrate, have been considered to obtain the overall GW sustainability index (0–100). These attributes are defined in linguistic variables by the experts working in water resources management. In this study, the water quality and quantity parameters have been defined using five attributes, namely very good (VG), good (G), fair (F), poor (P), and very poor (VP). In this case, as shown in Fig. 6, Gaussian membership functions are used to deal with these subjective variables.

Weights and the degree of importance of attributes have been separately defined using linguistic variables, namely Extremely Important (E), Very Important (V), Medium Importance (M), Low Importance (L), and Unimportant (U). As the particular attributes do not hold equal importance in terms of water use or the intended index, a degree of importance is assigned to each water quality and quantity attribute. In this case, as shown in Fig. 7, triangular membership functions are used to deal with the subjective variables.

The importance weight/degree for each attribute in the index acts as the weights for calculating different indices. In practice, the weights for each attribute are used in a crisp and non-fuzzy form. Thus, there is a need to calculate a representative weight for a particular fuzzy-based degree of importance. The centroid method for de-fuzzification is used to obtain the crisp weights for the linguistic importance scale. The representative weight (wi) for the above five importance classes can be calculated using Eq. (3). The obtained values for representative weights are given in Table 1.

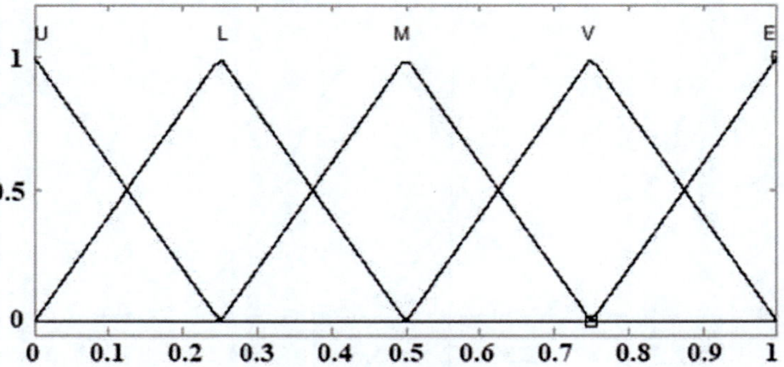

Fig. 7 Triangular membership function used for describing degrees of importance

Table 1 Linguistic terms used with their importance values to define quality attributes

Importance terms	Weight (w_i)
Unimportant (U)	0.0833
Low Importance (L)	0.2500
Medium Importance (M)	0.5000
Very Important (V)	0.7500
Extremely Important (E)	0.9167

$$w_i = \frac{\int_{w=0}^{1} w.\mu_{wi}(w)dw}{\int_{w=0}^{1} \mu_{wi}(w)dw} \tag{3}$$

Based on the values of the various attributes at various geo-grids of the spatial regions of Rajasthan, the decision-makers assign the weights and degree of goodness of quality and quantity at various stations. Equation (4) evaluates the final GW index with different combinations of attributes using Yager's max-min model. The quality index set $\widetilde{QI_n}$ for the nth station is defined as suggested by Singh et al. [27]:

$$\widetilde{QI_n} = \bigcap_{i=1,2,\ldots..I} S_{i,n}^{w_{i,n}} = \min_{i=1,2\ldots I}\left\{\left(\mu_{S_{i,n}}(S)\right)^{w_{i,n}}\right\} \tag{4}$$

where $S_{i,n}$ is the assigned score by the decision-maker for a given ith attribute at the nth sampling station selected in the region of study. Once the particular GW index set $\widetilde{QI_n}$ is computed, the quality index set with the maximum membership value is determined at the decision point in terms of GW index, QI_n, using Yager's algorithm as given in Eq. (5):

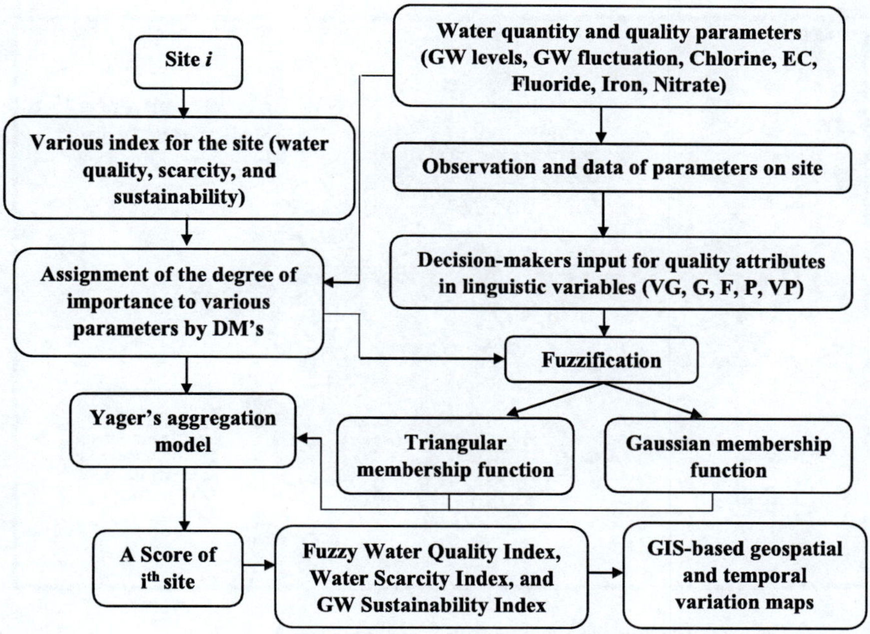

Fig. 8 Fuzzy decision support system used for mapping of the GW scarcity index, GW quality index, and GW sustainability assessment of arid region of Rajasthan

$$QI_n = \left\{ s | \mu_{Q_{i,n}}(s) = Max(QI_n) | \right\} \tag{5}$$

Thus the various sustainability-based indices at selected sampling locations or various geo-grids cells across the study area of Rajasthan are calculated and mapped using GIS-based geospatial and classification tools. A detailed methodology for the overall evaluation and mapping of the GW scarcity index, GW quality index, and GW sustainability index is shown in Fig. 8.

The data for the water quality attributes (concentration range and spatial distribution of Chloride, EC, Fluoride, Iron, and Nitrate) for the study region of Rajasthan are shown in Figs. 9, 10, 11, 12, 13.

It can be observed from the above figures that due to highly diverse sets of variables in overall groundwater quality attributes, it is essential to downscale the combined data and define a single comprehensible and comparable index. Hence, the study develops fuzzy-based downscaling techniques that help decision-makers to understand the overall GW quantity, quality, and sustainability in the large study area.

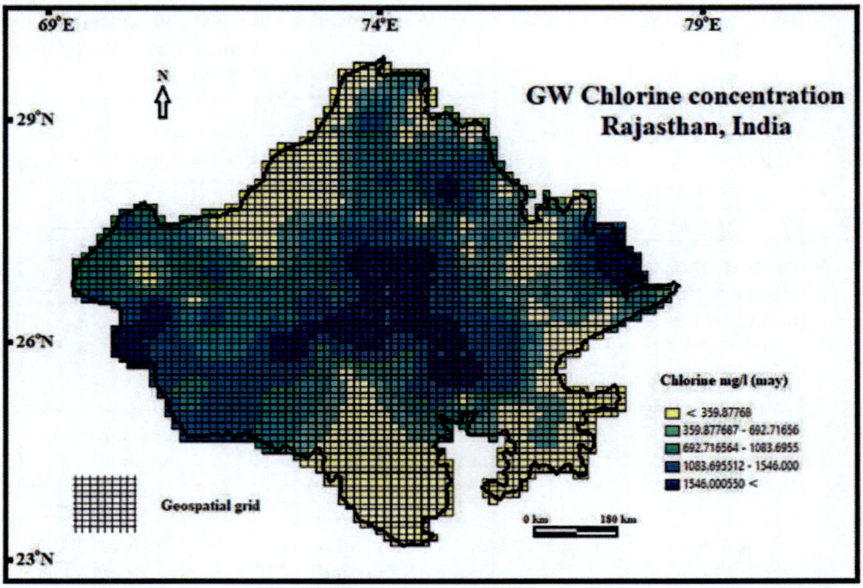

Fig. 9 Variation of Chloride in Groundwater of Rajasthan during May

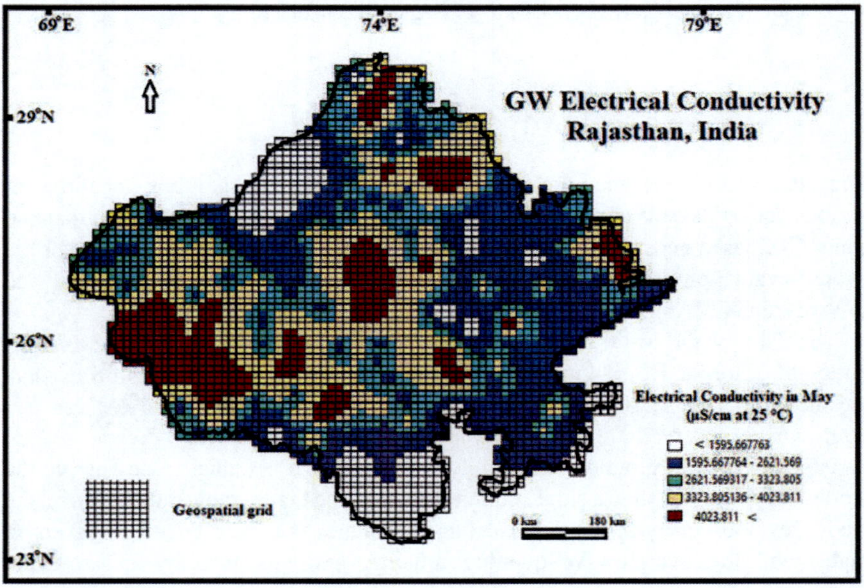

Fig. 10 Variation of EC in Groundwater of Rajasthan during May

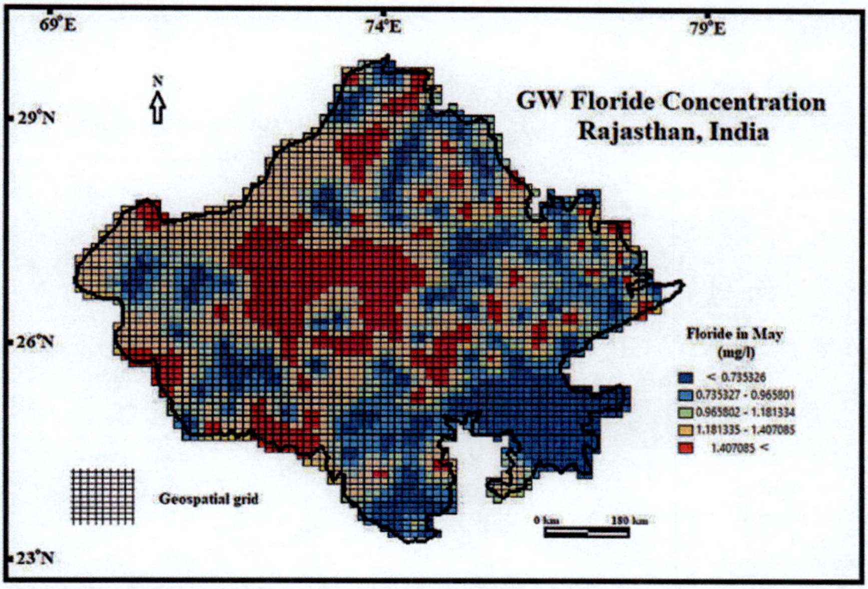

Fig. 11 Variation of Fluoride in Groundwater of Rajasthan during May

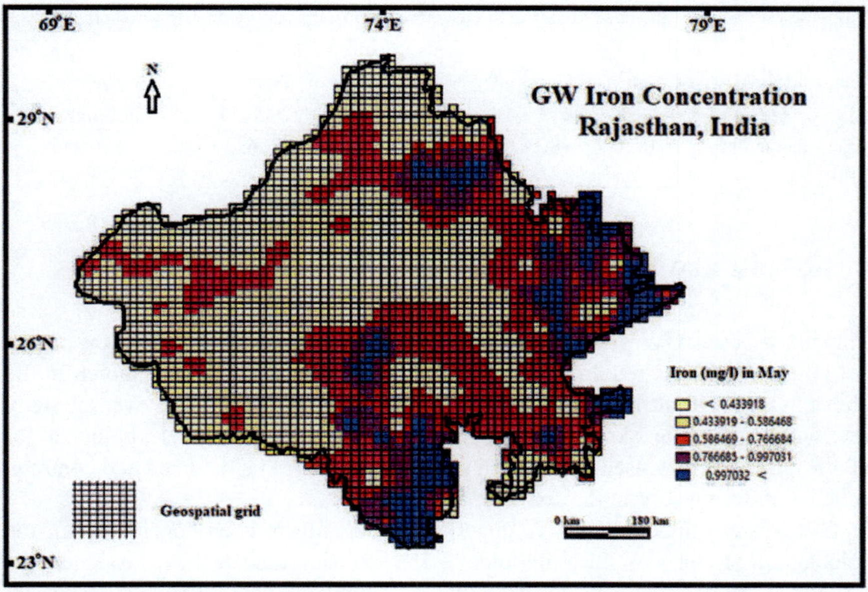

Fig. 12 Variation of Iron in Groundwater of Rajasthan during May

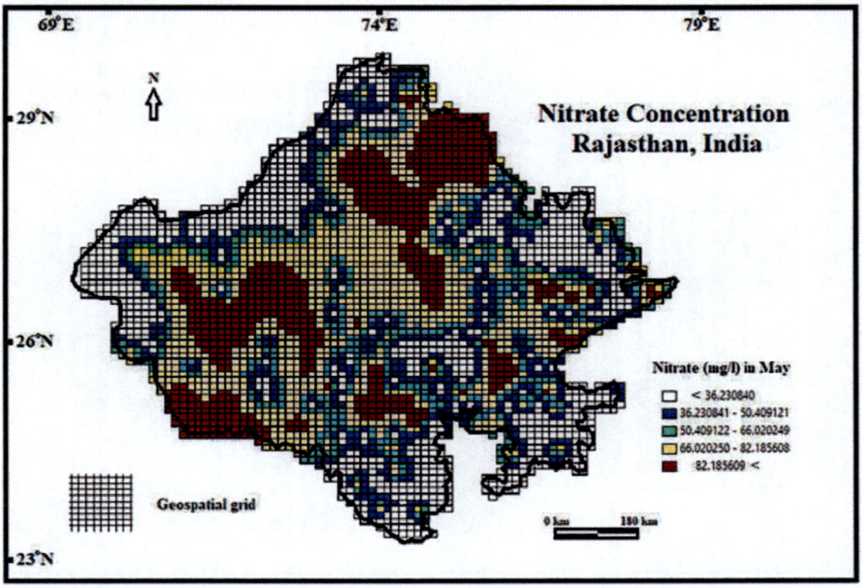

Fig. 13 Variation of Nitrate in Groundwater of Rajasthan during May

Table 2 GW scarcity index attributes defined on the linguistic scale by decision-makers

Index attribute	Very Good	Good	Fair	Poor	Very Poor
GW Level (m)	0–11.56	11.57–20.5	20.6–29.4	29.5–39.1	39.2 and above
GW Fluctuation (m)	> +1.92	+0.14 to +1.91	−1.35 to +0.13	−2.69 to −1.36	−2.70 and below

4 Results and Discussions

The fuzzy-based GW sustainability framework is briefly described in the above section, and the data required for the fuzzy index calculation is also shown in the above figures. The attribute data for the GW is geospatially distributed over the study area within a specific 1 km × 1 km geogrid. Each geogrid is assigned a unique id for all the data. The id is used to establish a specific region's standardized and common calculation framework in the study area.

GW Scarcity Index: The GW quantity attributes have been considered for the calculation of the GW Scarcity index. The decision-makers have assigned the linguistic weights for GW levels and GW fluctuation as Extremely Important (E) and Very Important (V). The representative weights for the above attributes come out to be 0.91 and 0.75, respectively. The linguistic scale to classify the data using a fuzzy set theory-based model is defined by the decision-makers in Table 2.

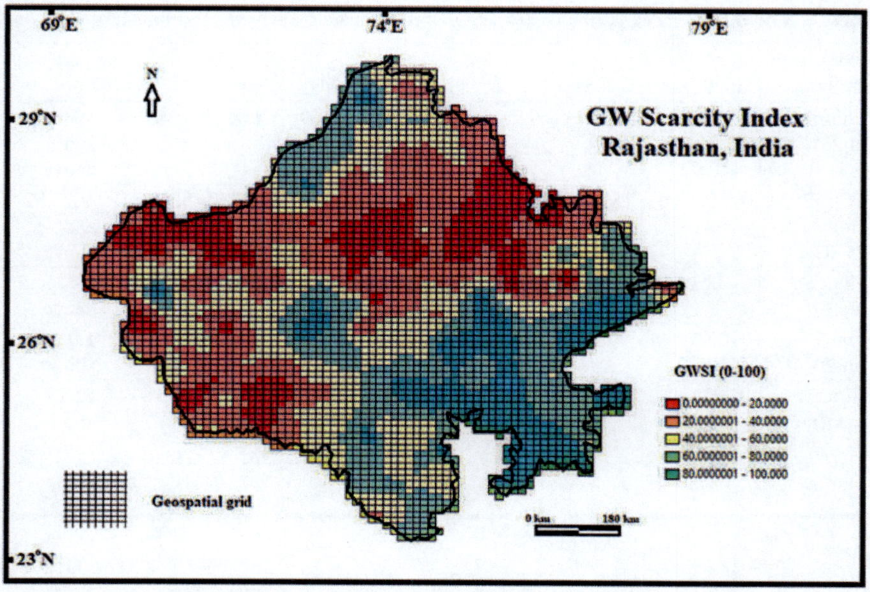

Fig. 14 Groundwater Scarcity Index for Rajasthan, India

The fuzzy inference model is developed using Eqs. (4) and (5) in the MATLAB FIS modeling tool. The input of the attributes in terms of the linguistic variables are used in the fuzzy-based index model, and a crisp number, i.e., the GW scarcity index ranging from 0–100, is obtained for each geogrid cell for the study area. With 100 being the GW abundant and 0 being the most GW scarce region in the study area. The overall geospatial distribution of the GW scarcity index is shown in Fig. 14.

GW quality index: Like the GW quantity attributes, five GW quality attributes have been considered to calculate the GW quality index. The decision-makers have assigned concentrations of Chloride, EC, Fluoride, Iron, and Nitrates the linguistic weights of Medium Important (M), Low Important (L), Extremely Important (E), Very Important (V), and Very Important (V), respectively. The representative weight for the above attributes comes out to be 0.50, 0.25, 0.91, 0.75, and 0.75, respectively. The data classes on a linguistic scale are defined by considering the opinion of the decision-makers, as given in Table 3.

The input of the attributes in terms of the linguistic variables have been used in the fuzzy-based index model, and a crisp number, i.e., the GW quality index ranging from 0–100, is obtained for each geogrid cell for the study area. With 80–100 being the GW region of very good water quality status and 0–20 being the most critically affected region in terms of degraded water quality in the selected study. The overall geospatial distribution of the GW quality index is shown in Fig. 15.

Similar to the above two indices, the overall groundwater sustainability index is obtained by combining the two quantity and five quality attributes. The sustainability

Table 3 GW quality index attributes defined on the linguistic scale by decision-makers

Index attribute	Very Good	Good	Fair	Poor	Very Poor
Chlorine (mg/l)	0–359.8	359.9–692.71	692.72–1083.69	1083.7–1546	1546.1 and above
EC (μS/cm) at 25 °C	0–1595.66	1595.67–2621.56	2621.57–3323.80	3323.81–4023.81	4023.82 and above
Fluoride (mg/l)	0–0.73	0.74–0.96	0.97–1.18	1.19–1.40	1.41 and above
Iron (mg/l)	0–0.43	0.44–0.58	0.59–0.76	0.77–0.99	1.0 and above
Nitrates (mg/l)	0–36.23	36.24–50.40	50.41–66.02	66.03–82.18	82.19 and above

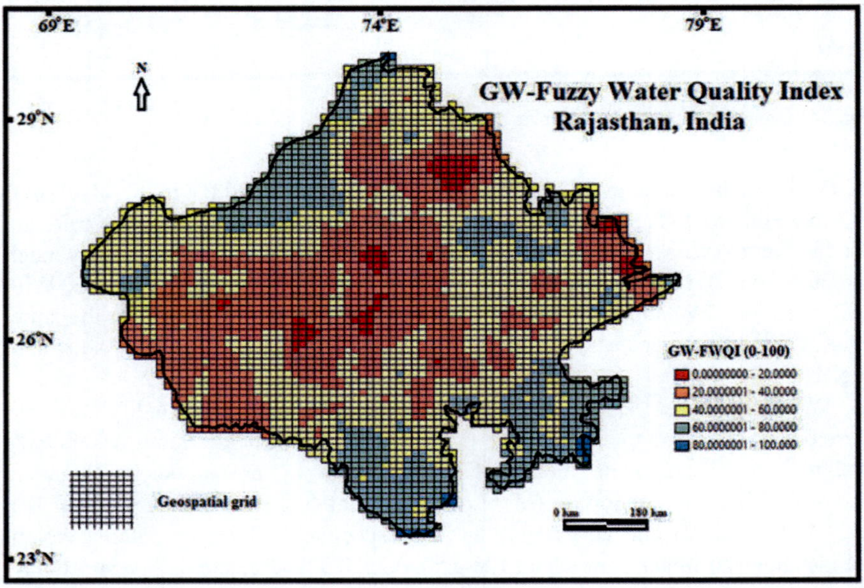

Fig. 15 Groundwater Quality Index for Rajasthan, India

aspects of GW resources in Rajasthan is finally obtained in terms of four priority classes. With a sustainability index value of 0–25 being the most critical and 75–100 as the most sustainable status of GW in Rajasthan.

From Fig. 14, it can be seen that the districts of Alwar, Jhunjhunu, Sikar, Churu, Nagaur, Jalor, and several blocks of western Jaisalmer lie in the region of severe GW scarcity. From Fig. 14, a similar interpretation can be drawn for water quality, and it has shown the critically affected districts as Churu, Bharatpur, Dhaulpur, Jodhpur,

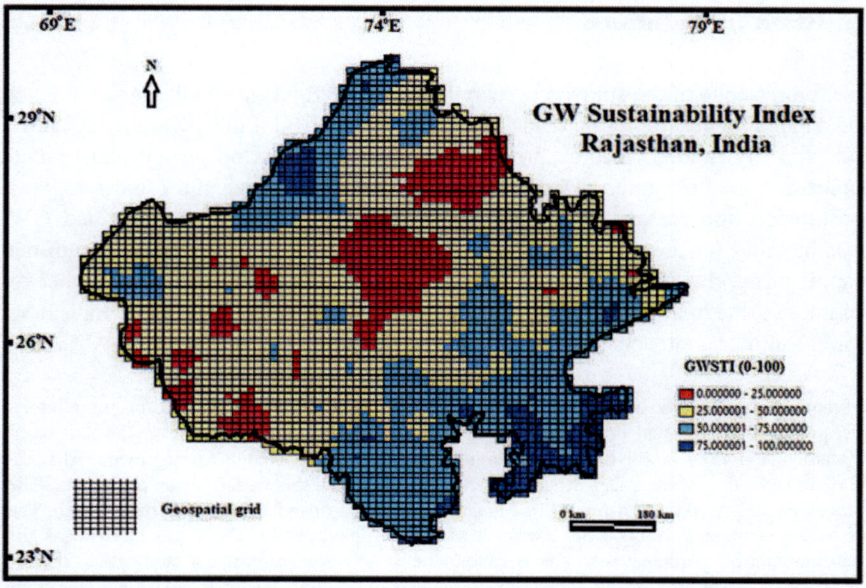

Fig. 16 Overall Groundwater Sustainability Index for Rajasthan, India

Barmer, Bhilwara, Ajmer, and Nagaur. The results show that the sustainability of GW in Rajasthan can be assessed from two contexts of scarcity and quality and one combined context of overall sustainability. It is clear from Fig. 16 that Churu, Nagaur, Jalor, Sikar, Jaisalmer, and Barmer lie in critical regions of GW sustainability and immediate actions need to be taken for these regions for ensuring the long-term sustainability of GW resources in these regions.

5 Conclusions

The study demonstrated how the fuzzy approach helps to define the sustainability indicators under uncertain environments. The fuzzy set theory has helped in handling linguistic variables with variation in GW quantity and quality attributes. It is not very easy to compare and incorporate non-comparable attributes in a mathematical model, but the fuzzy set theory gave a flexible and robust computational framework for the study. Integrating GIS in the fuzzy model helped standardize the input data to perform calculations for selected regions accurately. The final integrated model was able to downscale the vast array of data and obtain priority maps of the study area that decision-makers can readily use for sustainable groundwater management. The dual approach of quality and quantity for solving the sustainability assessment problem for GW in Rajasthan resulted in an in-depth insight into the problem at the regional and state levels.

6 Recommendations

The future scope of the study is to map the region of Rajasthan at the regional level and a much higher resolution, which can only be achieved if the governing agencies and privately funded research projects collaborate to develop an updated unified database from the region. Decision-makers should prioritize the critical regions identified in the present study in terms of GW Scarcity, GW Quality, and GW Sustainability for developing water infrastructures in the near future. Although the mapping and classification technique used in the study uses a limited number of parameters, the indicators can be modified and adapted for any number of indicators using a similar approach.

Acknowledgments The authors would like to acknowledge BITS Pilani, India for providing its infrastructure that assisted this research. The authors' special thanks are due to the Advanced Research Laboratory in Environmental Engineering and Faecal Sludge Management (ARLEE-FSM) of Civil Engineering Department, BITS Pilani, at which this research was carried out. The authors are also thankful to Aditya Birla Finance Ltd., for the financial support of the research. The references mentioned in this study are acknowledged whole-heartily. The authors also thank the contributors who participated in the questionnaire survey to perform this study. The authors acknowledge the efforts of the anonymous reviewers and editors for their valuable suggestions and efforts. The authors would also like to show their gratitude to Rajasthan State Pollution Control Board, Jaipur and Central Pollution Control Board, New Delhi (India), for making available the groundwater data for Rajasthan as an open-source used in the study.

References

1. Kumar A, Krishna AP (2018) Assessment of groundwater potential zones in coal mining impacted hard-rock terrain of India by integrating geospatial and analytic hierarchy process (AHP) approach. Geocarto Int 33(2):105–129
2. Singh AP, Khakoliya A, Tavanshetti S, Yadav J (2019) Groundwater quality assessment using GIS and fuzzy logic – a case study of Jhunjhunu District. Pollut Res 38(3):140–147
3. Das M, Kumar A, Mohapatra M, Muduli SD (2010) Evaluation of drinking quality of groundwater through multivariate techniques in urban area. Environ Monit Assess 166(4):149–157
4. Tiwari K, Goyal R, Sarkar A (2017) GIS-based spatial distribution of groundwater quality and regional suitability evaluation for drinking water. Environ Process 4(3):645–662
5. Vasanthavigar M, Srinivasamoorthy K, Vijayaragavan K, Rajiv Ganthi R, Chidambaram S, Anandhan P et al (2010) Application of water quality index for groundwater quality assessment: Thirumanimuttar sub-basin, Tamilnadu, India. Environ Monit Assess 171(1–4):595–609
6. Li P, Wu J, Qian H (2016) Hydrochemical appraisal of groundwater quality for drinking and irrigation purposes and the major influencing factors: a case study in and around Hua County, China. Arab J Geosci 9(1):1–17
7. Ahada CPS, Suthar S (2017) Hydrochemistry of groundwater in North Rajasthan, India: chemical and multivariate analysis. Environ Earth Sci 76(5):203
8. Getahun E, Keefer L (2016) Integrated modeling system for evaluating water quality benefits of agricultural watershed management practices: case study in the Midwest. Sustain Water Qual Ecol 8:14–29

9. Bhakar P, Singh AP, Mittal RK (2022) Assessment of groundwater suitability using remote sensing and GIS: a case study of Western Rajasthan, India. Arab J Geosci 15:41. https://doi.org/10.1007/s12517-021-09272-9
10. Kamra SK, Lal K, Singh OP, Boonstra J (2002) Effect of pumping on temporal changes in groundwater quality. Agric Water Manag 56(2):169–178
11. Kumar RP, Ranjan RK, Ramanathan AL, Singh SK, Srivastava PK (2015) Geochemical modeling to evaluate the mangrove forest water. Arab J Geosci 8(7):4687–4702
12. Ravindra K, Garg VK (2006) Distribution of fluoride in groundwater and its suitability assessment for drinking purpose. Int J Environ Health Res 16(2):163–166
13. Machiwal D, Islam A, Kamble T (2019) Trends and probabilistic stability index for evaluating groundwater quality: the case of quaternary alluvial and quartzite aquifer system of India. J Environ Manag 237:457–475
14. Bhakar P, Singh AP (2018a) Life cycle assessment of groundwater supply system in a hyper-arid region of India. Procedia CIRP 69(May):603–608
15. Bhakar P, Singh AP (2018b) Groundwater quality assessment in a hyper-arid region of Rajasthan, India. Nat Resour Res 28(2):505–522
16. Chintalapudi P, Pujari P, Khadse G, Sanam R, Labhasetwar P (2017) Groundwater quality assessment in emerging industrial cluster of alluvial aquifer near Jaipur, India. Environ Earth Sci 76(1):1–14
17. Singh AP, Bhakar P (2020) Development of groundwater sustainability index: a case study of western arid region of Rajasthan, India. Environ Dev Sustain 23:1844–1868. https://doi.org/10.1007/s10668-020-00654-9
18. Praveena SM, Lin CY, Aris AZ, Abdullah MH (2010) Groundwater assessment at Manukan Island, Sabah: multi-displinary approaches. Nat Resour Res 19(4):279–291
19. Singh CK, Mukherjee S (2015) Aqueous geochemistry of fluoride enriched groundwater in arid part of Western India. Environ Sci Pollut Res 22(4):2668–2678
20. Machiwal D, Cloutier V, Güler C, Kazakis N (2018) A review of GIS-integrated statistical techniques for groundwater quality evaluation and protection. Environ Earth Sci 77(19):1–30
21. Mogaji KA, Lim HS (2018) Development of groundwater favourability map using GIS-based driven data mining models: an approach for effective groundwater resource management. Geocarto Int 33(4):397–422
22. Srinivas R, Bhakar P, Singh AP (2015) Groundwater quality assessment in some selected area of Rajasthan, India using fuzzy multi-criteria decision making tool. Aquat Proc 4:1023–1030
23. Schmoll O, Howard G, Chilton G, Chorus I (2006) Protecting ground water for health-managing the quality of drinking-water sources. World Health Organization (WHO). Retrieved from http://www.who.int/water_sanitation_health/publications. Accessed on November 15, 2018
24. Ross TJ (2008) Fuzzy logic with engineering applications, 2nd edn. Wiley, New Delhi
25. Sakawa M (1993) Fuzzy sets and interactive multi objective optimization. Plenum Press, New York
26. Singh AP (2008) An integrated fuzzy approach to assess water resources potential in a watershed. ICFAI J Comput Math 1(1):7–23
27. Singh AP, Dhadse K, Ahalawat J (2019) Managing water quality of a river using an integrated geographically weighted regression technique with fuzzy decision making model. Environ Monit Assess 191(6). https://doi.org/10.1007/s10661-019-7487-z
28. Singh AP, Ghosh SK, Sharma P (2007) Water quality management of a stretch of river Yamuna: an interactive fuzzy multi-objective approach. Water Resour Manag 21:515–532

Estimation of Groundwater Quality Parameters for Drinking Purpose using IDW, GIS and Statistical Analysis Methods: A Case Study of Basaltic Rock in Mahesh River Basin, Akola and Buldhana Districts (MS), India

Chaitanya B. Pande and Kanak N. Moharir

Abstract Assessment of groundwater quality parameters is carried out using IDW, GIS and statistical analysis methods for the drinking purpose in the hard-basaltic rock area, little part of younger and older alluvium soil in the study area. The groundwater quality parameters have displayed a little high pH landscape and greatest numbers of alkalinity and electrical conductivity in area. Saline water is a very largest concentration values identified at near about teen villages. Some area of study area has been infection due to unidentified sediment, more ash, soil bricks and powerhouse plant located in basaltic hard rocks area. Total thirty-five open and bore wells were selected in the present area, which wells more suitable to used of drinking and irrigation in basaltic rock area. These observation wells were located through Global positioning system (GPS) instrument with the reference of toposheets. The groundwater parameters are very much highly affected due to basaltic rock minerals and unidentified sediment. Therefore, the groundwater physicochemical parameters were evaluated and these groundwater quality parameters were representation in the form of graphically and spatially distribution. To assessment of groundwater parameters representations have been computing the absorption of chemical factors to relating water quality standard values of drinking and irrigation uses in basaltic hard rock area. As per the results of Piper diagram can be found hardness of carbonate with the most of water are under in the salinity zone. The water quality of irrigation factors like sodium adsorption ratio, salinity hazard, sodium percentage and residual sodium carbonate were measured as per standards techniques. The water samples excluding at maximum places fell in good to excellent classification for agriculture uses in the area. The various maps of water quality were prepared using spatial

C. B. Pande (✉)
AICRP for Dryland Agriculture, Dr. Panjabrao Deshmukh Krishi Vidyapeeth, Akola, Maharashtra, India

K. N. Moharir
Department of Geology, Sant Gadge Baba Amravati University, Amravati, Maharashtra, India

S. Yadav et al. (eds.), *Water Quality, Assessment and Management in India*, Earth and Environmental Sciences Library, https://doi.org/10.1007/978-3-030-95687-5_16

interpolation method. These parameters are having poor groundwater quality due to highly pH values and This high pH values are directly impact on the human health as well as agricultural crops in the river basin area. The results can be helpful to development of sustainable water resources and agriculture yield under the basaltic hard rock area of Maharashtra at India.

Keywords GIS · Groundwater quality · Statistical analysis · Basaltic rock

1 Introduction

In India, approximately 80% of the rural communities and 50% of the urban people use the soil for internal agriculture determination. A 33% of the aquifers development in the region is not appropriate to human and irrigation uses. A significant number of anthropogenic pollutants have emerged over the past few years as severe groundwater activity in the aquifer [1–3]. "India is fast moving towards a crisis of groundwater overuse and contamination" [4]. Groundwater is the essential natural resource for agriculture development, human health, sustainable development and capable of environments in the India [5, 6]; further, it's procedures a basic natural resource to groundwater planning. "The hydro-geochemistry configuration of groundwater is controlled by hydrogeology, rock types, geochemical minerals, and anthropogenic activities, soils and crops, precipitation, geological structure" [7]. Water quality is an individual significant feature entire in the groundwater quality analysis. Here has marvelous requirement of fresh groundwater from human and industrial purposes. The chemical elements in groundwater such as (Ca^{2+}, Mg^{2+}, Cl^{-}, HCO_3^{-}, Na^{+}, K^{+} and SO_4^{-2}) of water composition represent an essential part in categorizing and measuring of groundwater quality. The estimation of water quality parameter has been necessary uses of high influenced of pH and ion concentrations for which groundwater suitability of the drinking, farmed and manufacturing drives [8, 9].

Consequently, water quality parameters are user in the physicochemical factors and these elements are primarily approaches by geological types, natural hazards, human-made activity, climatic change situations and anthropogenic happenings of any area in India [10–12].

Furthermore, the usage of insecticides and fertilizers increases the farming manufacture, but this outcomes harmful effect on water quality parameters in groundwater [13–16] According to [17] India day by days more harmful anthropogenic activities such as industrial wastes, open defecation, and untreated sewerage disposal has been included in the fresh groundwater and surface water quality. Due to the application of poor or hazardous quality water the agriculture land/soil is affected and damages the crop yield in several ways [18]. Its impact of tremendous variations in the groundwater quality parameters have been affected formulation, water cycle and social elements of the underground water recharging. in any country.

Groundwater quality factors had calculated by using several kinds of statistical scattering figures as Piper diagram [19–23] to recognize the spatial and temporal changes in the quality of the ground water taking place in the water table. The result is the stripe plot diagram is currently used to represent the statistical consequence of different parameters of groundwater quality. Groundwater evaluation was depending on a laboratory experiment; however, it was very easy to integrate several databases by the development of satellite technology and (GIS) [17, 24]. Groundwater analyses are studied of basin area and to identify suitable area to development of agriculture and consumption uses. In order to investigate the groundwater quality factors, thirty-five water data were gathered from the basin area. Samples of groundwater quality are verified for the estimation of major anions, which including chloride, nitrate, sulphate and phosphate; important cations like as sodium, potassium, calcium and magnesium; pH, electric conductivity (EC) and total dissolved solids (TDS). The water quality maps were made from groundwater quality data in the GIS software. The agriculture parameters of groundwater quality information are computed using different factors namely Kelly's ratio (KR) with magnesium hazard (MH) for irrigation purposes absent of rainwater under drought condition [25–28]. The suitable and unsuitable zones are found with the help of water quality maps. Piper's diagram, Giggenbach Triangle and Durov plots were helpful for understanding of water quality in the basaltic rock region.

2 Study Area

Study area is situated in between 76°46′11″E and longitude 20°40′36″N latitude at the districts of Akola and Buldhana of Maharashtra in India. The total basin area calculated was 328.25 sq. km^2. The elevation is approximately 598 m above MSL. Total 90% study area are deposited under unidentified sediment area. In study area observed was very high runoff zone and recharge zone [29]. The annual rainfall 750–900 mm was observed as per IMD metrological stations. In this study maximum area is rainfed zone with agriculture land depended on rainwater and groundwater. Suppose some time rainwater is coming as per time that time maximum agriculture land based on the groundwater. Hence groundwater quality analysis is more important factors in the development, planning and management of study area.

3 Geology, Climate and Drainage System

Geologically, in basaltic in Mahesh river consists of Ajanta and Chikhali formations of basaltic hard rocks [14]. In this study area, Alluvium, un-identified sediment area was found using field verification, satellite image, and geological map. The large section is bounded by basaltic rocks and the north side enclosed by little alluvium patches. Geomorphological, the Mahesh river basin is categorized into the mesa, butte, moderately dissected, pediment and, pediplain with alluvial plain [30]. The floodplain area was not noticed in the river basin and pediment, and pediplain were found along the southern and northern side of the basin area. Two types of soil namely black and red were recovered from the satellite image and soil map of Mahesh river basin. Drainage type is significant indications of the hydrology system. The stream network type has measured from toposheets, and satellite data are easily show to geomorphology landform and rock types with propose soil features and stream situation in Mahesh river basin [31].

4 Methodology

The groundwater samples from the study area were obtained after the process or scented in [32, 33]. In the course of the pre-post seasons, groundwater sampling was done. In the study area, a total of 35 groundwater samples were received. Groundwater locations were noted using a GPS device, and they are presented in Fig. 1. In pre-reinforced and dried polyethylene bottles, water samples were composed. The groundwater samples collected in the field area were analyzed for electrical conductivity (EC), pH, the standard procedures for the implementation are total solids dissolved (TDS), major cations and anions [4, 18, 32, 34]. The approaches to sampling are choose using aims and the area's geology, drainage and hydrogeology. During the pre-post monsoons, floor quality maps were prepared using techniques of spatial interpolation. Table 1 show the chemical quality data for pre-post monsoons (Table 1). Different methods and charts for the study and interpretation of the groundwater analysis data have been used.

5 Statistical Analysis

The measure for correlation is the easy correlation coefficient in those variables, which denotes that the single variable is sufficient to compute its other variable [35, 36]. This coefficient is used when the dependent (x) is affected by individual (y) or conversely the correlation between variables is determined between both variables [37, 38]. Correlation matrix of eight factors during pre-post monsoon period. The expression of the correlation coefficient (r) is expressed as

$$\mathrm{r} = \frac{\sum_{i=1}^{n} xi\, yi - \frac{\sum_{i=1}^{n} xi \sum_{i=1}^{n} yi}{n}}{\sqrt{\left[\sum_{i=1}^{n} x_i^2 - \frac{(\sum_{i=1}^{n} xi)2}{n}\right]}\sqrt{\left[\sum_{i=1}^{n} y_i^2 - \frac{(\sum_{i=1}^{n} yi)2}{n}\right]}}$$

Where, x and y are the variables, and n is the number of water elements. The r values range from -1 to +1; r = +1 displays the toughest positive linear correlation and r = 1 reveals strongest negative linear correlation [39].

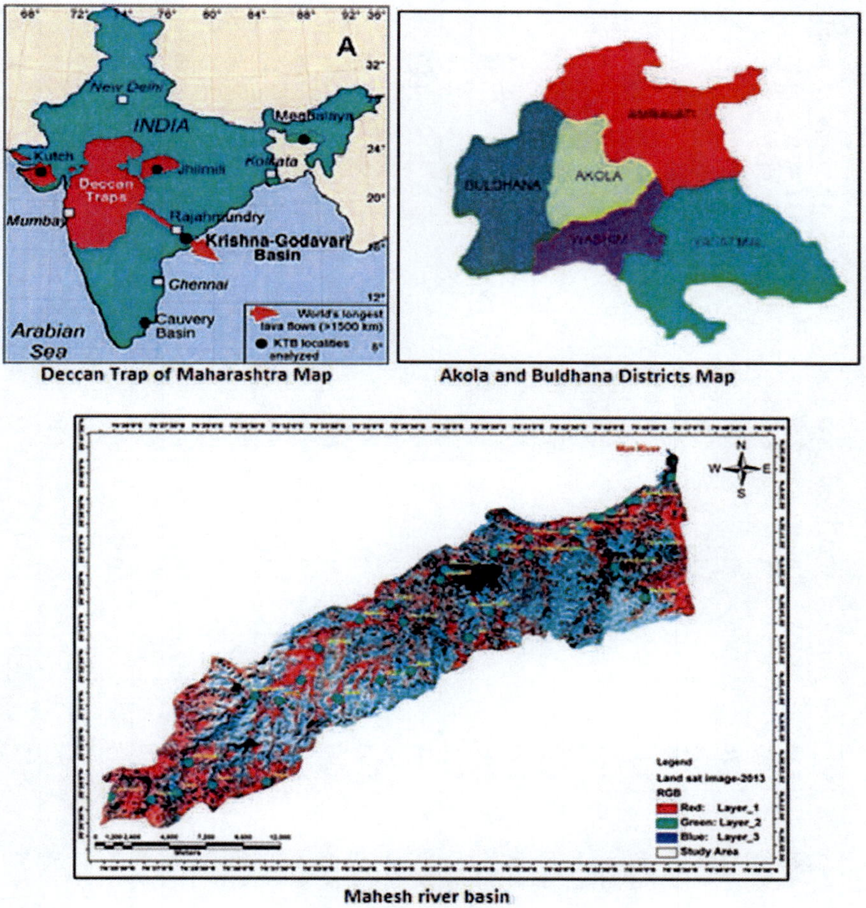

Fig. 1 Location map of the study area

6 Spatial Interpolation Techniques

The spatial interpolation method available in the software of Arc-GIS 10.3 has utilized to the analysis of the groundwater quality mappings at the un-sampled sites of the dry land zone area. Thematic maps have been generated through the IDW interpolation method, and that adopts water values that are adjoining to one another were added in the Mahesh basin area. The surrounding location has used for pre-post-post monsoon water quality prediction in the arid region to measure groundwater quality values. This study of most popularly used IDW interpolation methods has been compared to water quality values using the mapping of groundwater quality parameters and statistical analysis in the pre-post monsoon samples information.

Table 1 Water quality data in pre-monsoon

S. N.	Village Name	Ec μS/cm	pH	TDS meq/l	Ca meq/l	Mg meq/l	Cl meq/l	CO_3 meq/l	HCO_3 meq/l	SO_4 meq/l	Na meq/l	K meq/l
1	Balapur	879	7.6	378.35	4.44	3.13	1.55	0.13	2.61	0.65	1.48	1.02
2	UmraLasura	476	6.9	501.81	6.09	3.95	2.57	0	5.87	1.33	2.91	0.31
3	Sambhapur	1396	7.2	492.5	4.09	5.76	12.86	0	7.47	0.52	3.7	0.1
4	Shendri	867	7.8	392.66	4.14	3.71	7	0.79	7.51	0.85	1.83	0.64
5	HingnaUmra	1587	7.5	451.91	5.09	3.95	24.03	0.13	5.26	2.27	2.09	0.38
6	Ambikapur	934	7.2	364.73	3.59	3.7	3.47	0	5.83	1.12	1.83	0.1
7	Koregaon	986	7.4	426.91	3.89	4.65	1.35	0.23	2.41	0.83	2.09	0.38
8	Hingna	678	6.5	597.82	4.14	7.84	12.86	0.54	5.83	0.92	0.52	0.43
9	Nilegaon	745	7.6	447.87	4.09	4.87	11.02	0.18	5.26	1.6	1.96	0.26
10	Ramnagar	960	7.1	464.53	5.59	3.7	25.22	0.38	7.47	0.52	2.09	1.02
11	Wihigaon	789	6.5	309.7	2.25	3.95	11.71	0	4.23	2.91	3.87	0.38
12	Kherdi	869	6.1	509.99	5.09	5.11	11.71	0.31	5.87	3.14	3.39	0.26
13	UmraAtali	845	6.2	675.1	7.09	6.42	16.62	0	1.07	2.27	1.74	1.05
14	Naidevi	896	6.8	357.6	3.19	3.96	7.31	0	1.39	1.6	2.35	0.23
15	Lokhanda	845	6.3	387.54	3.79	3.96	15.63	0.18	4.23	3.93	1.52	0.26
16	Pala	845	6.9	587.14	7.78	3.96	1.95	0.23	1.39	3.14	3.87	1.92
17	Sirala	978	6.8	411.99	4.29	3.95	16.31	0	5.83	3.93	1.96	0.64
18	Ganeshpura	876	6.8	342.19	2.54	4.3	16	0.31	1.39	2.52	1.48	1.1
19	Wairagad	968	7.1	519.97	5.29	5.11	9.99	0	1.39	2.08	1.48	1.36
20	Undri	987	7.3	402.51	4.09	3.96	11.62	0.1	1.39	3.93	1.96	1.15
21	Dasala	987	6.3	379.7	3.89	3.7	12.86	0	8.36	3.93	1.96	0.38
22	Kinhi	812	7.1	514.96	4.54	5.76	9.06	0	1.39	2.52	1.48	0.23
23	Pimpri	945	7.1	412.21	4.54	3.7	3.55	0.31	5.87	2.58	1.96	0.26
24	Nirod	645	6.5	526.76	6.59	3.95	9.17	0	2.61	3.08	2.35	0.23
25	Gharod	605	6.3	476.86	5.59	3.95	4.32	0.13	1.39	3.93	1.96	0.38
26	Akoli	645	7.3	487.53	4.64	5.11	11.57	0.23	4.11	0.83	1.48	0.36

27	Atali	521	6.9	668.29	5.14	8.23	13.68	0.1	5.26	2.52	1.52	0.31
28	Patunda	548	7	620.88	4.19	8.23	12.86	0	1.39	2.94	3.31	0.26
29	Pedka	451	6.3	554.49	6.14	4.95	5.02	0	5.87	0.52	1.74	0.38
30	Kadmapur	354	6.8	307.34	2.45	3.7	11.62	0	7.47	2.58	2.83	0.36
31	Palsi Bk.	345	7	545.64	6.29	4.63	3.39	0.05	5.83	1.33	1.96	0.61
32	PalshiKh.	456	6.9	373.36	4.34	3.13	4.49	0	4.11	0.65	2.83	0.36
33	UmraLasura	345	6.3	469.66	4.44	4.95	1.27	0	5.87	3.08	1.74	0.23
34	Takarkhed-1	387	6.5	409.05	3.64	4.54	12.72	0	1.07	3.19	1.48	0.31
35	Takarkhed-2	412	6.9	409.78	3.24	4.95	20.9	0	1.97	0.83	1.48	0.26

6.1 *Inverse Distance Weighted*

Predictions by this deterministic method are based on a linear grouping of nearly situated ranges [40]. Defines IDW as

$$Z = \frac{\sum_{i=1}^{n} W_i Z_i}{\sum_{I=1}^{n} W_i}$$

$$W = \frac{1}{d_i^p}$$

Where z is interpolated number, z_i is called range, n is whole value helpful in interpolation, di is the distance called and interpolated numbers and p is power factor where weight reductions as distance rises from the interpolated numbers. Outcomes of IDW Interpolation approaches are very particular by weighting power [41]. This interpolated technique has usually used, and defaulting power in the ArcGIS software, i.e., the inverse of the distance has higher to second power point.

7 Results

7.1 *Groundwater Quality for Drinking Purpose*

In this chapters, the hydrogeochemistry analysis has referred to as the geochemical properties and values which control the performance of dissolved chemical components in the groundwater. The dissolved components are investigated as ions, molecules or solids. The substantial product of rock weathering and changes in time and space has linked to the chemical composition of ground water. Therefore, its utility for internal, manufacturing and irrigation uses determines its variation in the levels of concentrations of different hydrogeochemical constituents. This field shows, along with the calculated parameters for hydrogeochemical data, the minimum, highest and average values of the pre-post monsoon ground water quality data is presented in Table 2.

8 Groundwater Chemistry

Table 2 shows that the pH is between 6.1 and 7.8, 6–7.9 and the EC is between 345 to 1587, 348 to 1456 µS/cm in the values of river basin area. Calcium is changed from 25 to 173 mg/l and Mg from 27 to 133 mg/l; sodium is from 6 to 86 mg/l, and potassium from 2 to 73 mg/l is varied. The total value of dissolved solids is from

Table 2 Water quality parameters (min, max, and avg.) compared with BIS (10,500 Limits 2012) and WHO (2014) standards

	Pre-monsoon			Post-Monsoon				
Parameters	Min	Max	Avg.	Min	Max	Avg.	BIS Stds. (2012)	WHO Stds. (2011)
EC	345	1587	767.54	348	1456	809.09	–	1500
pH	6.1	7.8	6.88	6	7.9	7.07	6.5–8.5	7.5
TDS	307.34	675.1	462.27	266.9	717.05	470.44	500	500
Ca	2.25	7.78	4.58	2.05	6.79	4.76	75	75
Mg	3.13	8.23	4.67	2.96	9.15	4.66	30	30
Cl	1.27	25.12	10.21	0.92	22.15	7.71	250	250
CO_3	0	0.79	0.13	0	0.64	0.18	–	–
HCO_3	1.07	8.36	4.18	1.64	8.97	3.92	–	300
SO_4	0.52	3.93	2.13	0.69	4.21	2.27	150	500
Na	0.52	3.87	2.12	0.52	3.87	1.83	–	200
K	0.1	1.92	0.51	0.23	1.99	0.51	10	20
Kelly Ratio	0.04	0.62	0.25	0.06	0.42	0.20	–	–
Magnesium hazard	33.73	60.26	50.60	37.47	68.59	50.85	–	–

787 to 1456 mg/l. The study varies from 8 to 94 mg/l in sulphate concentration. The concentrations of bicarbonate and carbonate range in ground water vary between 130 and 509 mg/l and between 38 and 222 mg/l. Particles of chloride vary between 21 and 207 mg/l/. The alkalinity varies between 36 to 144 mg/l and the hardness ranges between 120 and 596 mg/l. Tables 1 and 3 is shown the largest ions of abundance in groundwater.

9 Correlation Analysis

Tables 4 and 5 show the analysis of the correlation matrix of groundwater parameters for the pre-post monsoon. A similar trend is reflected in the correlation between the parameters of water quality in post-monsoon seasons. Significant ($r > 0.68$) to slightly less significant ($r < 0.68$) correlation was found among the analyzed of water quality parameters.

10 Physical Parameters

10.1 pH

The study of pH is an essential factor that determines the bicarbonate and carbonate process in the freshwater. PH is assumed a biodiversity component and is caused by

Table 3 Water quality data in post-monsoon

S. N.	Village Names	Ec µS/cm	pH	TDS meq/I	Ca meq/I	Mg meq/I	Cl meq/I	CO_3 meq/I	HCO_3 meq/I	SO_4 meq/I	Na meq/I	K meq/I
1	Balapur	1004	7.6	377.61	4.59	2.96	2.59	0	2.46	2.56	2.91	0.26
2	UmraLasura	697	6.8	464	5.99	3.29	2.65	0	6.05	1.77	3.87	0.36
3	Sambhapur	1345	7.6	538.23	4.99	5.77	11.58	0	2.02	1.17	2.35	0.36
4	Shendri	829	7.2	390.62	4.09	3.72	9.36	0	2.41	0.96	0.52	0.38
5	HingnaUmra	1456	7.8	419.09	5.09	3.29	22.15	0.43	8.52	1.92	2.04	1.15
6	Ambikapur	1000	7.8	394.67	4.19	3.7	3.6	0	6.2	0.85	2.09	0.38
7	Koregaon	857	7.9	471.01	4.74	4.68	1.61	0.36	2	1.52	1.78	0.46
8	Hingna	876	7.2	617.4	4.44	7.91	7.63	0.05	5.26	1.35	2.04	0.36
9	Nilegaon	789	7.5	445.17	4.79	4.11	9.25	0.31	2.52	0.69	0.52	0.31
10	Ramnagar	987	7.3	497.63	6.79	3.17	21.46	0.43	2.02	1.17	1.09	0.36
11	Wihigaon	854	6.8	266.9	2.05	3.29	6.94	0.23	2.59	2.54	1.78	0.26
12	Kherdi	888	6.2	606.67	6.79	5.35	10.03	0.38	8.97	2.75	1.74	0.23
13	UmraAtali	842	6.7	589.81	6.04	5.76	10.31	0	1.98	1.92	2.44	0.89
14	Naidevi	802	7	381.93	3.09	4.54	5.76	0.38	1.64	1.64	1.96	0.38
15	Lokhanda	926	7	471.75	4.89	4.54	10.2	0.31	2.59	3.02	1.22	0.38
16	Pala	926	7.1	564.07	6.74	4.54	1.73	0.36	2.05	2.75	1.78	1.15
17	Sirala	867	6.3	414.1	4.99	3.29	14.41	0.23	6.2	3.02	1.74	0.38
18	Ganeshpura	880	6.9	491.75	3.24	6.59	13.14	0.13	1.64	4.21	0.96	1.15
19	Wairagad	997	7.2	539.3	5.44	5.35	6.19	0.31	1.64	1.89	2.91	1.23
20	Undri	1289	7.9	429.34	4.04	4.54	9.36	0.08	2.05	3.02	1.74	1.99
21	Dasala	1024	6.8	355.41	3.94	3.17	11.87	0	4.23	3.02	2.91	0.61
22	Kinhi	902	7.1	518.27	4.59	5.77	9.39	0	1.64	2.56	1.48	0.36
23	Pimpri	989	7.1	419.68	4.59	3.8	1.87	0.64	4.16	1.64	1.31	0.38
24	Nirod	682	6.6	439.05	5.49	3.29	2.88	0	2.02	3.83	0.96	0.38
25	Gharod	682	6.3	419.09	5.09	3.29	1.3	0	2.05	3.02	1.09	0.36
26	Akoli	702	7.6	516.85	4.99	5.35	11.87	0.36	8.9	1.52	1.48	0.82

27	Atali	536	6.2	717.05	5.19	9.15	9.94	0.13	2.52	4.21	1.22	0.26
28	Patunda	463	7.5	667.15	4.19	9.15	3.34	0	2.05	4.16	3.09	0.23
29	Pedka	483	6.6	553.71	6.14	4.94	3.6	0.31	8.97	1.17	2.44	0.36
30	Kadmapur	348	7	275.57	2.35	3.17	11.52	0	2.02	1.64	1.22	0.23
31	Palsi Bk.	378	7.5	521.09	6.29	4.13	2.42	0.23	5.26	1.77	1.74	0.54
32	PalshiKhurd	623	6.9	370.12	4.44	2.96	3.54	0.38	8.9	2.56	1.22	0.23
33	UmraLasura	387	6	491.33	4.89	4.94	0.92	0.31	6.05	3.27	2.44	0.38
34	Takarkhed-1	421	7.2	415.92	3.79	4.53	3.6	0	1.98	2.52	0.96	0.26
35	Takarkhed-2	587	7.2	413.99	3.34	4.94	11.87	0	5.82	1.52	2.91	0.23

Table 4 Details of pre-monsoon correlation coefficient matrix

Parameters	EC	pH	TDS	*Ca*	Mg	Cl	CO_3	HCO_3	SO_4	Na	K
EC	1	–	–	–	–	–	–	–	–	–	
pH	**0.361**	1	–	–	–	–	–	–	–	–	
TDS	−0.097	−0.187	1	–	–	–	–	–	–	–	
Ca	0.192	−0.131	0.736	1	–	–	–	–	–	–	
Mg	−0.117	−0.147	**0.752**	0.107	1	–	–	–	–	–	
Cl	0.808	−0.058	0.023	−0.185	0.214	1	–	–	–	–	
Co_3	**0.772**	**0.295**	−0.283	−0.006	0.006	0.061	1	–	–	–	
HCO_3	0.173	0.049	−0.143	−0.087	−0.125	0.067	**0.251**	1	–	–	
So_5	0.430	−0.480	−0.039	−0.015	−0.043	0.113	−0.223	−0.226	1	–	
Na	0.464	−0.054	−0.052	0.078	−0.153	−0.134	−0.219	0.098	0.117	1	
K	0.624	0.143	0.157	**0.390**	−0.150	0.020	0.210	−0.340	0.085	0.018	1

Table 5 Details of post-monsoon correlation coefficient matrix

Parameters	EC	pH	TDS	*Ca*	Mg	Cl	CO_3	HCO_3	SO_4	Na	K
EC	1	–	–	–	–	–	–	–	–	–	
pH	0.423	1	–	–	–	–	–	–	–	–	
TDS	0.078	−0.132	1	–	–	–	–	–	–	–	
Ca	0.078	−0.152	**0.601**	1	–	–	–	–	–	–	
Mg	−0.170	−0.057	0.818	**0.032**	1	–	–	–	–	–	
Cl	0.446	0.120	−0.018	−0.028	−0.003	1	–	–	–	–	
Co_3	0.148	−0.003	0.040	0.288	−0.157	0.083	1	–	–	–	
HCO_3	−0.082	−0.134	**0.034**	0.257	−0.141	0.084	**0.335**	1	–	–	
So_4	−0.167	−0.460	0.220	−0.065	**0.321**	−0.108	−0.173	−0.200	1	–	
Na	0.072	**0.048**	0.156	0.103	0.122	−0.164	−0.215	0.212	−0.023	1	
K	**0.497**	0.320	0.087	0.104	**0.033**	0.216	**0.111**	−0.109	**0.131**	**0.057**	1

interface between many substances in liquid solution as well as various biological occurrences. Low pH is one of the common problems of groundwater. The underlying reasons for the lowest pH are totaling of acid rainwater. Further ions like nitrates and sulfates observed in groundwater can show lower pH. The adverse effects of acidic water are many. Especially acidic water can lead to corrosion of the pipe and may cause iron, lead or copper to be released into the tap water. A low pH can decolorate and taste the water bitterly. As most of them grow in a very smaller and critical pH range, pH is indication of presence of biological life. pH is record tributaries unaffected in the human pH values in between 6.5 to 8. Soils of the western area, which is alkaline is normal provider to high pH situations. People donate higher pH primarily in the form among the most frequently used fertilizer (nutrient rinse) which denotes the growth of algae and pH improvement. The low pH of aquatic organisms can be especially harmful. The low pH impacts the features of marine life (physiological, biological) by decreasing enzymatic activity and water system miles in usefulness. The pH ranges show 7.5 to 7.9 under in the excellent water quality. pH samples have collected in the river basin area as per the norms of standards to water drinking. pH ranges were varied in between 6.1–7.8 and 6–7.9 respectively, six is low value obtained from the central part of the basin and the higher value of 7.9 was recorded in the upper part of the basin show a little acidic and alkaline nature of quality of the groundwater. For Pre-monsoon, minimum pH (6.1) was observed in the Kherdi and Umra Atali villages while maximum pH (7.46) values were found in the Balapur, Shendri and Nilegaon villages (Fig. 2a and b). For Pre-monsoon, minimum pH (6) values were seen in the Khardi, Atali and Umaraatali villages while maximum pH (7.9) values were found in the Undri, Ambikapur and Koregaon villages.

10.2 *Electrical Conductivity (EC)*

In this study show EC depend on the ionic concentration, various kinds of ions and temperature in the groundwater [34]. The EC changes from 345 to 1587 and 348 to 1456 μS/cm in the pre-post monsoon seasons. The maximum EC values have observed in the HingnaUmra and the low values of EC are reported in Kadmapur, Palsi Bk. and Atali during the pre-post monsoon seasons (Fig. 3a and b). Therefore, thirty-five water samples were examined according to WHO values and classification (1500 μs/cm) [42].

10.3 *Total Dissolved Solids (TDS)*

The total dissolved solids (TDS) are calculated by (meq/l) and are melted in groundwater resources. The different types of Ions are donated to suspended items in groundwater of study area. Resource departments in so many occurrences are

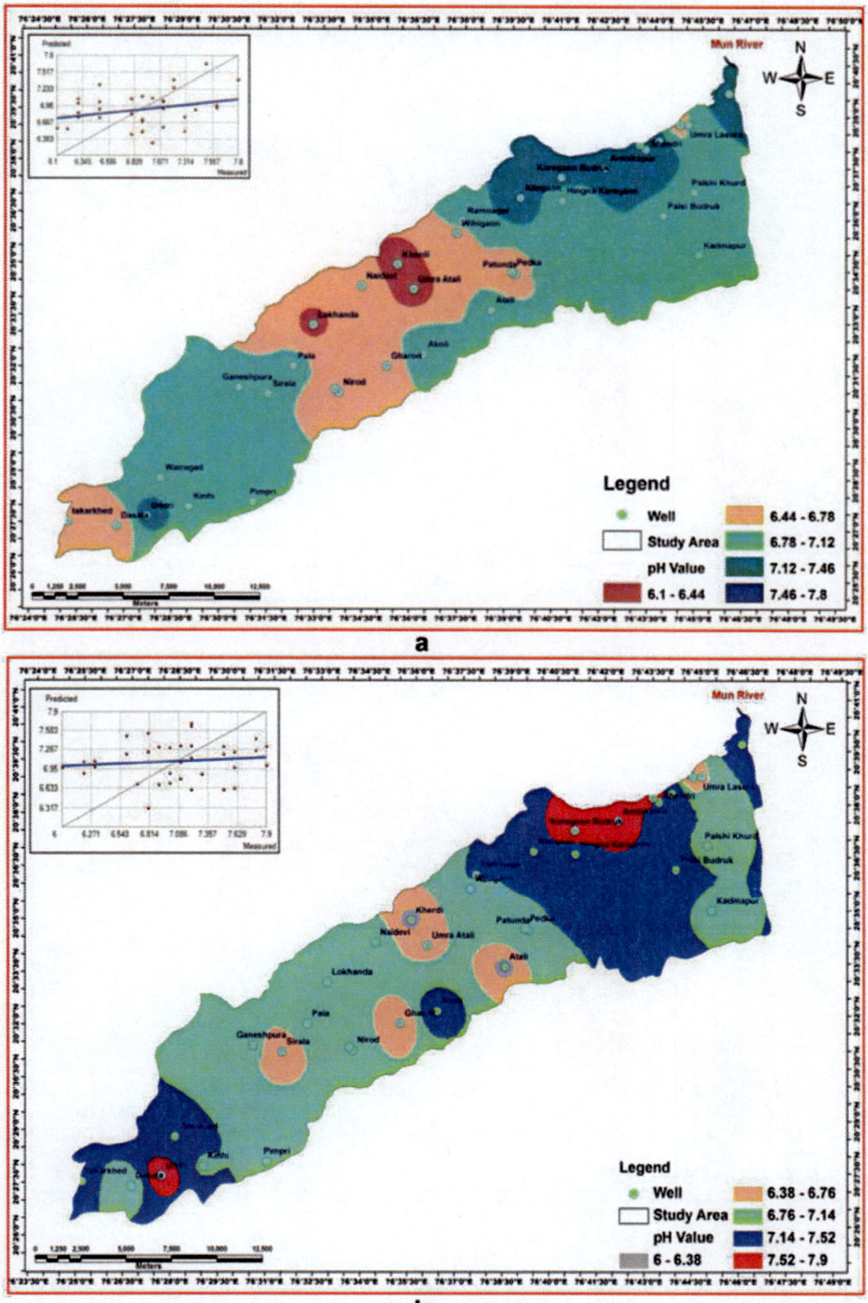

Fig. 2 (**a**) Water Quality map of pH in pre-monsoon. (**b**) Water Quality map of pH in post monsoon

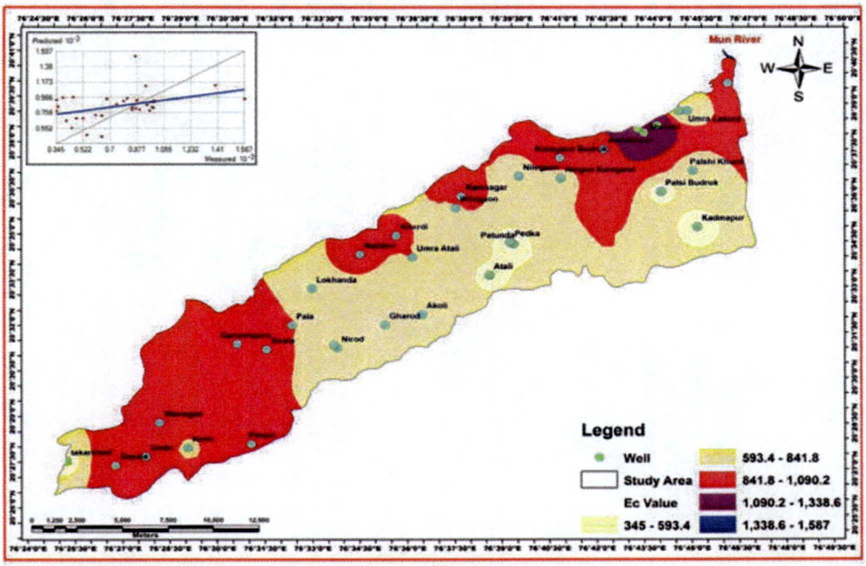

a

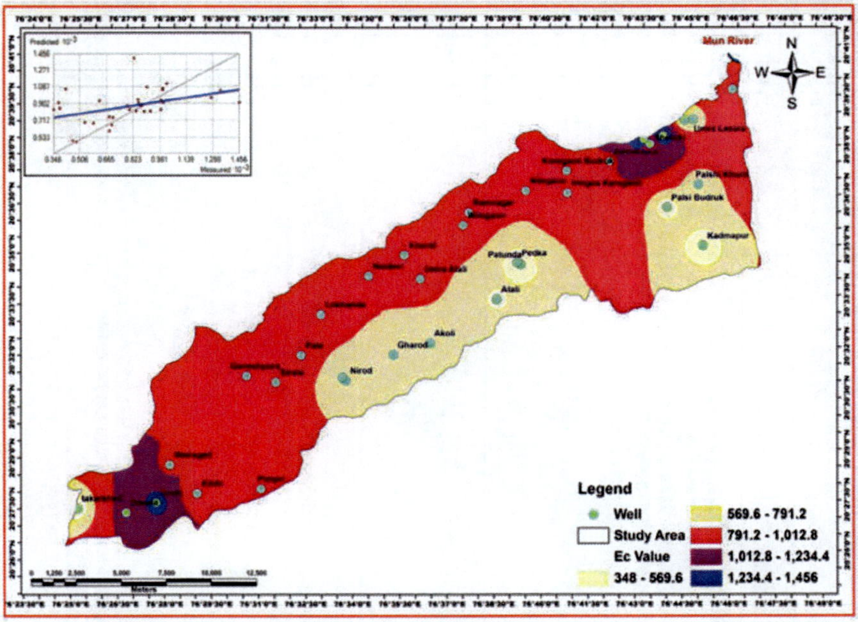

b

Fig. 3 (**a**) Water Quality map of Ec in pre monsoon. (**b**) Water Quality map of pH in post monsoon

Table 6 The concentration (in ppm.) of TDS classified water

TDS	Water Quality
<1000	Fresh
1000–10,000	Brackish
10,000–1,00,000	Saline
>1,00,000	Brine

using the determinate TDS and salinity exchangeable because all these ions are normally in the salt factor techniques. This is an essential factor for drinking water because higher TDS values may output in a 'salty' taste to the groundwater. TDS analysis, salt and suspended particles are included under the water. Salts contain chemical mixtures in natural groundwaters, including anions like carbonates, chlorides, sulphates and nitrate and cations such as potassium (K), magnesia (Mg), calcium (Ca) and sodium (Na). The research shows the environmental conditions, in proportions which create a balanced solution, these compounds are currently in river basin area. It's the concentration in water of molecular and colloidal dispersion of non-volatile factors. Due to the movement of water surface and subsurface, it includes a range of different quantities of inorganic dissolved chemicals. The TDS-classified water concentration (in ppm) seen in Table 6.

It is an obtained by concentration of Na, Cl which calculates the conductivity which in significantly influence TDS. The ppm range of TDS is from 250 to 0.55. There's a value of 0.90 [35] is provide the connection of the TDS groundwater ranges as 100–5000 micro moho's/cm at 250c. 1 ppm. TDS = 1.56 EC x 106. TDS value is 307.34 to 675.1 and 266.9 to 717.05 in the pre-post periods of 2014 (Fig. 4a and b).

10.4 Chemical Parameters

In this research continuing groundwater water dissolved solids contain mostly of inorganic salts like bicarbonate, phosphate, chlorides, carbonates, and sulphates etc. and a low value of carbon-based material and liquified gases element computed with the help of water quality parameters investigation.

10.5 Calcium (Ca)

Ca is a cation in water and is one of the soil alkaline metals dispersed in the ground's outside. With the occurrence of CO_2, calcium bicarbonates usually can be produced at a heavy atmosphere of about 20 ppm and a high pressure of 100 ppm. In water approaching from limestone area the absorption may be larger. Calcium values are equalised with chemical analysis in the water sample of basin area. Pre-post Ca value observations in the basin of the Mahesh river ranges from 2.25 to 7.78, from 2.05 to

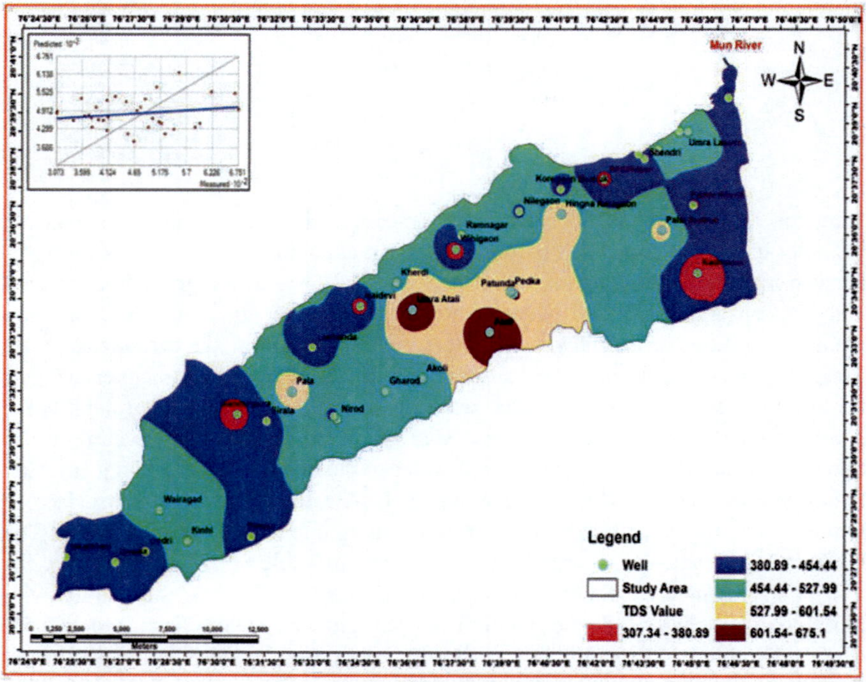

a

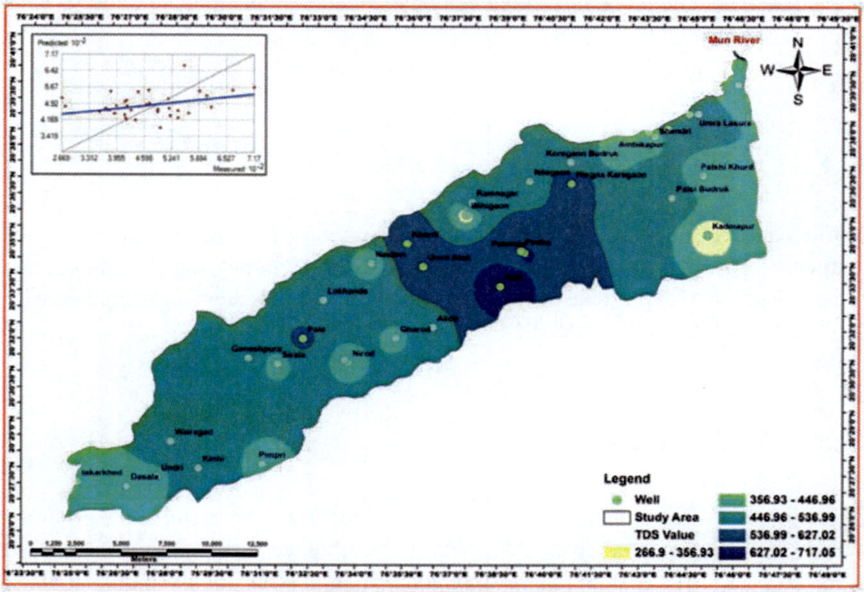

b

Fig. 4 (**a**) Water Quality map of TDS in pre monsoon. (**b**) Water Quality map of TDS in post monsoon

6.87 respectively in pre–post monsoons. For the pre-monsoon higher values were observed of Umra Atali, Pala and Nirod villages and then lower values were observed of Kadampur, Nilegaon, Koregaon and Ganeshpur villages, while respectively in the post-monsoon larger values were reported of Kherdi, UmaraAtali, and Palsi Budruk while smaller numbers were observed in Fig. 5a and b.

10.6 Chloride (Cl)

In the present research Chloride is the component halogen group and the geochemical scenario determine the concentration in all-natural water. The natural chloride-containing water has supported the adsorption of soil and chloride that included roach. This surface may lead to local pollution of the surface area, with 250 mg/litreous chloride concentration and ions of sodium. In the absence of sodium ion also the chloride concentration is up to 1000 mg/liter, salty taste can be unavailable in water. The Cl ranges are shown in between 1.27 to 25.22, 0.92 to 22.15 during the pre-post period of river area (Fig. 6a and b).

10.7 Magnesium (Mg)

Mg and Ca are the both elements mostly accountable for inflexibility water. The necessary limit of magnesium is 30 mg/Lit. in normal groundwater. The Mg values were observed in between 3.13 to 8.23 and 2.96 to 9.15 for pre-post periods respectively. The representation of Mg maximum value (9.07) was reported in Atali and Pedka villages while minimum amount (2.88) has found in Balapur, Dasala, Ramanagr, Undri, and Kadampur villages during pre-monsoon in the Mahesh river basin. The observation of Mg maximum value (9.15) was observed in Atali and Pedka villages while minimum amount (2.96) were found in Balapur, Dasala, Palshi Bk. villages in post-monsoon (Fig. 7a and b).

10.8 Sodium (Na)

Sodium has been unconfined due to weathering of plagioclase feldspar and amphiboles. Na values is 0.52 to 3.87 and 0.52 to 3.87 during pre-post monsoon. Sodium distribution map shows the greater, it is towards to eastern and lower ranges reported with South-West in area. The concentration of Ca, Mg, Na, SO4 and their distribution plotted in piper trilinear plots of the study area. The observation of the Na higher value (3.87) was identified in Umra Lasura, Palshi, Kadmapur, Shendri, Pedka villages while minimum value (0.52) has observed in Hingna Karegaon villages in pre-monsoon period, while minimum value (0.52) has detected in the Takarkhed village for post season of the Mahesh river basin (Fig. 8a and b).

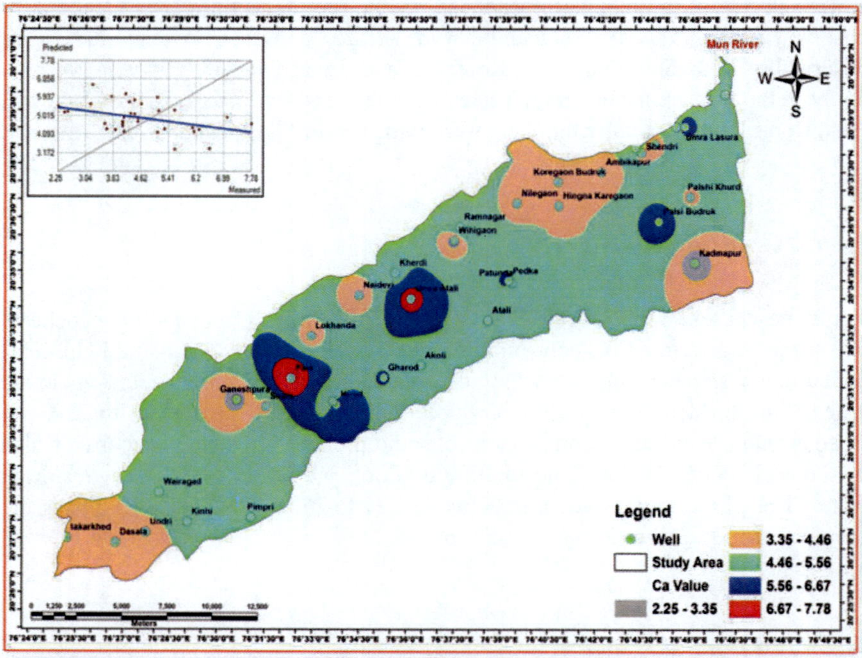

a

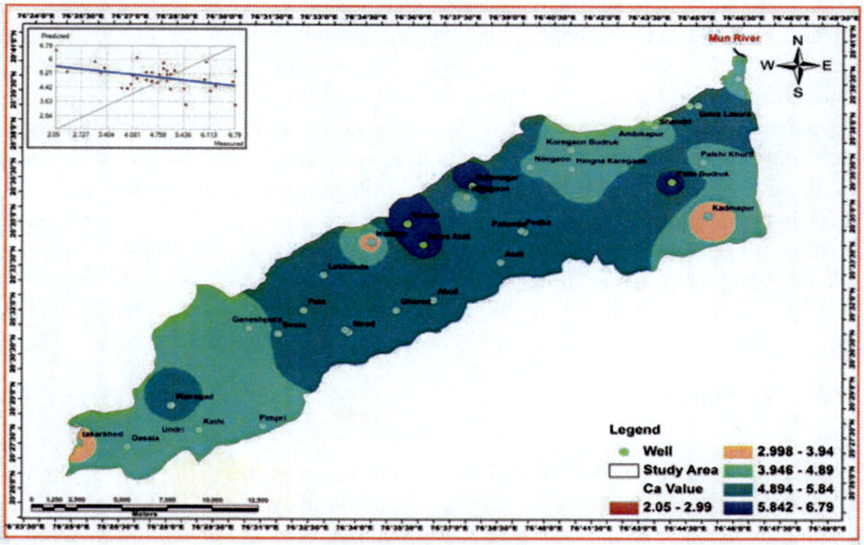

b

Fig. 5 (**a**) Water Quality map of Ca in pre monsoon. (**b**) Water Quality map of Ca in post monsoon

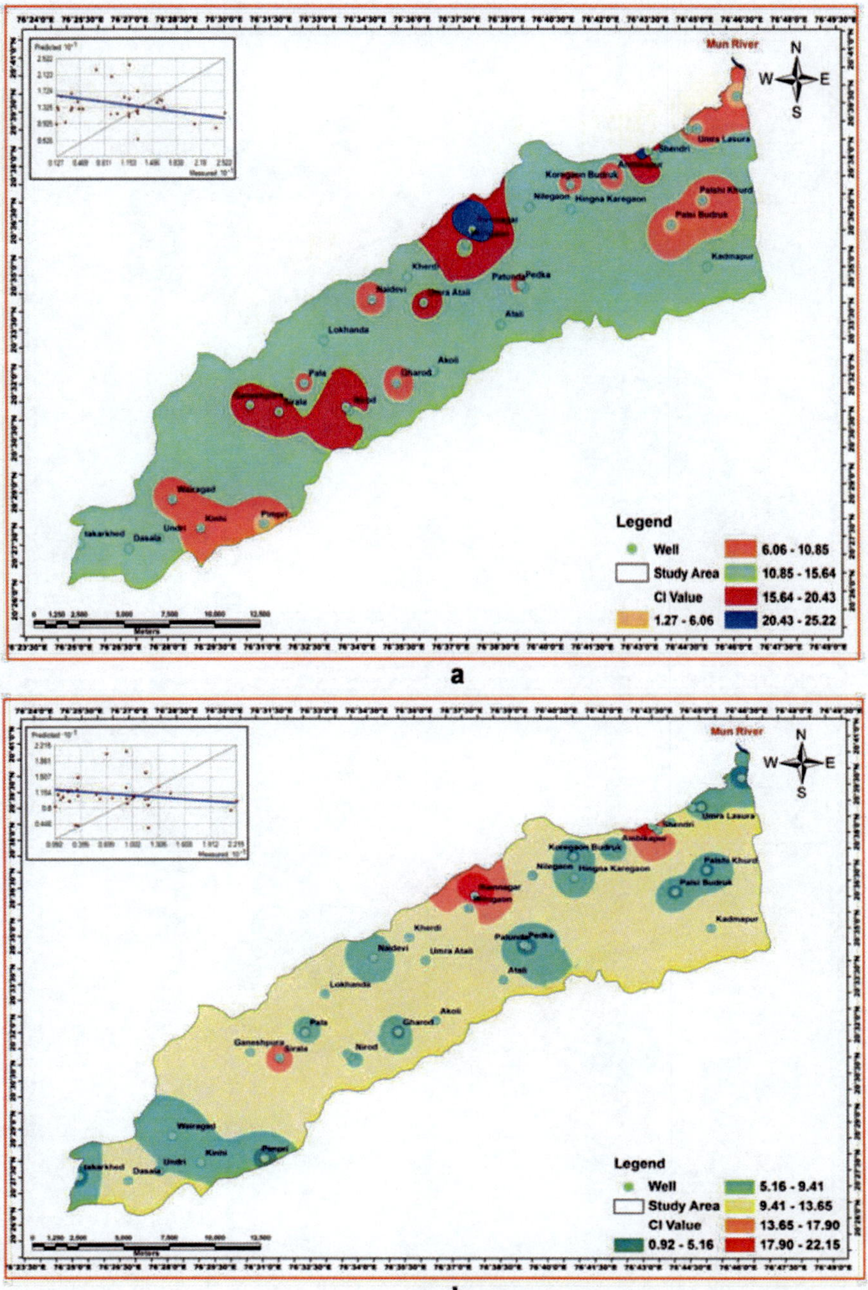

Fig. 6 (**a**) Water Quality map of Cl in pre monsoon. (**b**) Water Quality map of Cl in post monsoon

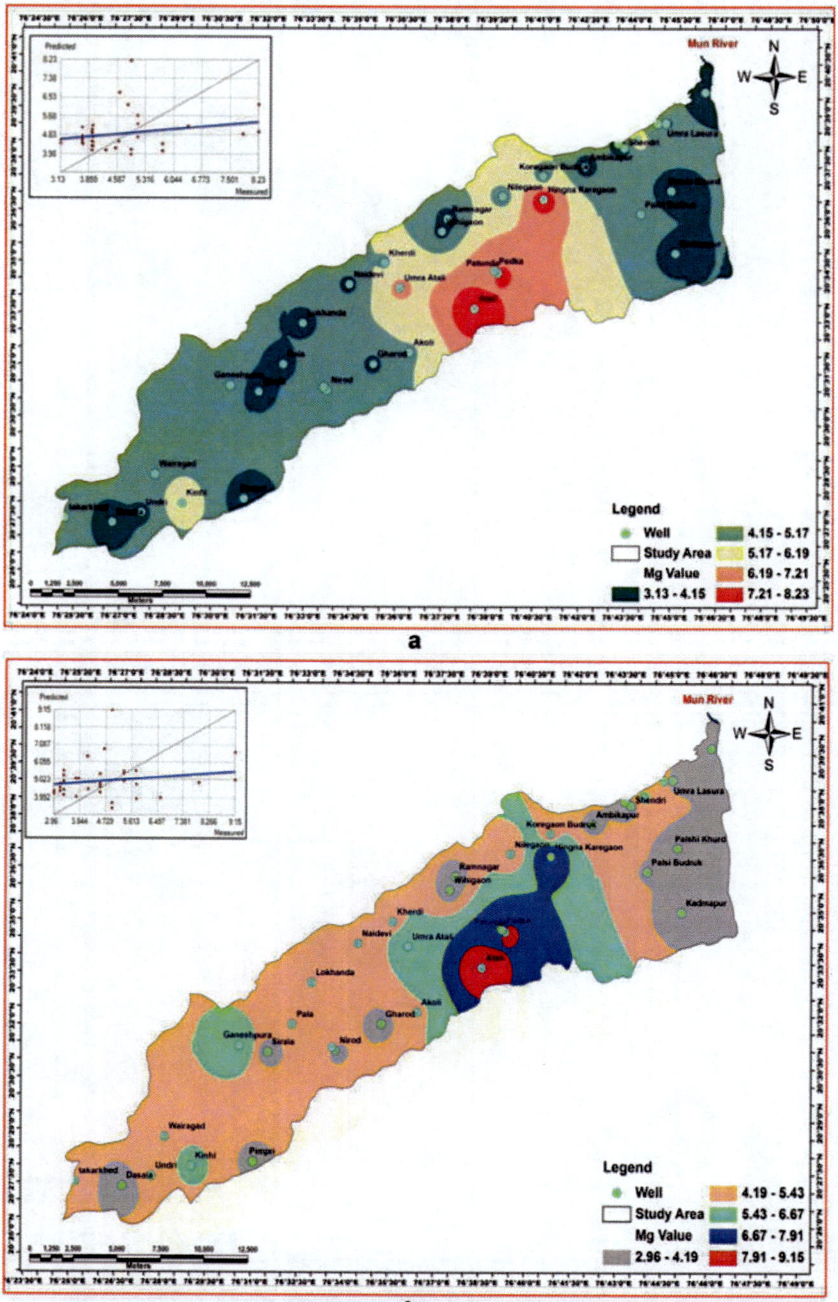

Fig. 7 (**a**) Water Quality map of Mg in pre monsoon. (**b**) Water Quality map of Mg in post monsoon

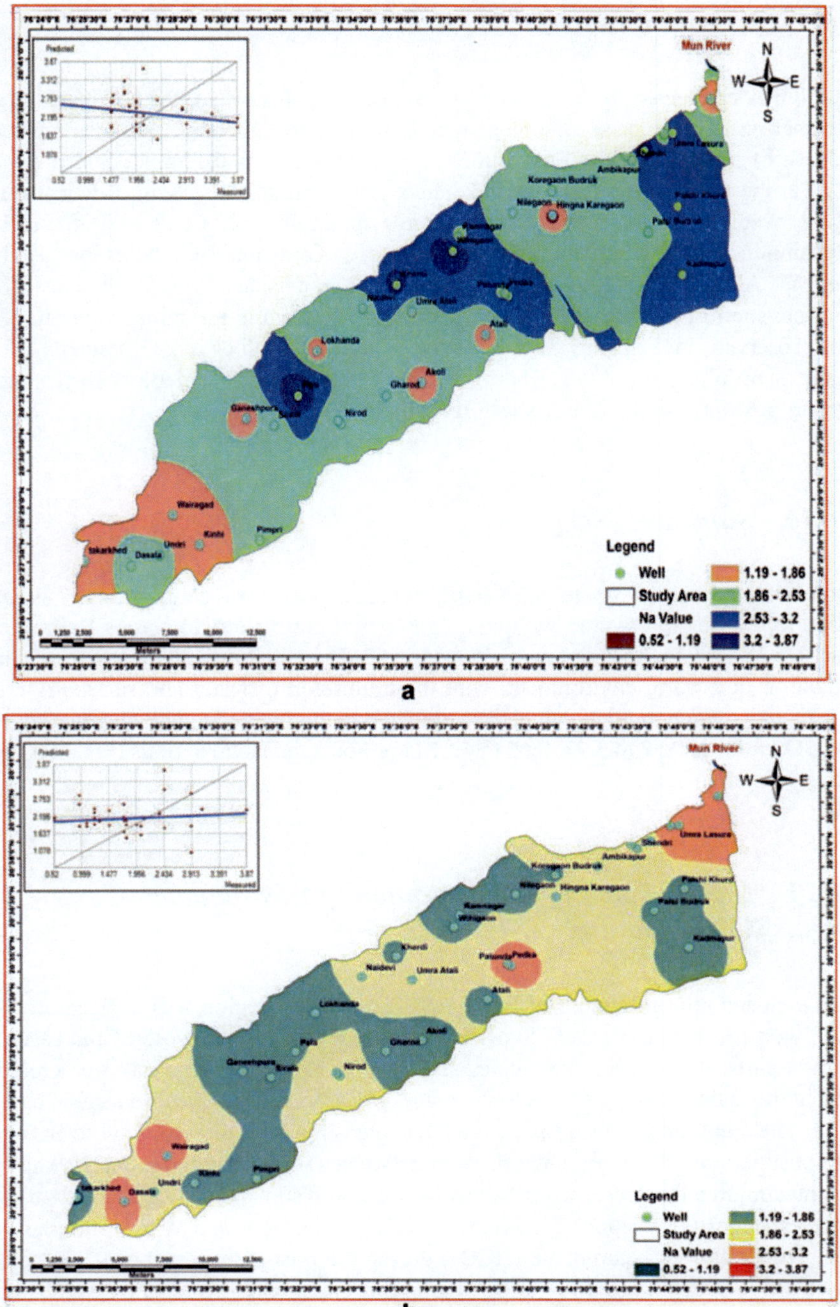

Fig. 8 (**a**) Water Quality map of Na in pre monsoon. (**b**) Water Quality map of Na in post monsoon

10.9 Potassium (K)

Sodium is connected to the study of potassium in many ways. In the ground, the weathering of orthoclase, microcline and strain level igneous rocks has created release of potassium. The Potassium value is observed as 0.1 to 1.2 and 0.23 to 1.99 in the pre-post-monsoon period. The representation of the K maximum value (1.92) were observed in Pala villages while minimum value (0.1) was found in Kadampur, Palshi Kh., Umra Lasura, Shendri, Koregaon Bk., Nilegaon, Atali, Pedka, Akoli, Kherdi, Naidevi, Gharod, Nirod and Pimpri villages for pre-monsoon of the area. While the Dasala villages while minimum value (0.23) has observed in Balapur, Umra lasura, Kadmapur, Nilegaon, Koregaon Bk., Ramnagar, Wihigaon, Atali, Kherdi, Naidevi, Nirod, and Palshi Bk. villages, etc. during post-monsoon in the Mahesh river basin (Fig. 9a and b).

10.10 Sulphates (SO_4)

At concentrations of up to 4.50 mg/l, sulphates occur in natural water. When 1000 mg/lit has discovered with some geological formations in waters Rainwater has significantly higher sulphate stages, specifically in regions with high levels of air pollution in a warm environment, with the infiltration of water and surface rivers, which can quickly release sulphates from the weathering zone. The observation of the SO_4 value is 0.52 to 3.93, 0.69 to 4.21 in pre-post monsoon periods (Fig. 10a and b).

10.11 Classification of Concentration of Bicarbonate + Carbonate Ions

The concentration minimum of HCO_3 within the study region was analyzed to be 1.03 meq/1 in a clustering of "Super carbonated Water". In the western and eastern central parts of the basin, carbonates and bicarbonates were reported to be greater and in the main lower area to be lower values of carbonate and bicarbonates in basin area. The study of the bicarbonate (HCO_3) values are 1.07 to 8.36, 1.64 to 8.97 in pre-post periods. The representation of bicarbonate (HCO_3) the pre-monsoon shows the maximum values were identified in Dasala and Kadmapur villages and then the smaller values were reported in Kinhi, Naidevi, Ganeshpur and Wairagad villages. The observation of bicarbonate (HCO_3) during the post-monsoon shows the maximum value was observed in Atali, Kherdi and Palshi BK. while lower value was observed in Kinhi village. The observations of bicarbonate (HCO_3) post monsoon maximum area is in moderate values (Fig. 11a and b).

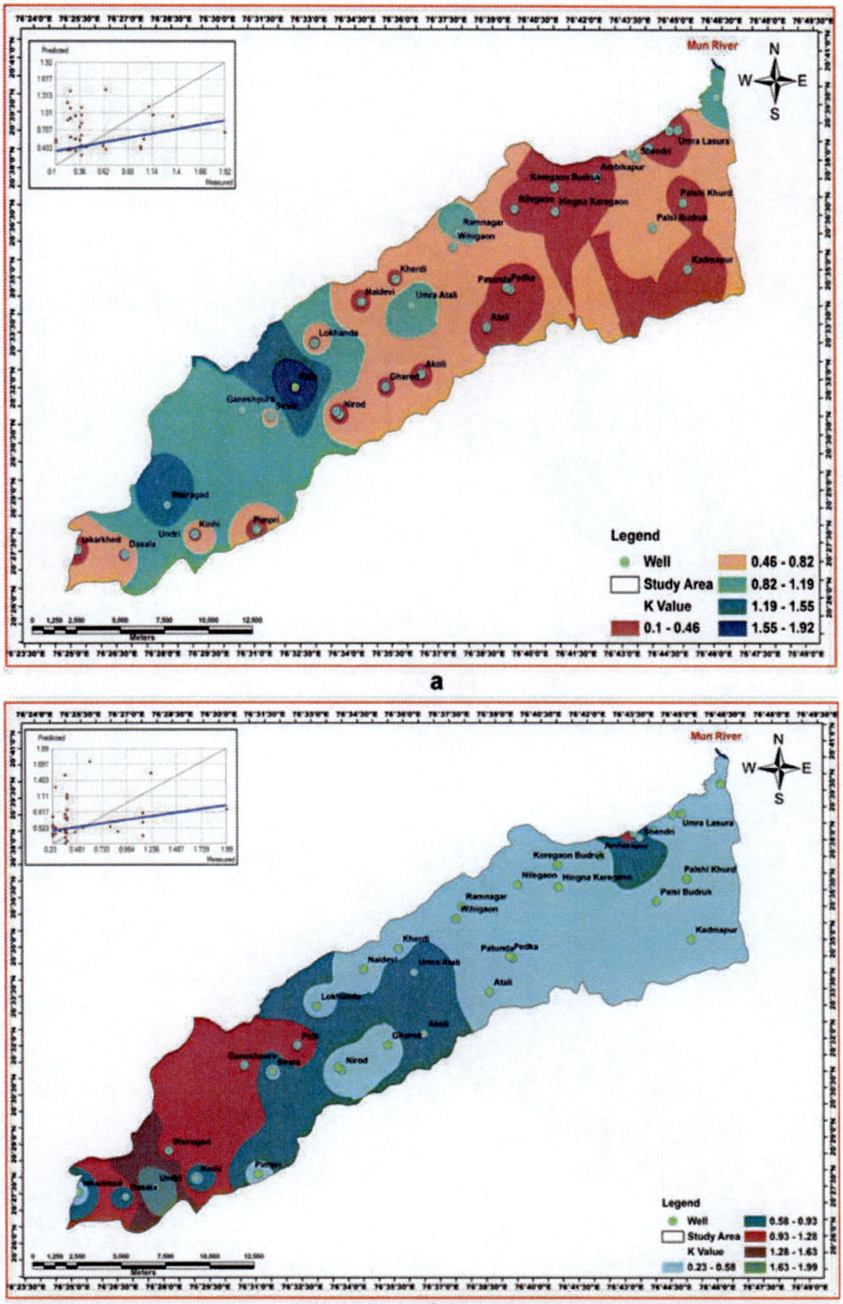

Fig. 9 (**a**) Water Quality map of K in pre monsoon. (**b**) Water Quality map of K in post monsoon

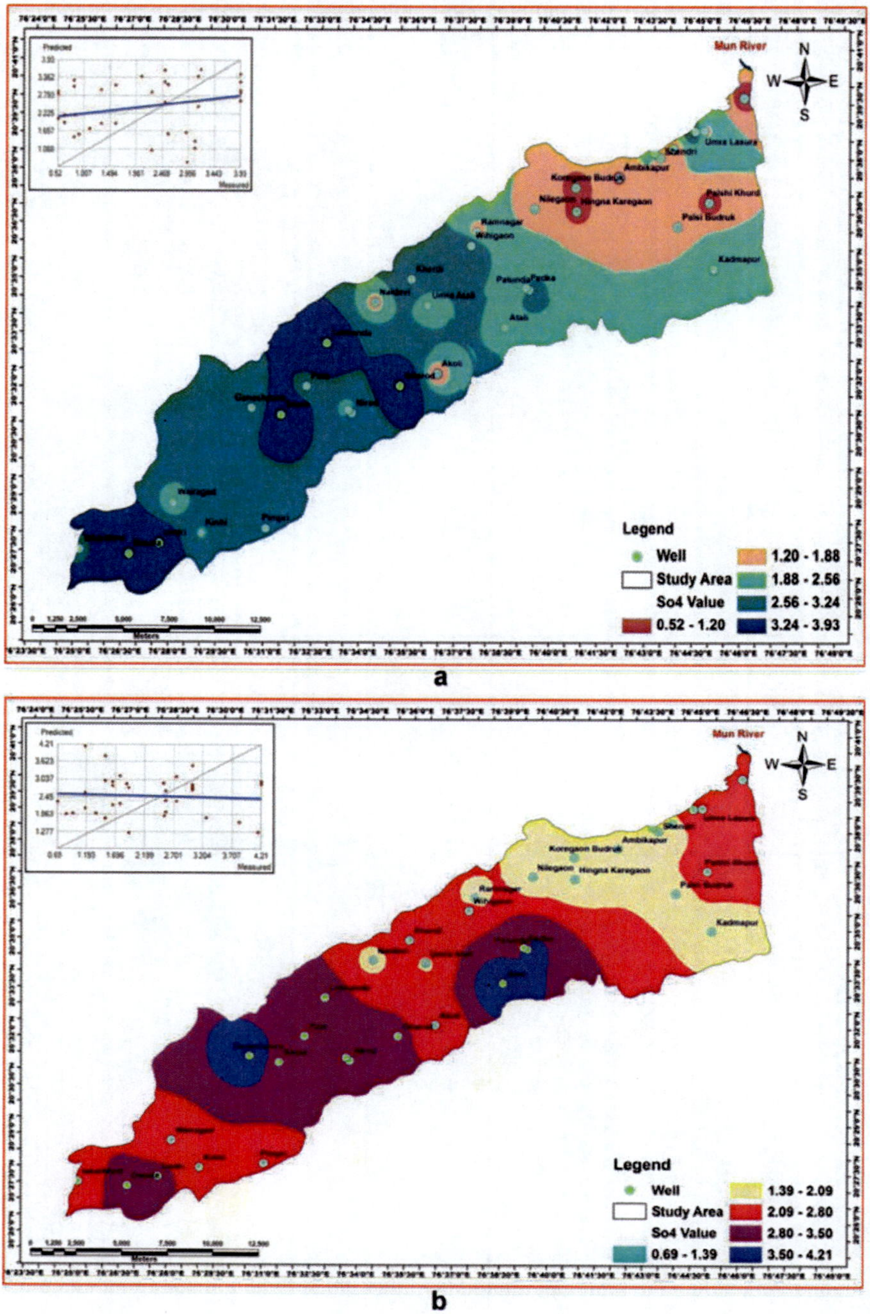

Fig. 10 (**a**) Water Quality map of SO_4 in pre monsoon. (**b**) Water Quality map of SO_4 in post monsoon

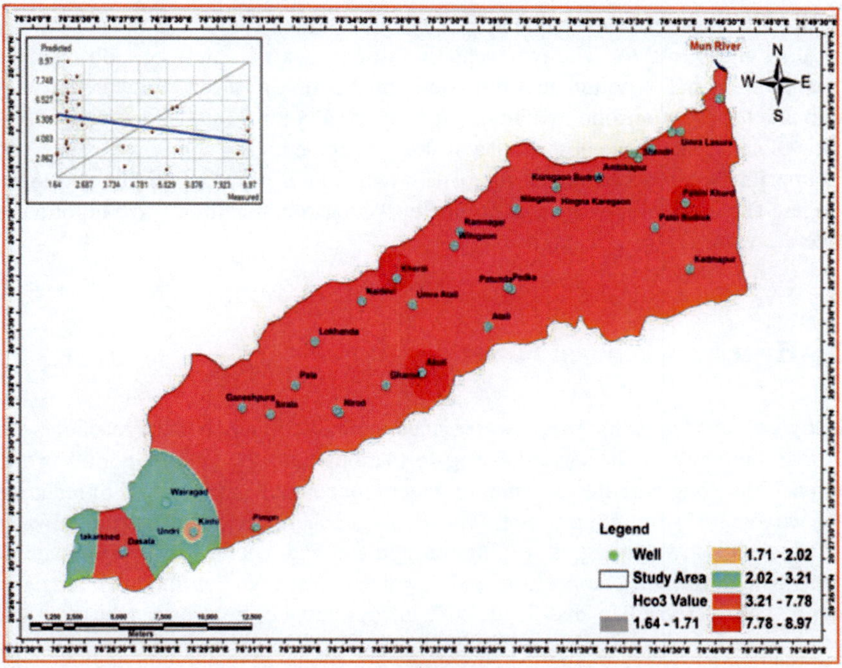

a

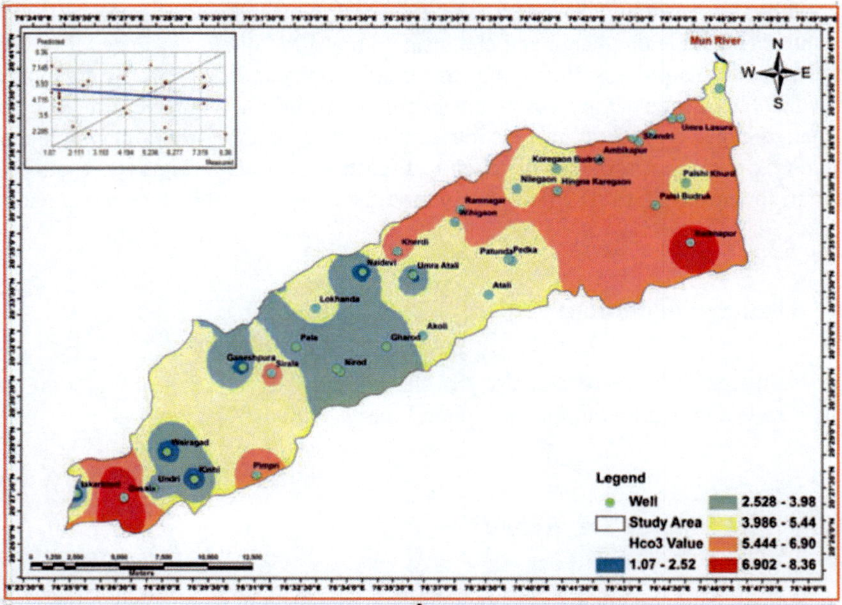

b

Fig. 11 (**a**) Water Quality map of HCO_3 in pre monsoon. (**b**) Water Quality map of HCO_3 in post monsoon

The observation of the carbonate (CO_3) values in the study area range between 0 to 0.79 and 0 to 0.64. The representation of carbonate in pre-monsoon displays greater values has reported in the eastern region of the basin. The observation carbonate the pre-monsoon maximum area reported in moderate values. The observation of carbonate in the post-monsoon shows that the higher values were observed in Pimpri village. While lower value was found in Undri, Ganeshpur and Atali villages. The observation carbonate of the post-monsoon maximum area reported in moderate values (Fig. 12a and b).

11 Hydrogeochemical Facies

The hydro-geochemically groundwater is a cumulative function of its reaction with soil, the lithology of the water-bearing formations, and the storage period in the composition along with the direction of water movement from the place of recharge to discharge [43] of groundwater. The hydrogeochemical cycle of groundwater started when rainwater begins seeping through soil and reaches the deeper stage of the earth with the soil composition and dissolves a variety of different cations and anions. The increased in dissolved CO2 in groundwater contributed by soil and micro-organisms increases the solubility power of the groundwater increases [44].

During the movement of groundwater through the aquifer, a series of dissolution or precipitation reaction takes place, and it depends on the chemical composition of groundwater on the one hand, and chemical composition of aquifer material on the other side, in this process, the aquifer material is weathered along the groundwater flow path [20]. Some of the clays which are formed by weathering have strong cation exchange capacity and hence reaction resulting due to this process takes place. Therefore, through a series of the chemical reaction between groundwater and the aquifer, the groundwater is continuously modified along its subsurface travel path.

12 Hydrogeochemistry

The evaluation of hydrogeochemistry of the groundwater is carried by plotting of anions and cations present in meq/1 in the Piper's Trilinear diagram [45].

12.1 The Piper's Trilinear Diagram

The Piper's trilinear graphs is showing the groundwater quality situation. The Fig. 13 is present water locations decrease in area of 1, 4 and 6 fields recommend alkaline ground exceeded alkali's and strong acid exceeds loose acid respectively. The overall chemistry of the water samples in cation anion pairs is helpful to

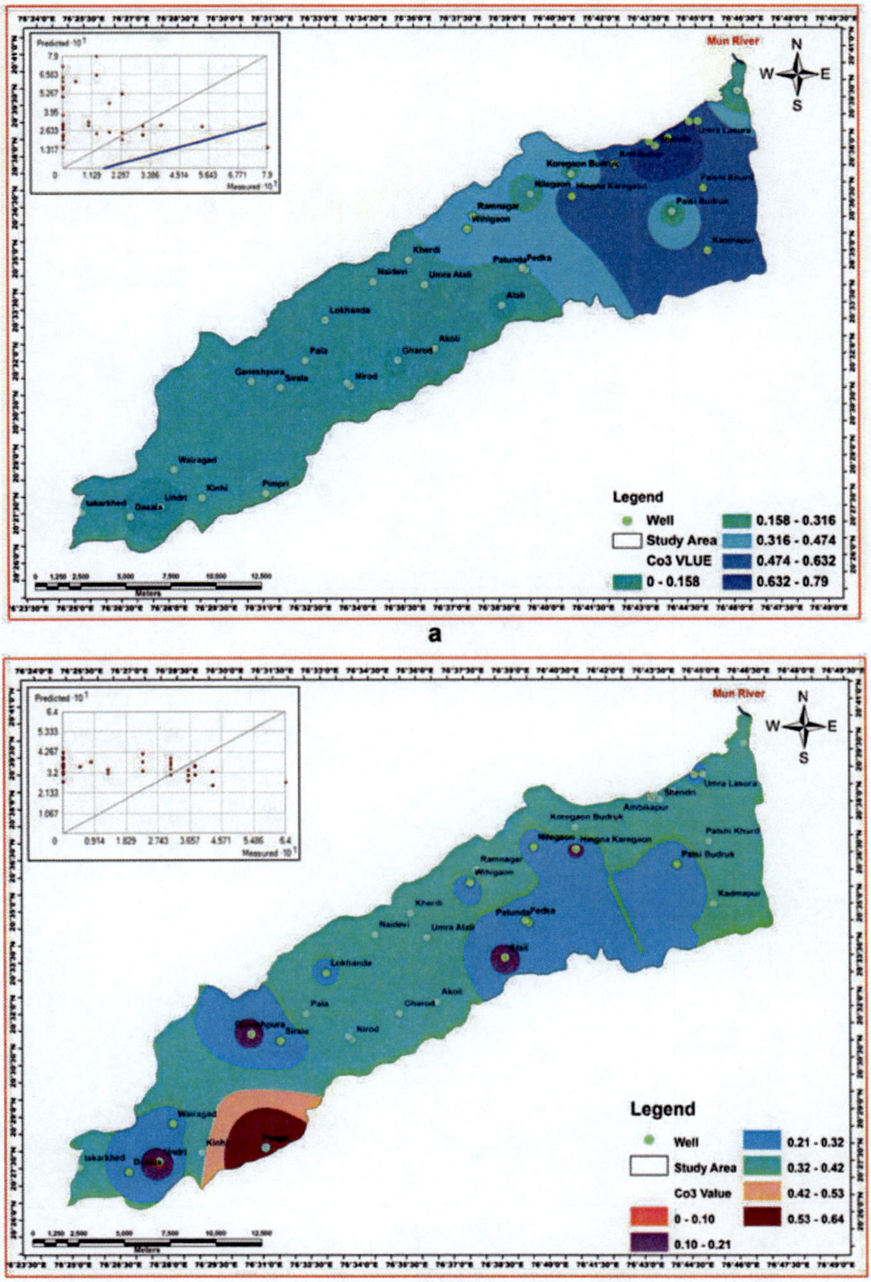

Fig. 12 (**a**) Water Quality map of CO_3 in pre monsoon. (**b**) Water Quality map of CO_3 in post monsoon

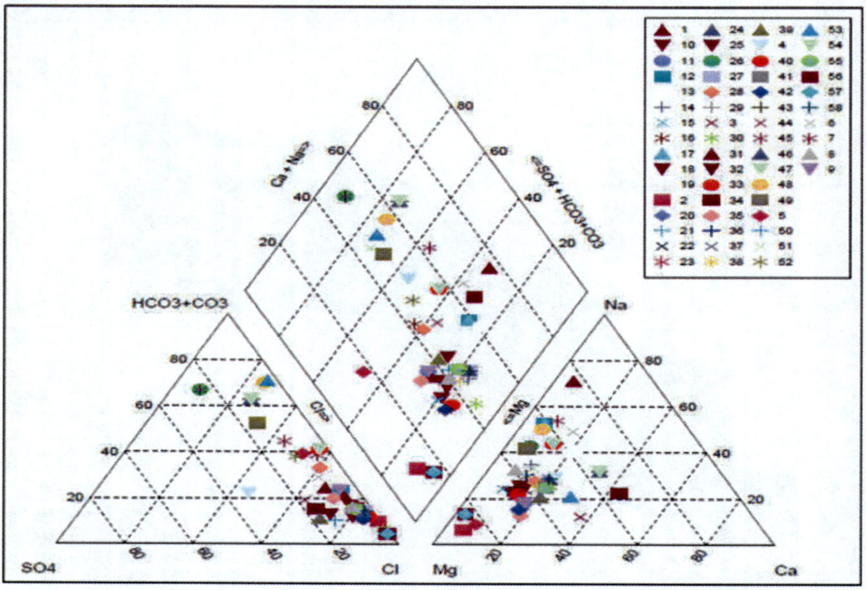

Pre-Monsoon

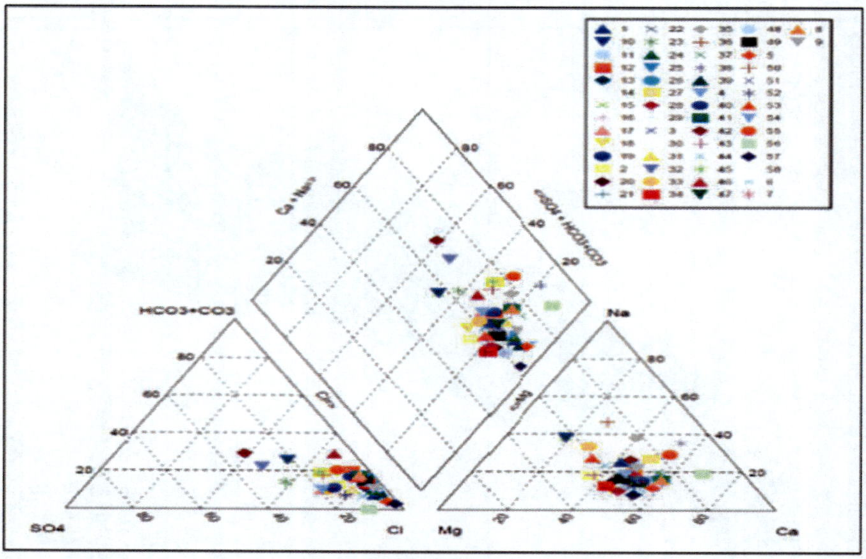

Post-Monsoon

Fig. 13 Piper Trilinear diagram

understand. The Piper diagram shows commonalities and dissimilarities between observation wells, since those with equal enthusiasm tend to be categorised around each other. The Giggenbach Diagram includes two triangular and one diamond-shaped intermediary ground (Fig. 14).

13 Durov Diagrams

It is designing the most of the ions as fractions of milli-equals in the both base triangles. Study of overall cations and anions have been fixed 100% equivalent and the information about the locations in both triangles are estimated onto a square grid parallel to every triangle to the third axis. This graph showed useful properties and links for huge numbers of specimens. This diagrams very systematically represented the groundwater quality during pre-post monsoons (Fig. 15).

14 Kelly's Ratio (KR)

It is a union, or <1, symbolic of outstanding ground water quality for irrigation while higher, it has shown improper farming due to alkaline hazards [7]. By using the following equation, the ratio of Kelly was computed.

$$KR = \frac{Na}{Ca + Mg}$$

where whole ionic concentrations are expressed in milliequivalents per litre.The excess level of Na+ ion in groundwater, which is unwanted for agricultural purpose, demonstrates a more major KR than "1." In the research area, the KR value less than 1, excluding thirty-five locations, demonstrates that groundwater is appropriate for drinking and irrigation uses.

15 Magnesium Hazard

Raghunath [46] explained that the Mg^{2+} and Ca^{2+} ions are significant for the development of the vegetations, but largest concentrations of Mg^{2+} in water effect on earth and yield manufacture. A greater level of Mg^{2+} is typically occurrence of exchangeable Na^{+} in flooded. For requiring magnesium hazard (MH) for agriculture water [47].

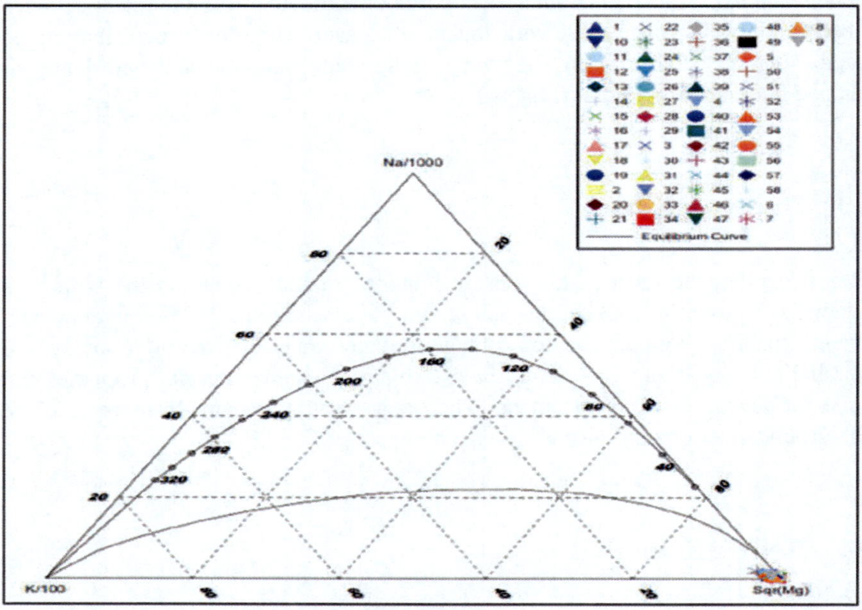

Pre-Monsoon

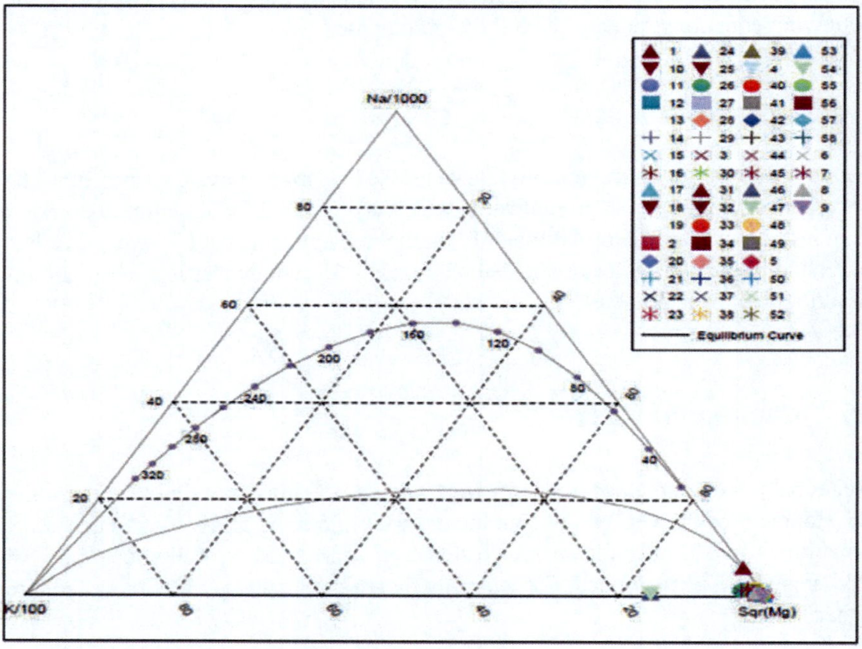

Post-Monsoon

Fig. 14 Giggenbach Triangle diagram

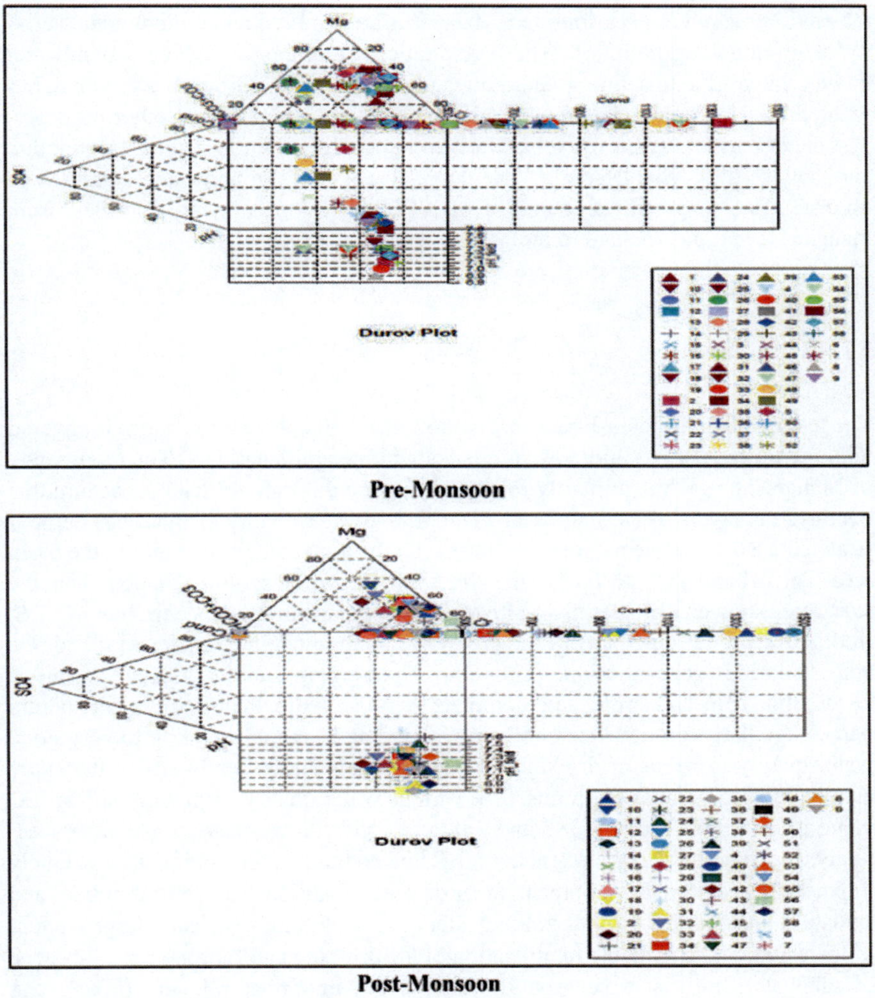

Fig. 15 Durov Plots diagrams

16 Discussion

There are fewer necessary changes in groundwater quality values were reported among the 35 water quality sampling sites in pre-post monsoon periods, while shows monitoring locations in Mahesh river basin. The groundwater water quality parameters of slight changes were reported due to seasonality and climatic condition. According to the results of water quality parameters analysis and different water has used categories, large amounts demonstrate that water table must be preserved previous to use for potable or irrigation purposes or for other uses. In order to finalize the environmental impact, long term databases are required and human-made activity of groundwater pollution. The precipitation variations the whole groundwater quality values due to saline, weathering and erosion affect. Groundwater which has greater values requirements the cure before drinking, irrigation or other purposes. The major part is covered by saline zone, and it is affected for agriculture production and soil fertility characteristic. The largest range of groundwater quality was observed due to greater amounts of pH, TDS, nitrate, hardness, fluorides, iron, manganese and bicarbonate in the groundwater at the present area [48].

17 Conclusion

The feature of the Maharashtra is that approximately 80 percent of its area is covered with the horizontal basaltic, volcanic rocks with insignificant dips. The topography of Maharashtra is thus primarily related to erosion through a fluvial procedure that overlays the layers of Deccan basalt. The groundwater quality analysis has demonstrated the a little saline nature of the basin and the high alkalinity ranges in the basin area. The pH and TDS profiles of the two river-side pre-post monsoon periods in the study area showed a greater rise on lower drainage stage in the concentration of TDS. In this study area some parts of the groundwater are moderate impact on the nitrate and Chloride concentration. The ion representing non-carbonate hardness, i.e. alkaline soils and strong acid dominate the total hydrochemistry. As per sodium percentage, large groundwater wells are belonging to the outstanding to very good water quality to needs of drinking and agricultural. The groundwater of the study area is good to permissible limits. The various water quality maps have shown EC value ranges from 345 to 1587 and 348 and 1598 for pre-post monsoons respectively. In irrigation water classification EC and sodium concentration are extremely important agriculture development under dryland condition. The various indices and statistical analysis and various groundwater quality plotting's indicates large amount of the water quality suitable to utilization of the drinking and farming needs absence of rainwater and that time groundwater playing important role in drinking and irrigation purposes. However, before agriculture, a combination of low and high salinity water is also recommended in the study region to decrease the salinity danger. Magnesium hazard is presenting 75% of groundwater wells showed the magnesium ratio of above 45%. Kelly's ratio is excluding at large sites are very good to excellent group to drinking and agriculture purpose. The Piper's diagram, Giggenbach Triangle and Durov plots are very helpful for understood of groundwater quality and which location is suitable or not suitable is easily show in the different plotting. All these analysis of the hydro-geochemical characteristics during different seasons were spatially studied so that it will be simply understood and helpful to human society and groundwater development and management.

18 Recommendations

1. The selected groundwater quality parameters is so important to human and agriculture uses in the basaltic rock region.
2. The regularly observing water quality should be needed to development, and management of human, animal and sustainable agriculture.
3. Results can be utilized to development, and planning of policy in whole area.
4. These results may utilization to sustainable watershed planning and agriculture water management in the semi-arid regions.

Acknowledgement We are grateful thanks towards AICRP for Dryland Agriculture, Dr. Panjabrao Deshmukh Krishi Vidyapeeth, Akola, Maharashtra, India and Department of Geology, Sant Gadge Baba Amravati University, Amravati for providing the necessary support for this research.

References

1. Chakraborti D, Das B, Murrill MT (2011) Examining India's groundwater quality management. Environ Sci Technol 2011(45):27–33
2. Mishra AP et al (2022) Assessment of water quality index using Analytic Hierarchy Process (AHP) and GIS: a case study of a struggling Asan River. Int J Environ Anal Chem:1–13. https://doi.org/10.1080/03067319.2022.2032015
3. Panneerselvam B, Muniraj K, Pande C et al (2021) Geochemical evaluation and human health risk assessment of nitrate-contaminated groundwater in an industrial area of South India. Environ Sci Pollut Res. https://doi.org/10.1007/s11356-021-17281-0
4. Subramani T, Elango L, Damodarasamy SR (2005) Groundwater quality and its suitability for drinking and agricultural use in Chithar River basin, Tamil Nadu, India. J Environ Geol 47: 1099–1110
5. Hem JD (1985) Study and interpretation of the chemical characteristics of natural water. USGS water supply paper 2254:117–120
6. Shepard D (1968) A two-dimensional interpolation function for irregularly-spaced data. Proceedings of the 1968 ACM National Conference: 517–524
7. Moharir K, Pande C, Patil S (2017) Inverse modeling of Aquifer parameters in basaltic rock with the help of pumping test method using MODFLOW software. Geosci Front Elsevier J 5:1–13
8. Panneerselvam B, Muniraj K, Duraisamy K et al (2022) An integrated approach to explore the suitability of nitrate-contaminated groundwater for drinking purposes in a semiarid region of India. Environ Geochem Health. https://doi.org/10.1007/s10653-022-01237-5
9. Elbeltagi A, Pande CB, Kouadri S et al (2022) Applications of various data-driven models for the prediction of groundwater quality index in the Akot basin, Maharashtra, India. Environ Sci Pollut Res 29:17591–17605. https://doi.org/10.1007/s11356-021-17064-7
10. Piper AM (1953) A graphic procedure in the geochemical interpretation of water analysis. USGS Ground Water Note 12:63
11. Mukate S, Panaskar D, Wagh V, Muley A, Jangam C, Pawar R (2018) Impact of anthropogenic inputs on water quality in Chincholi industrial area of Solapur, Maharashtra, India. Groundw Sustain Dev 7:359–371

12. Singh AK, Raj B, Tiwari AK, Mahato MK (2013) Evaluation of hydrogeochemical processes and groundwater quality in the Jhansi district of Bundelkhand region, India. Environ Earth Sci 70(3):1225–1247
13. Carpenter SR, Caraco NF, Correll DL, Howarth RW, Sharpley AN, Smith VH (1998) Nonpoint pollution of surface waters with phosphorus and nitrogen. Ecol Appl 8(3):559–568
14. Gurav Chandrakant (2017) Application of remote sensing, geology and geomorphological studies for mass wasting zone analysis in Jotiba-Panhala Hill Range Area, Kolhapur District, Maharashtra, India IOSR Journal of Applied Geology and Geophysics (IOSR-JAGG) e-ISSN: 2321-0990, p-ISSN: 2321-0982.Volume 5, Issue 2 Ver. I (Mar. – Apr. 2017), pp 29–37
15. Kanak moharir (2019) Spatial interpolation approach-based appraisal of groundwater quality of arid regions. J Water Supply Res Technol AQUA. https://doi.org/10.2166/aqua.2019.026
16. Karanth KR (1987) Groundwater assessment, development and management. Tate McGraw Hill, New Delhi, p 720
17. Ravikumar P, Somashekar R, Angami M (2011) Hydrochemistry and evaluation of groundwater suitability for irrigation and drinking purposes in the Markandeya River basin, Belgaum District, Karnataka State, India. Environ Monit Assess 173(1):459–487
18. Mueller TG, Pierce FJ, Schabenberger O, Warncke DD (2001) Map quality for site-specific fertility management. Soil Sci Soc Am J 65(5):1547–1558
19. Dehghanzadeh R, Safavy Hir N, Shamsy SJ, Taghipour H (2015) Integrated assessment of spatial and temporal variations of groundwater quality in the eastern area of Urmia salt Lake basin using multivariate statistical analysis. Water Resour Manag 29(4):1351–1364
20. Garrels RM (1967) Genesis of some ground waters from igneous rocks, Researches in Geochmistry, vol 2. Wiley, New York, pp 405–420
21. Ramakrishnaiah CR, Sadashivaiah C, Ranganna G (2009) Assessment of water quality index for the groundwater in Tumkur Taluk, Karnataka State, India. J Chem 6(2):523–530
22. Sadat-Noori SM (2014) Groundwater quality assessment using the Water Quality Index and GIS in Saveh-Nobaran aquifer. Iran Environ Earth Sci 71(9). https://doi.org/10.1007/s12665-013-2770-8
23. Singh AK, Mondal GC, Kumar S, Singh TB, Tewary BK, Sinha A (2008) Major ion chemistry, weathering processes and water quality assessment in upper catchment of Damodar River basin, India. Environ Geol 54:745–758
24. Mukherjee I, Singh UK, Singh RP, Kumari D, Jha PK, Mehta P (2020) Characterization of heavy metal pollution in an anthropogenically and geologically influenced semi-arid region of east India and assessment of ecological and human health risks. Sci Total Environ 705:135801. https://doi.org/10.1016/j.scitotenv.2019.135801
25. Boateng TK, Opoku F, Acquaah SO, Akoto O (2016) Groundwater quality assessment using statistical approach and water quality index in Ejisu-Juaben Municipality. Ghana Environ Earth Sci 75(6):489–502
26. CGWB (2010) Ground water quality in shallow aquifers of India. Central Ground Water Board, Faridabad
27. Chaturvedi A, Bhattacharjee S, Singh AK, Kumar V (2018) A new approach for indexing groundwater heavy metal pollution. Ecol Indic 87:323–331
28. Egbueri JC (2018) Assessment of the quality of groundwaters proximal to dumpsites in Awka and Nnewi metropolises: a comparative approach. Int J Ener Water Res 2(1–4):33–48
29. Pande CB, Moharir KN, Pande R (2018a) Assessment of Morphometric and Hypsometric study for watershed development using spatial technology–A Case Study of Wardha river basin in the Maharashtra, India. Int J River Basin Manag 19:43–53. https://doi.org/10.1080/15715124.2018.1505737
30. Pande CB, Khadri SFR, Moharir KN, Patode RS (2018b) Assessment of groundwater potential zonation of Mahesh River basin Akola and Buldhana districts, Maharashtra, India using remote sensing and GIS techniques. Sust Water Resour Manag 4:1–15. https://doi.org/10.1007/s40899-017-0193-5

31. Humphreys WF, Watts CHS, Cooper SJB, Leijs R (2009) Groundwater estuaries of salt lakes: buried pools of endemic biodiversity on the western plateau, Australia. Hydrobiologia 626:79–95. https://doi.org/10.1007/S10750-009-9738-4
32. APHA (2012) Standard methods for the examination of water and wastewater, 22nd edn. American Public Health Association, New York
33. WHO (2011) Guidelines for drinking-water quality, 4th edn
34. Handa BK (1964) Modified classification procedure for rating irrigation waters. Soil Sci 98: 264–269
35. Davis SN, De Wiest RJM (1967) Hydrogeology, 2nd edn. Wiley, New York, p 463
36. Pande CB, Moharir KN, Singh SK et al (2020) Groundwater evaluation for drinking purposes using statistical index: study of Akola and Buldhana districts of Maharashtra, India. Environ Dev Sustain 22:7453–7471. https://doi.org/10.1007/s10668-019-00531-0
37. Steube C, Richter S, Griebler C (2009) First attempts towards an integrative concept for the ecological assessment of groundwater ecosystems. Hydrogeol J 7:23–35
38. Bhutiani R, Kulkarni DB, Khanna DR, Gautam A (2017) Geochemical distribution and environmental risk assessment of heavy metals in groundwater of an industrial area and its surroundings, Haridwar, India. Ener Ecol Environ 2(2):155–167
39. Batabyal AK, Chakraborty S (2015) Hydro-geochemistry and water quality index in the assessment of groundwater quality for drinking uses. Water Environ Res 7:907–617
40. Ramkumar T, Venkarramanan S, Anithamary I, Ibrahim SM (2012) Evaluation of hydrogeochemical parameters and quality assessment of the groundwater in Kottur blocks, Tiruvarur district, Tamilnadu, India. Arabian J Geosci 6:101–108
41. Kolpin D (1997) Agricultural chemicals in groundwater of the Midwestern United States: relations to land use. J Environ Qual 26:1025–1037
42. Shanmugasundharam A, Kalpana G, Mahapatra SR, Sudharson ER, Jayaprakash M (2015) Assessment of groundwater quality in Krishnagiri and Vellore districts in Tamil Nadu, India. Appl Water Sci 7:1869–1879
43. Griffith AJ (2001) Geographic techniques and recent applications of remote sensing to landscape-water quality studies. Water Air Soil Poll 138:181–197
44. Feth JH, Roberson CE, Polzer WL (1964) Sources of mineral constituents in water from granitic rock, Sierra Nevada, California and Nevada. U.S. Geological Survey Water Supply Paper 1535-I
45. Matson P, Parton J, Power A, Swift M (1997) Agricultural intensification and ecosystem properties. Science 277:504–509
46. Raghunath HM (1987) Groundwater, 2nd edn. Wiley Eastern Limited, New Delhi, p 563
47. Doneen LD (1964) Notes on water quality in agriculture. Water science and engineering. University of California, Davis
48. Muhammad Arshad (2017) Irrigation water quality. https://www.researchgate.net/publication/320531819_Irrigation_Water_Quality

Evaluation of Hydro-meteorological Conditions and Water Resource Prospects in East Bokaro Coalfield, Damodar Basin, India

Mukesh Kumar Mahato, Prasoon Kumar Singh, Abhay Kumar Singh, and Gurdeep Singh

Abstract Evaluation of the natural groundwater recharge has been done in EB coalfield of Damodar basin. The hydro-meteorology, water table, surface planes, geo-morphological factors influences the groundwater table fluctuation. The average annual rainfall (AAR) over last 35 years has been recorded 1366 mm with maximum rainfall of 2544 mm in 2001 and minimum rainfall of 733 mm in 2004. The groundwater level fluctuation in the coalfield area ranged from 0.15 to 2.87 mbgl. The water level hydrographs indicate a slight decline in the water level. The insignificant declining trend seems to be endorsed due to over water draft by the local residents. The rainfall is the key source of groundwater recharge by filtration through soil surface. In addition to rainfall, the coal mine water pumping from the local coal mining areas and open water sources, such as abundant mine pits, contribute to the groundwater recharge. The EB coalfield is highly distressed, permeability of geological component is spatially inconsistent and depends on stratigraphic, cracking, and depth attenuation. By using the rainfall infiltration and the groundwater table fluctuation method, the total annual replenishable recharge (TARR) was anticipated to be 48.6 Mm^3/year and 62.3 Mm^3/year, respectively. According to the rainfall infiltration method, the net annual availability of groundwater is 13.7 Mm^3. The GEC-1997 guideline was used to evaluate the stage of groundwater development, which belongs to the secure class (i.e. 67%). However, there was an urgent need for water resource management in the coalfield area.

Keywords Hydro-meteorology · Groundwater availability · Water table fluctuation · Recharge prospective · GIS · East Bokaro

---

M. K. Mahato (✉) · A. K. Singh
Water Resource Management, CSIR-Central Institute of Mining and Fuel Research, Dhanbad, Jharkhand, India

P. K. Singh · G. Singh
Indian Institute of Technology (Indian School of Mines), Dhanbad, Jharkhand, India

S. Yadav et al. (eds.), *Water Quality, Assessment and Management in India*, Earth and Environmental Sciences Library, https://doi.org/10.1007/978-3-030-95687-5_17

1 Introduction

Rapid population growth and the resulting demand for easily accessible water resources are well-known facts [1, 2]. As a result of this demand, many regions around the surface of the earth are facing water scarcity and increased water stress [3, 4]. Rainfall and water level fluctuations must be studied. Because of its significance in determining the water level, the correlation between these has been widely used in recent decades. The environmental degradation and ecological disruption have posed a threat to basic human sustainability [5]. The discharge of such untreated and undertreated effluent leads to serious ecological degradation in general, as well as water contamination in particular. As a result of pollution, the amount of available water, both surface and sub-surface has decreased. Because of the extensive mining in the area, the groundwater quality may affect, owing to leaching of mining effluents [5]. In many countries across the world, groundwater resources are essential to sustaining livelihood. In India, groundwater is a valuable resource for meeting the water requirements of agriculture, industrial, and households. Groundwater is a natural water resource that is both dynamic and replenishable, but its availability in hard rock terrain is limited [5]. In India, about 30% population lives in cities and 90% lives in rural areas, they depend entirely on surface or groundwater and access to safe drinking water remain a critical issue [6, 7]. Water shortage is one of the most critical problems in several areas of Jharkhand, West Bengal, Orissa, and several other states of India in recent years [8, 9]. Every year, the groundwater table depletes due to an overdraft of groundwater that exceeds the replenishable recharge [10–13]. Considering this in view the groundwater availability assessment is the obligation for coal mining areas.

Hydro-meteorological conditions of an area have subantantial impact on the fluctuation of groundwater table [12]. The rainfall is the prime source of groundwater replenishment by percolation through deep aquifers. A huge percentage of the rainfall drains out as surface run-off into the local drainage systems, which then discharge load into the areas master drainage system [11]. During the monsoon season, the south west rainfall is responsible for majority of India's rainfall, and other water-related persuit, whether natural or human-made like agriculture with groundwater extraction [14]. Overexploitation and contamination are serious threat to groundwater resources in the current situation. Understanding groundwater fluctuation and flow is therefore important and in some cases, overexploitation is contaminating aquifers; unscientific development of groundwater flow and hydrogeochemical properties intensifying the problem [15]. To explore the prospective region of groundwater recharge is exceedingly essential for the maintenance of water quality and groundwater management [3, 16]. Aquifer recharge in and around of coal based industries such as coal washery, power plants and coal mining activities has had a minor impact on the geology, geomorphology, and soil structure [17]. There are a variety of techniques for determining water recharge-potentials. Large aquifer properties including transmissivities, storage coefficients, and other parameters are difficult to evaluate due to lack of resolution [18]. Therefore, the present study is imperative in determining the status of the variations in rainfall potency, groundwater quantity, and water resource prospective in East Bokaro coalfield of Damodar basin of India.

2 Study Area

The present study was undertaken in the East Bokaro (EB) coalfield of Damodar basin, which is located in the Bokaro district of Jharkhand State. It covers about 237 sq. km and spread between 23° 45′ N to 23° 50′ N latitude and 85° 30′ E to 86° 03′ E longitude. The coalfield is located in the Barkakana-Gomoh section of the eastern railways, which runs through the coalfield in an east-west direction. The EB coalfield is highly undulating and hilly all over the area, known for the part of Chhotanagpur Plateau.. The overall slope of the sites is facing east, which controlled the all tributaries of Damodar River. This district has a humid and subtropical climate with hot dry summer (March to May), and cold winter (November to February). The humidity is found to be higher during July to September with average annual humidity is approximately 60%. The summer temperature rises up to 46 °C and winter temperature goes down to 4 °C.

2.1 Geology of East Bokaro Coalfield

The EB coalfield has numerous thick and thin regional Barakar coal seams, with thin interlarded regional Raniganj coal seams and a consistent deposition from Talchir to Supra-Panchet formations [19, 20]. The geological feature is illustrated in Fig. 1. The lithology comprised of coarse-grained ferruginous sandstone, pebble sandstone, and red clay, as well as a little local coal seams within the coal measures. Tillite, greenish sandstone, and needle shale can be contained in very few scattered Talchir formation exposures. The Barakar formation is found in the eastern part of the basin, and it contains coarse-grained arkosic sandstone, fine-grained laminated sandstone, grey shale, carbonaceous shale, and coal seams include flaggy, fine grained ferruginous sandstone along with micaceous sandy and black shales together with siderite band. They are displayed in the arc-shaped outcrops of the barren measures and are exposed in the central and western parts. The Raniganj formations are medium to coarse grained calcareous sandstone along with fine grainular green sandstone with grey and carbonaceous shales, and thin coal seams. The Panchet formations revealed similar arc-shaped pattern surrounding green micaceous sandstone with buff fine gainnular sandstone, red and green shale, and thin coal seams [21, 22]. The Supra Panchet formation is identified in the western most part of the coalfield, on Lugu Hill, at a higher elevation.

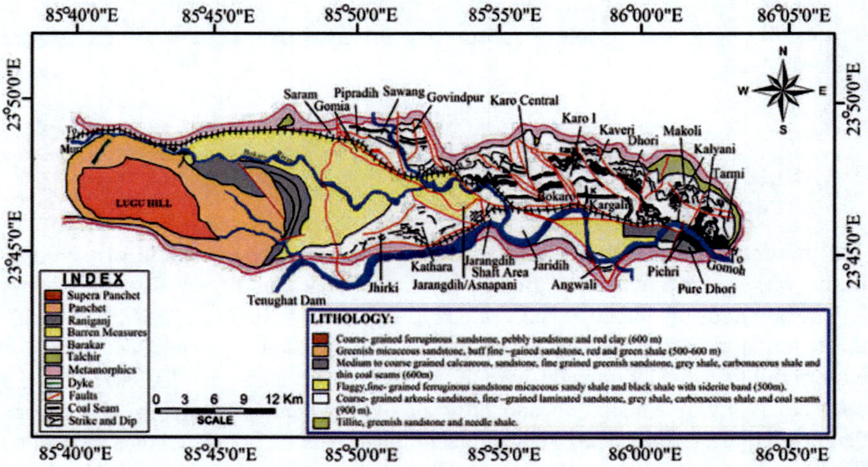

Fig. 1 Geological map of East Bokaro coalfield; modified from [12]

2.2 *Hydrogeology of East Bokaro Coalfield*

EB coalfield is situated on the southern bank of Damodar River. The elevation in the coalfield ranges between 195 m to 215 m above mean sea level. When the water level drops down to about 190 m during the summer, the Damodar River can be crossed on foot. However, during the monsoon season, the Damodar River floods a part of the coalfield. The slope of the ground is from south to north in the coalfield area. The drainage of EB coalfield is mainly controlled by the Damodar River, which flows from west to east along the northern frontier [23]. The highest flood level of the Damodar River was 205.7 m in 1976, above mean sea level (Source: CMPDIL, Ranchi). There are also many other nalas that drain, including Khanjo nadi, Tisri nala, Dhakal jor and Godo nala. The coalfield area has a mostly dendritic drainage pattern. Above the Kargali coal seam, an alluvial formation of clay loose sand, weathered sandstone, poorly cemented thin shaly sandstone, acts like an unconfined aquifer. The lower formations, which are comprised of compact fine to medium-grained sandstone with lamination and intercalation with thin shale and carbonaceous shale bands with secondary porosity, are semi-confined and have lower potential. The deeper aquifers act as an unconfined aquifer at the outcrop region. Groundwater moves laterally through the sandstone inter-granular pore spaces in the sandstone aquifer (Table 1).

Table 1 Hydro-geological units of East Bokaro coalfield

Hydro-geological unit	Formations	Average thickness (m)
Unconfined aquifer	Loose alluvium soil, weathered sandstone followed by sandstones and shale.	37.0
Aquiclude	Seam Kargali (To be worked)	9.6
Semi confined aquifer	Carb. Shale, Intercalation of shale and sandstone	39.0
Aquiclude	Seam Bermo (To be worked)	9.3
Semi confined aquifer	Intercalation of shale and sandstone	36.0
Aquiclude	Seam Karo VIII/IX/X (To be worked)	22.4
Semi confined aquifer	Generally gray shale and intercalation of shale and sandstone	8.5
Aquiclude	Seam Karo VI/VII (To be worked)	27.6

2.3 *Aquifer Parameters*

The entire coalfield area was not exposed to a detailed aquifer performance test. However, aquifer performance test were performed in the nearby Makoli of EB coalfield, which is located 1.8 km north of the opencast project, and the following aquifer parameters for the formations were reported:

Transmissivity (T): 21.8 m^2/d
Permeability (k): 0.38 m/d (1.0 m/d for unconfined aquifer)
Storage coefficient: 9.6 x 10^{-5}
Wells yield: 50–300 L/min

However, Lower aquifer permeability (Formation such as alternating bands of sandstone and shale and carbonaceous shale) lying above the working seam, on the other hand, is usually <0.38 m/d. The major source of groundwater replenishment in the EB coalfield area is rainfall. Groundwater in the study area is also influenced by seepages from canals, lakes, and other bodies of surface water. Because of the vide variability of geology, topography, drainage, and activity in the study area, the hydro-geological condition is complicated [24]. In general the EB coalfield can divide into two hydro-geological units:

i. Fissured formation
ii. Porus formation or unconsolidated formation

Based on the degree of consolidation, the fissured formation can be further divided into two types based on the degree of consolidation:

(i) **Consolidated formation:** Groundwater occurs in the area, but it is confined or semi-confined. Fractures and lineaments have been found to be helpful in establishing groundwater movement and storage.

(ii) **Semi consolidated formation:** This formation contains groundwater that is confined to semi-confined. These can be located in the-centre of the study area.

2.4 Unconsolidated Formation

It is in a low laying area that is covered in recent alluvium, which was deposited mainly by Damodar, Konar, and Bokaro Rivers. The groundwater is found in these areas when the water Table (W.L) is low. The topography of the plains of the river Damodar and EB area is gently undulating mostly towards southern zone of the coalfield. The lower Barakar comprises of compact and massive sandstone and resulted comparatively rugged terrain in the northern part. Lugu hill, the most prominent topographical feature rises to nearly 950 m. Bokaro, Konar and Damodar are the three prominent rivers flowing in the central, eastern and southern regions, respectively through the coalfield.

2.5 Terrain Analysis

The digital elevation model (DEM) depicts the digital ground surface terrain that can be used to classify areas that need cutting and filling as well as estimate the volume of earthwork needed to level the undulating topography of the coalfield area. The DEM can be used to determine the hydrological characteristics the study area such as flow direction, flow accumulation, aspect and slope [25].

2.6 Hill Shade or Digital Terrain Model

Hill shade is a hypothetical illumination surface raster that depicts the surface illumination ratio. It improves the terrain elevation visualization capabilities [26]. In the Hill shade theme, it is also possible to see the terrain undulation. The hill shade map of the EB coalfield revels that there is little undulation in the west northern part of the area, while the rest of the area is flat (Fig. 2).

2.7 Slope of the Study Area

The slope is defined as the maximum rate of change between any two cells (in a surface raster or DEM) [25]. A slope value is assigned to each cell in the output raster. A lower slope value represents flat terrain, while a slope value denotes steep terrain. The slope of the EB coalfield area ranged from 0 to 60 degree (Fig. 3). While

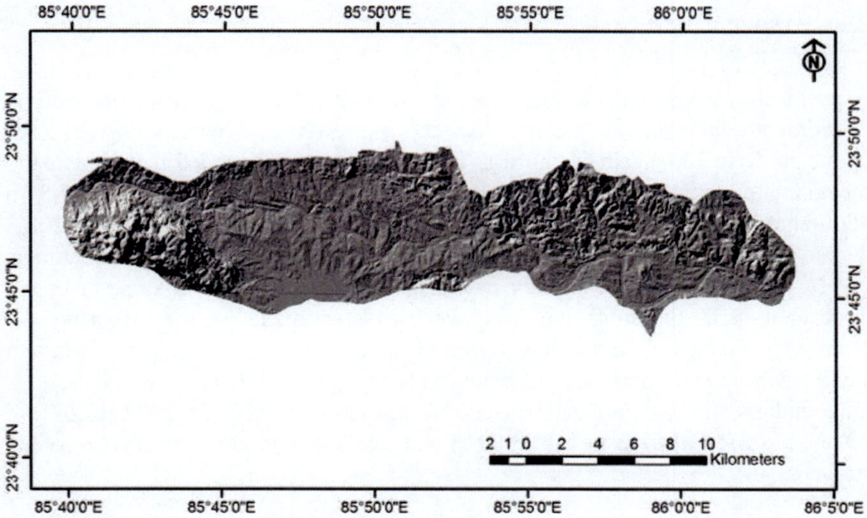

Fig. 2 Hill shade or Digital terrain model of East Bokaro coalfield

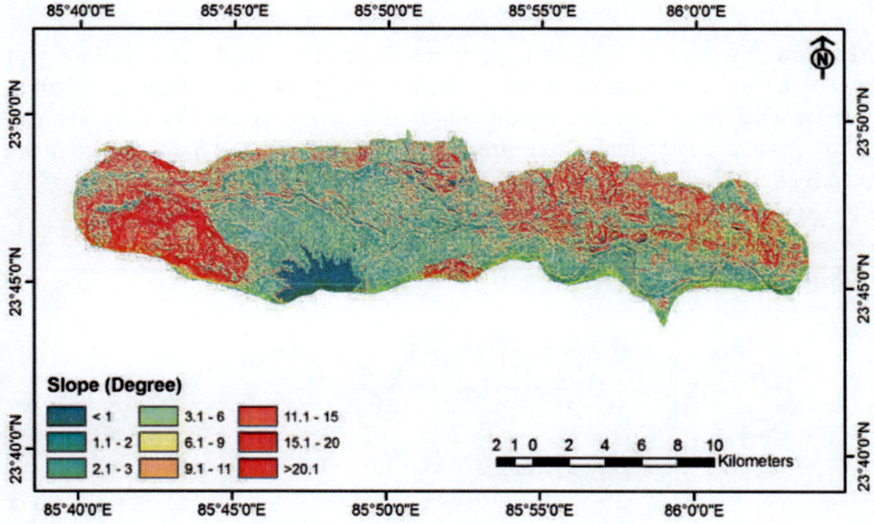

Fig. 3 Slope map of the East Bokaro coalfield

the average slope value of the mining area was 0.67 degree, it indicates flat topography of the area. Slight undulation can be observed erratically in area (mostly in the north) with maximum slope of ranging from 1 to 10 degree. The highest slope is seen in southern margin of EB coalfield in a narrow band (and on the margin of the drain).

2.8 Aspect

Aspect is the direction that a slope faces of EB coalfield. It indicates the down slope direction at a location on a surface. Aspect is expressed clockwise in degrees from 0 (due north) to 360 (again due north). Aspect image is colour coded according to the prevailing direction of the slope at each pixel. 0 to 22.5 degree and 337.5 to 360 degree symbolize north, 67.5 to 112.5 degree symbolize east, 157.5 to 202.5 degree stand for south and 247.5 to 292.5 degree represent west direction and minus one (−1) value is used to identify flat surfaces. In this area, most slope faces are directed towards the south and southwest, as seen in the aspect map (Fig. 4). Northwest directed slope are mostly found in southern part of the plant area and small patches are seen in the northern part of the region. Moreover, other directions are found in entire area as small patches. North and middle (26° 24′ 15″ N to 26° 24′ 40″ N and 81° 7′ 30″ E to 81° 8′ 00″ E) of this coalfield area contains some flat surfaces.

2.9 Flow Direction

The flow direction process in a (sink free or hydrologically corrected) digital elevation model determines in which neighboring pixel some water in a central pixel will naturally flow. D-8 algorithm is used to determine the flow direction map of this plant area. Two major flow directions (south and southeast) can be seen in the map (Fig. 5). North to southeast flow direction can be noticed in southern part of the

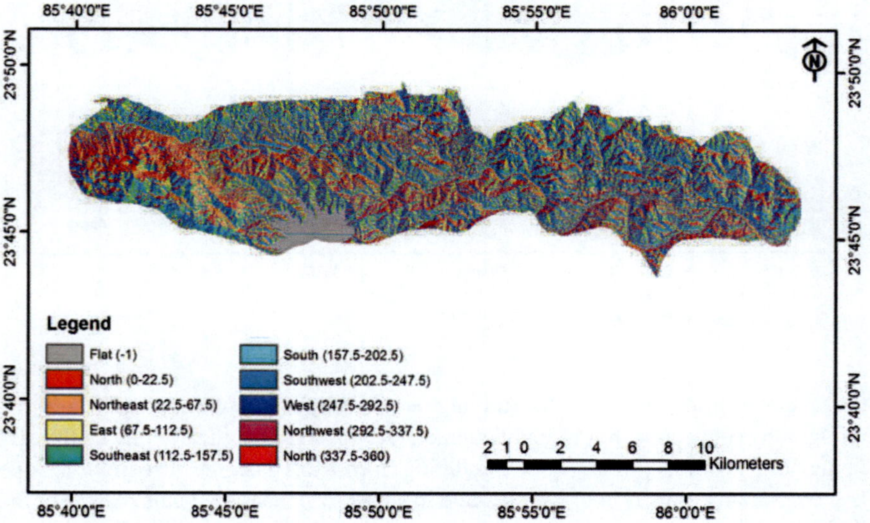

Fig. 4 Aspect map of East Bokaro coalfield

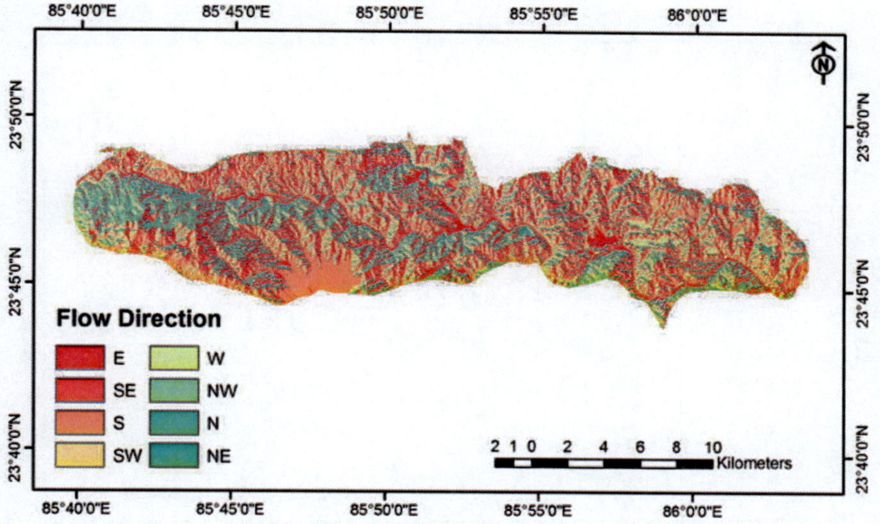

Fig. 5 Flow direction map of East Bokaro coalfield

plant area and other types of flow direction can be distinguished as small patches in entire coalfield area.

2.10 Geomorphology

The EB coalfield area is a part of Chhotanagpur plateau with undulating land and hilly terrain spreading all over the surface area. The overall slope of the area is towards east and regulated the Damodar River tributaries distribution. The average elevation varies from 200–350 m above MSL higher with maximum distinct hill block is Gomia. The north-western zones are distinguished by hilly arrays. Denudation hills form over metamorphic rocks and are found in northeastern/northwestern parts of the Bokaro district as relatively high hills [27] and the soils are mainly residual. Maximum temperature and heavy rainfall distribution resulted deposition of lateritic soils from rocks of archaean metamorphic structure in vast land area especially lower Gondwana rock in the west-central and east-central areas [27].

2.11 Drainage Pattern

The drainage pattern distribution revealed the typical surface as well as subsurface formation in the EB coalfield area. The higher the drainage density, the higher the runoff: hence lower the drainage density, the higher the probability of groundwater

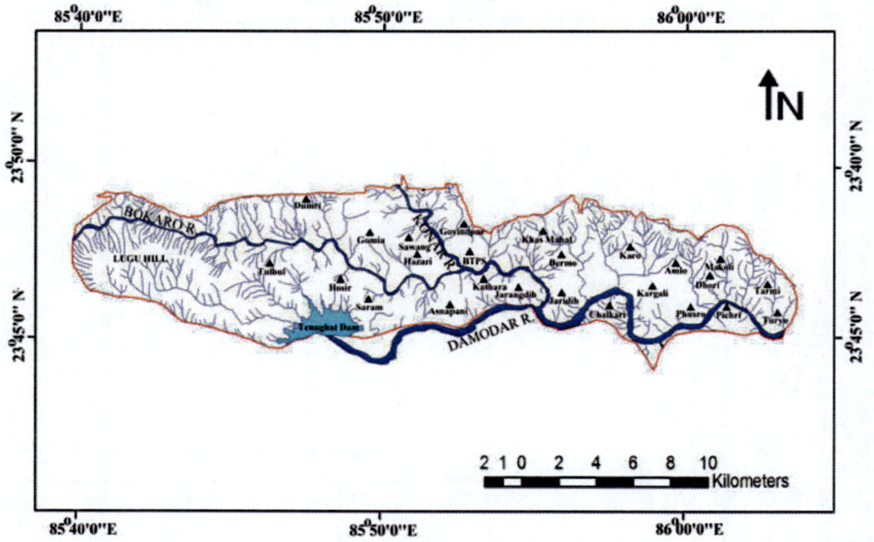

Fig. 6 Drainage map of East Bokaro coalfield

recharge. The EB coalfield area has a dentritic drainage pattern. Lineaments are weaker zones that have been formed due to crustal movements of the earth. They may represent fault zones, geologic contacts, fractures, shears, or major joints. They are the proven secondary aquifers in a hard rock area. The groundwater is only found in fractured and interconnected joints. Following the mining of shallow weathered aquifers, targeting of groundwater is focused towards the exploration of these heterogenetic hydro-fracture areas. The most prominent Damodar River flows in West-East direction in central part of the area (Fig. 6). The major tributaries of Damodar are Konar and Bokaro Rivers. The drainage system is mostly restricted to vulnerable zones viz. joints, fractures, and faults.

3 Methodology

Secondary data was collected from the Geological Survey of India (GSI), Kolkata. Qualitative data of groundwater resources and mine water discharge data were obtained from Central Ground Water Board (CGWB) were collected from Central coalfield limited, Ranchi, respectively [27]. Census and rainfall data were obtained from Bokaro district statistical office and Dhori weather station office, Phusro, respectively for demographic and meteorological studies.

3.1 Water Level Monitoring

Groundwater level monitoring by electronic water level Indicator (Model-K-11107) was monitored in thirty-two observation shallow dug wells during the pre-monsoon and post monsoon seasons in order to understand the behavior of groundwater regime in the EB coalfield area (Table 2 and Fig. 7).

3.2 Evaluation of Groundwater Prospects

The purpose of the groundwater resource assessment was to learn more about the yearly groundwater recharge and to classify assessment units based on their usage status and long-term water level patterns. The groundwater level fluctuation method is a widely used method for calculating groundwater recharge around the world. In addition common derived equations from various water study related projects are used for computation and validation of various recharge components [28]. Groundwater resources were estimated using the methods approved by the Groundwater Estimation Committee, 1997 [29].

3.3 Calculation of Groundwater Prospective

Water levels presumed to have impacted in the EB coalfield area. The coal mining areas are drastically disturbed and the water permeability are spatially varies depending upon lithology, fracturing, and depth attenuation in a confined geological units. The porous, easily jointed, and permeable sandstone assists in the formation of aquifers, whereas the shaly formations are aquitards, which may be cracked, weakly permeable, and leaky, and found to have weak porous barriers for upright water movement. The subsurface groundwater rejuvenation evident on the surface area was assessed using rainfall infiltration and water table fluctuation methodologies.

4 Results and Discussion

4.1 Normal Rainfall Pattern in East Bokaro Coalfield

The mean annual rainfall (1976–2012) of EB coalfield areas are accessible for data analysis and synthesis (Fig. 8). In the study area, the average annual rainfall was observed 1366 mm. The average rainfall varies significantly from year to year. The maximum average rainfall during 1976–2012 (with two gaps: 1978 and 1993), occurred during the year 2001, with 2543.8 mm.

Table 2 Monitoring of water level (mbgl) in East Bokaro coalfield

Monitoring Sites	Latitude (N)	Longitude (E)	Pre-monsoon (W.L)	Post-monsoon (W.L)	Fluctuation (W.L)	Well depth (m)	Diameter (m)
Pichri	23.7506	86.0206	3.94	3.79	0.15	12	3.17
Phusro	23.7598	86.0055	4.96	4.57	0.39	14	2.72
Tarmi	23.7605	86.0509	8.57	5.70	2.87	9	2.45
Karipani	23.7814	86.0400	4.31	3.64	0.67	12	2.12
Amlo	23.7800	86.0095	2.82	2.54	0.28	14	2.25
Dhori	23.7635	86.0074	4.02	3.66	0.36	12	3.15
Chalkari	23.7538	85.9577	4.36	3.49	0.87	9	2.00
Sarubera	23.7977	86.0167	4.34	3.93	0.41	11	1.35
Makoli	23.7796	86.0224	1.80	1.65	0.15	15	0.35
Subhas nagar	23.7805	85.9902	1.64	1.36	0.28	11	2.38
Kargali bazar	23.7704	85.9807	2.55	2.32	0.23	9	3.58
Kargali	23.7630	85.9758	4.60	2.00	2.60	10	3.10
Bermo	23.7648	85.9702	3.20	2.38	0.82	9	2.27
Sunday bazar	23.7855	85.9396	3.68	3.44	0.24	9	2.10
Angawali	23.7265	85.9818	4.30	3.78	0.52	21	2.39
Jaridih	23.7605	85.9275	6.95	6.63	0.32	14	3.58
Asnapani	23.7617	85.8910	5.05	4.56	0.49	11	1.85
Jhirki	23.7536	85.8469	4.16	3.63	0.53	12	1.70
Kurukpania	23.7904	85.9427	6.74	6.35	0.39	17	2.60
Khas mahal	23.7949	85.9371	4.50	3.08	1.42	41	2.20
Pilpilo	23.8080	85.9264	11.20	10.80	0.40	32	2.25
Lahariatanr	23.7903	85.8708	5.68	5.30	0.38	4	2.10
Govindpur	23.8002	85.8811	5.98	5.56	0.42	18	1.77
Kathara	23.7651	85.8723	6.72	6.31	0.41	9	2.12
Hazari	23.7834	85.8513	4.90	4.64	0.26	3	1.90
Sawang	23.7977	85.8367	5.20	4.03	1.17	13	1.72

Gomia	23.7964	85.8226	4.56	4.04	0.52	12	3.40
IEL Gomia	23.8102	85.8235	4.94	3.84	1.10	14	8.32
Hosir	23.7691	85.8013	5.82	4.82	1.00	30	4.56
Saram bazar	23.7605	85.8171	5.52	5.19	0.33	12	1.75
Tulbul	23.7732	85.7657	5.25	4.00	1.25	11	1.65
Dumri	23.8038	85.7643	5.17	4.75	0.42	15	3.38
Minimum			1.64	1.36	0.15	3.00	0.35
Maximum			11.20	10.80	2.87	41.00	8.32
Average			4.92	4.24	0.68	13.9	2.57

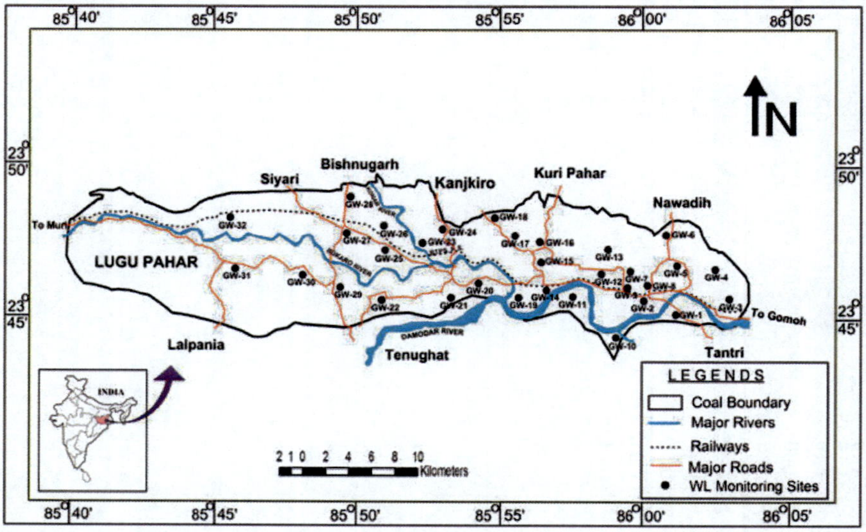

Fig. 7 Location map of water level monitoring sites in East Bokaro coalfield; modified from [12]

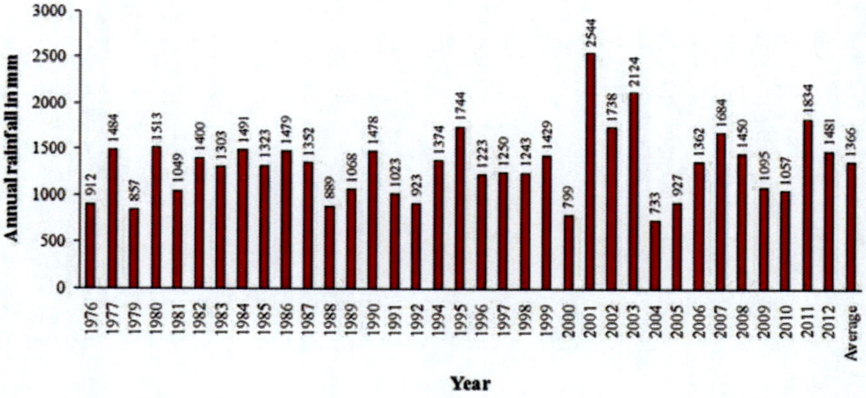

Fig. 8 Annual rainfall distribution of East Bokaro coalfield

According to the present study, average rainfall values ranged from 1001 and 1400 mm throughout a 28 years period with less than 1000 mm in 10 years and more than 1400 mm in 25 years [12]. From 1975 to 2012, the EB coalfield experienced the highest rainfall in single day. On July 17, 1975, the highest amount of rainfall in a single day was 239.7 mm [12]. The average peak rainfall on a day during 35 years was 89.5 mm, with a peak rainfall day of 239.7 mm in 1975 and a trough of 52.8 mm in 2004 [12]. The basic probability of rank method was used to calculate the probability percentage of rainfall in the study area. The mean annual rainfall found to be in descending order of magnitude and computations for possibility analysis

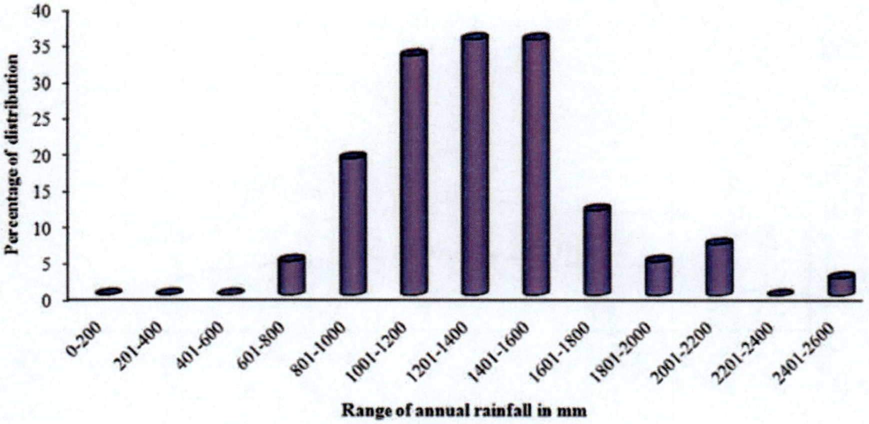

Fig. 9 Distribution percentage of annual rainfall in East Bokaro coalfield

performance. The AAR during 35 years was observed 1366 mm, with the maximum rainfall being 2544 mm in 2001 and minimum being 733 mm 2004 [12]. The probability of a minimum rainfall of 733 mm occurring is 99%, according to probability analysis. Similarly, the probability of 1366 mm rainfall was 50% while the probability of maximum rainfall of 2544 mm was only slightly more than 1% [12]. The rainfall varied between 1000 and 1400 mm were more frequent during the period (28). The range of annual rainfall distribution in percent is shown in Fig. 9.

4.2 *Temperature and Humidity Variations*

The meteorological parameters of the year 2012, indicated that the weather conditions are dry humid, sub-tropical with three distinct seasons namely dry hot summer season (March to May), rainy season (June to October) and cold winter (November to February). The coalfield area is located in tropical region with very hot summers and cool winters. However, the winter temperatures indicate that it is in the sub-tropical zone. Summer is usually from mid-March to mid-June with temperatures ranging from a maximum of 43.5 °C during daytime (08.30 am) to a minimum of 21.5 °C at night times (17.30 pm). In June, temperature can reach 43.5 °C on rare occasions. Winter persists from December to February, with of 25.8 °C during the day and minimum of 10.2 °C at night. The average annual relative humidity is about 68.5% (Fig. 10). A general analysis of rainfall occurrence, as observed in the rainfall stations indicates that the precipitation is mostly uncertain and unevenly distributed.

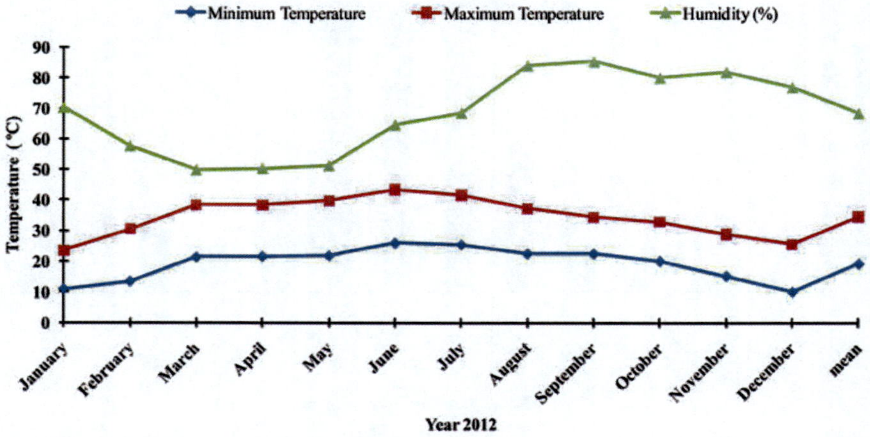

Fig. 10 Monthly variations of temperature and humidity in East Bokaro coalfield

4.3 Historical Water Levels

The study area does not have a permanent observation well. The pre-monsoon and post-monsoon historical water levels recorded by CMPDI ltd. Ranchi at the nearest permanent hydrograph stations like at Tarmi and Makoli (Dhori) for the last few years (2002 to 2012).

4.4 Water Level Trend

The water level data shows that the pre-monsoon season varies between 1.75 m (at Makoli) during 2002 to 9.95 m (at Tarmi) during 2011with an average of 5.54 m. while, the post-monsoon water levels vary from 1.60 m (Makoli) during 2004 to 7.80 m (at Tarmi) during 2009 with an average of 3.94 m. The water level fluctuation varies from 0.07 m (at Makoli) during 2011 to 4.95 m (at Tarmi) during 2011 with an average fluctuation of 1.61 m in the area. The pre-monsoon and post-monsoon water level hydrograph depict the trends of the EB coalfield area (Figs. 11 and 12).

The pre-monsoon water levels at Tarmi and Makoli shows a normal trend. The post-monsoon water level trend at Tarmi and Makoli monitoring stations shows a marginally decreasing. The marginal declining trend may be attributed to utilization by the local population. Overall, the increased population and coal mining had little impact on the local regime.

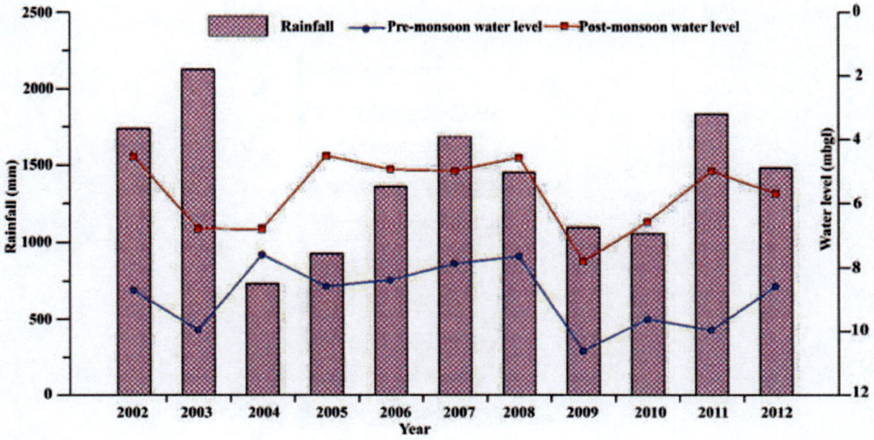

Fig. 11 Hydrograph of Tarmi monitoring site

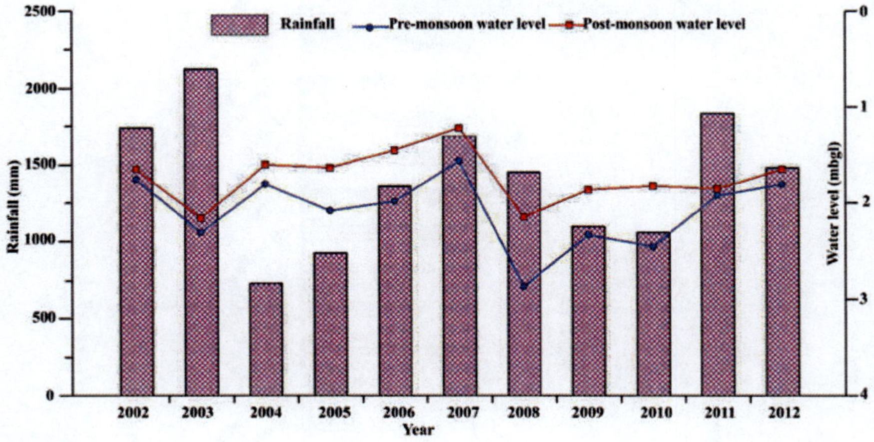

Fig. 12 Hydrograph of Makoli monitoring site

4.5 Water Table Behaviors

In the entire EB coalfield area, 32 wells were selected as observation points for monitoring the water levels in pre-monsoon as well as post-monsoon seasons during the year 2012. Table 2 and Fig. 13 show the fluctuation of groundwater level. The groundwater flow direction in the EB coal mining area was shown by the drainage pattern and water level monitoring.

The under groundwater system is primarily determined by topographic characteristics. Aquifer regime is mainly fractured controlled comprising of stratiform ore bodies due to metamorphic terrain prominent shear zone and other structural features

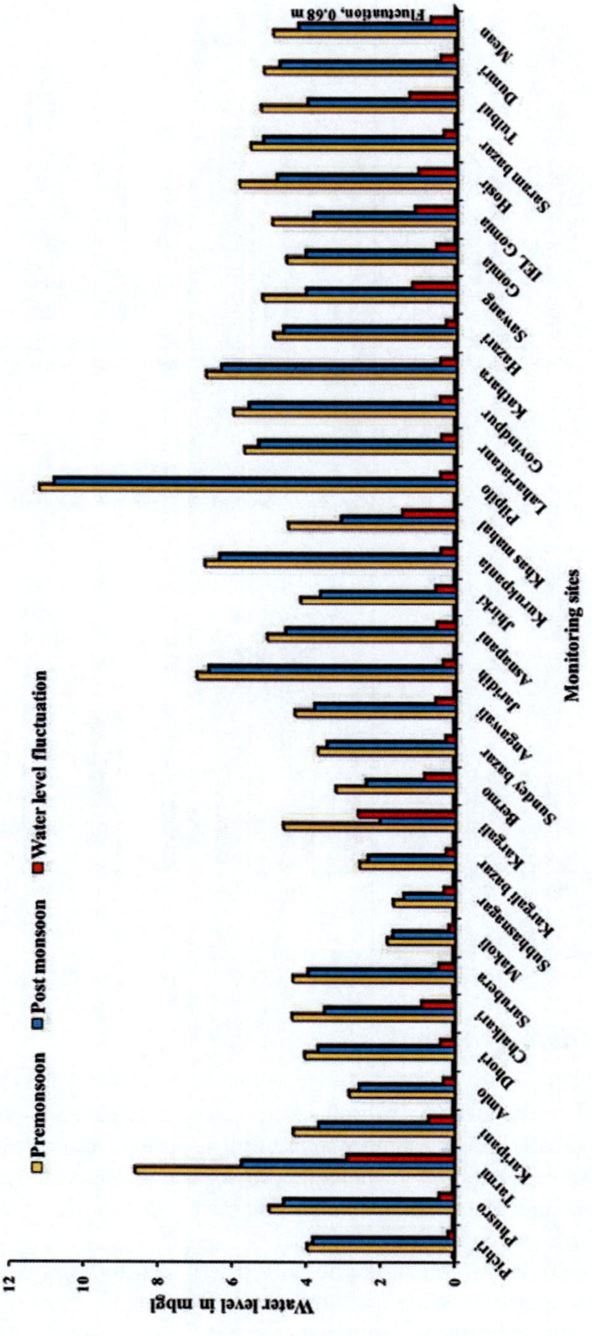

Fig. 13 water level fluctuation of East Bokaro coalfield

in the studied areas. The 32 water table monitored wells ranged from 1.64 to 11.2 m in pre-monsoon and 1.36 to10.8 m during post-monsoon. During the pre-monsoon and post monsoon season, the average water table was observed 4.92 m and 4.24 m, respectively. The well depth varied from 3–41 m. In general average water table variations found to be ranged between 0.15–2.87 m and the annual average water level fluctuation was 0.68 m [12]. The water level fluctuation in the EB coalfield area was highlighted in Fig. 13.

4.6 Assessment of Rainfall Pattern and Water Level Trend

The water quality is a critical issue for mankind since it is directly related to the human beings. Poor water quality has adversely impact on vegetative plant growth as well as human health. Groundwater quality data provide valuable hints towards geologichistory of rocks and suggest possible evidences of groundwater recharge, infiltration, and finally storage. The groundwater capacity to interact with surrounding environment and coherently spatial sharing of water flow are the two underlying causes of its active role in nature. Interaction and flow occur simultaneously at both spatial and temporal scales at changing rates and intensities. Therefore, the groundwater flows are created from the superficial land to porous deep depths of the earth's crust in different geologic duration. Between 1970 and 1990, most parts of the developing world saw a dramatic increase in groundwater extraction [30]. Evaluation, development, and management of groundwater resources depend upon the quantum of natural recharge. The quantity of natural recharge affects the evaluation, development, and management of groundwater resources. The quantity, duration, and intensity of rainfall, the depth of weathering, the specific yield of the formation, and the general slope of the terrain into drainage channels, among many other factors, affect the rise and fall of water levels. Precipitated water that enters the ground surface may get surface runoff and partially discharged into streams or partially infiltrate underground. It percolates into groundwater aquifers, gradually appearing as springs or stream or subsurface storage. When the soils moisture gets water and saturated water infilterate to recharges the groundwater. Nevertheless, during the dry season, it releases slowly as subsurface flow to join the stream as base flow. If water supplies are not always maintained by rainfall or by other resources, groundwater levels will slowly decline because of deep infiltration or evapotranspiration [31]. The recharge from adjacent surface watersheds can cause changes in groundwater levels. Pumping more groundwater than usual from a local aquifer or regional aquifers will significantly lower the water elevation in wells. Depending on the abstraction and recharge, the intensity of groundwater level fluctuation can vary between places. Water level estimation and monitoring in wells is a vital strategy for proper assessment and management.

4.7 Groundwater Prospects in East Bokaro Coalfield

Groundwater is a valuable, natural resource. As a result, assessing the potential of groundwater is not enough, but also necessary to manage it effectively to achieve long term benefits. The groundwater demarcation conservation area is needed to conserve as aquifer recharge zones from water contamination. While there are numerous methodologies to estimate the quantitative water recharge, only limited can be effectively implemented in the study areas.

Groundwater is withdrawn usually for domestic and irrigation purposes in the study area by open dug wells and small diameter hand operated tube wells. The tube wells are usually found deep (19–58 m) found compared to dug wells which have access to aquifer through weathered mantle. The coalfield areas are found in the hot-tropical reason, consequently high temperature regime is of long duration. The maximum temperature found to exceeds 46 °C daily. The extensive heat causes significant moisture depletion through evaporation by 60–65%. The net evaporation is significantly lower than the precipitation in rainy season, leading to in excess water loss due to either surface runoff or subsurface percolation and storage. The rainfall quantity, land topography and slope, land use distribution, soil characteristics, physiography, drainage pattern and hydro-geomorphology of the study sites catchment and sub-catchment all together influence surface runoff and subsurface storage of water. The study sites found to have gentle slope facing south and southeast. The hilly rain falls collected in low-lying areas through drainage and absorbed in the topsoil surface, finally reach to groundwater storage that flows towards the slope to transform into seasonal streams or Nalla. According to the GEC Report (1997), the infiltration factor for partially consolidated sandstones, weathered hard rock and fractured land should be 10–15% of normal rainfall, despite the fact that the infiltration factor of 20–25% can be utilized to evaluate the groundwater recharge prospects in sandy areas [32].

(i) *Calculation of TARR by rainfall infiltration factor method (*Table 3*):*

Other losses include natural discharges (i.e. 15% of recharge) = (−) 7.3 Mm^3
Annual net recharging of groundwater (TARR) = 41.3 Mm^3

(ii) *Calculation of TARR by groundwater table fluctuation method:*

Total annual replenishable recharge (TARR) = {Area × water table fluctuation (Average) × Specific yield}

Table 3 Total annual replenishable recharge (TARR) in Mm^3/year in the East Bokaro coalfield

TARR	= (Area × average rainfall × infiltration factor)
Total area of East Bokaro coalfield	= 237 km^2
Average rainfall	= 1366 mm
Infiltration factor	= 15%
TARR	= 237 km^2 × 1.366 m × 15%
	= 48.6 Mm^3

Table 4 Estimation of the availability of water in the East Bokaro coalfield

Parameters		Unit	Rainfall (%)	Values (Mm^3)
Total area of EB coalfield	237	km^2		
Total rainfall	1366	mm	100%	323.7
Evapotranspiration	860.6	mm	63%	203.9
Total surface runoff	300.5	mm	22%	71.2
Groundwater recharge	204.9	mm	15%	48.6

EB coalfield area = 237 km^2
Average water table fluctuation = 0.68 m
Specific yield = P*Rg/HW*(P-Rs)

The annual rainfall is P, the annual groundwater runoff is Rg, the annual surface runoff is Rs and the water table fluctuation is Hw.

= 1366 × 204.9/680 × (1366–300.5)
= 279893.4/724540 = 0.386

Table 4 represents the different parameters for estimating of water availability.

Total annual replenishable recharge (TAAR) calculation = (237 km^2 × 0.68 × 0.386)

TARR = 62.2 Mm^3

4.8 Annual Groundwater Draft Estimation

(a) *Water draft annually for irrigation purposes*

(i) Net irrigation draft is calculated using 54.5 km^2 of total irrigation land = 0.79 Mm^3
(ii) Irrigation return flow to the ground (−20%) = −0.16 Mm^3

Net irrigation use (i–ii) = 0.63 Mm^3

(b) *Water draft annually for domestic purposes*

(i) Annual domestic utilization = (Population × 70 Liter × 365 days)

= 872,821 × 70 liter × 365 days = 22.3 Mm^3

(ii) For livestock population (10% of the domestic uses) = 2.23 Mm^3

Total water draft (i + ii) = 24.5 Mm^3

(c) *Water draft annually from mine water discharges*

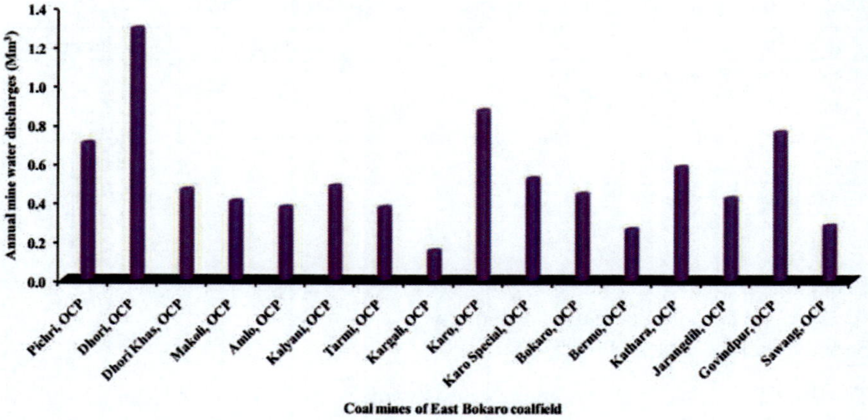

Fig. 14 Mine water discharges annually (in Mm^3) from East Bokaro coalfield

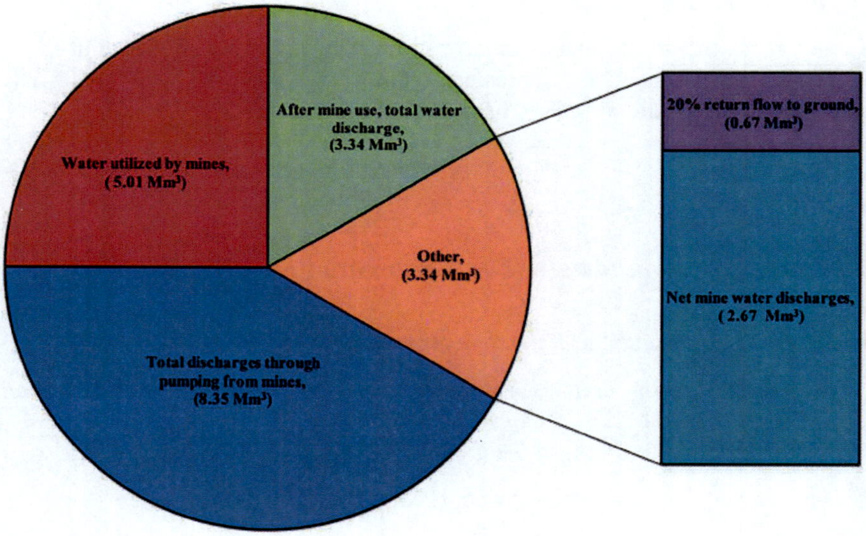

Fig. 15 Net mine water discharges annually (in Mm^3) from East Bokaro coalfield

The EB coalfield area is known to several opencast coal projects, including Pichri, Dhori, Dhori Khas, Makoli, Amlo, Kalyani, Tarmi, Kargali, Karo, Karo special, Kathara, Jarangdih, Bokaro, Govindpur and Sawang. The mine water discharges annually (in Mm^3) from each coal mines of East Bokaro coalfield depicted in Fig. 14. The highest mine water discharge calculated to be 1.29 Mm^3 per annum from Dhori open cast coal project. The net annual mine water discharges estimated to be 2.67 Mm^3 from East Bokaro coalfield shown in Fig. 15.

(i) The anticipated discharges = 3.34 Mm^3
(ii) Return flow 20% to ground = 0.67 Mm^3
(iii) Net annual discharges (i-ii) = 2.7 Mm^3

Net groundwater draft annually = (a + b + c)
= 0.63 + 24.5 + 2.7 = 27.5 Mm^3.

4.9 Net Annual Groundwater Availability Calculation

Net groundwater recharge by rainfall infiltration (annual) = 41.3 Mm^3
Net groundwater draft (annual) = 27.5 Mm^3
Net groundwater availability (annual) = (41.3–27.5) = 13.7 Mm^3

For calculating annual recharge prospective, we used a 15% average rainfall infiltration factor. The rainfall infiltration is predicted to provide 41.3 Mm^3 of total annual groundwater replenishable recharge (TARR) for the coalfield area.

In accordance with the Groundwater Assessment Report-1997 prepared by CGWB, irrigation annual draft in the study area was created [33].which was estimated to be 0.63 Mm^3. To calculate the domestic water consumption of a population (872821), @70 liters/day/person was measured. It was estimated that 22.3 Mm^3 of water will be taken for domestic uses. The water use for cattle was anticipated to be 10% of domestic consumption, with a total volume of 2.23 Mm^3 estimated. The total mine water releases into the coalfield area was estimated to be 3.34 Mm^3 through several active and abundant coal mines. Return flow to the groundwater system is anticipated to be roughly 20% of this (i.e. 0.67 Mm^3. The net mine water discharge, net groundwater recharge and net draft for EB coalfield were estimated to be 2.67 Mm^3, 41.3 Mm^3 and 27.5 Mm^3, respectively. Thus the total available groundwater resource is estimated to be 13.7 Mm^3. The water resource potentiality has been summarized in Table 5.

4.10 Groundwater Development Stages

There was only major coal mining and some coal-based industrial development activities are found in the selected study areas with no other industrial activities. As per GEC-1997 shown in Table 6, the total annual replenishable groundwater resource in the area is assessed as 41.2 Mm^3 as per rainfall infiltration factor method and total withdrawal from the area as 27.5 Mm^3 and the calculated stage of groundwater development as 67%, it falls within the category "Safe." The overall groundwater balance in the EB coalfield area is illustrated in Fig. 16.

Table 5 Summary of groundwater prospect estimation

(a)	Average of the water table	Pre-monsoon (May 2012)	4.92	mbgl
		Post-monsoon (November 2012)	4.24	mbgl
(b)	Total annual replenishable recharge (TARR)	By rainfall infiltration factor	41.3	Mm^3/year
		By groundwater table fluctuation	62.2	Mm^3/year
(c)	Annual draft excluding estimated draft through mine discharge		25.2	Mm^3/year
(d)	Estimated draft through mine discharge		2. 7	Mm^3/year
(e)	Net annual groundwater availability		13.7	Mm^3/year
	Stages of groundwater development		**67**	(%)

Table 6 Classification as per the stage of groundwater development

Groundwater level trend	Stage of groundwater development (%)	Classifications
Either pre-monsoon or post- monsoon Water levels do not show a falling trend	≤70	Safe
Both the trends during pre- and post-monsoon seasons do not show a falling trend	70–90	Safe
Either pre-monsoon or post- monsoon water levels show a falling trend	70–90	Semi-critical
Either pre-monsoon or post- monsoon water levels show a falling trend	>90	Critical
Both the trends during pre and post- monsoon seasons show a falling trend	<100	Critical
Both the trends during pre and post- monsoon seasons show a falling trend	>100	Over exploited

5 Conclusions

Groundwater recharge is mainly dependent on rainfall. The EB coalfield area receives an average annual rainfall of around 1366 mm with the maximum value estimated in 2001 being 2543.8 mm. In addition to rainfall, the present study sites experiences the mine water, water bodies and water logged coal mine pits release water contributing in recharge of groundwater, as return flow. However, insignificant depleting trend of groundwater level fluctuation was observed and may be attributed due to increase in consumption by the local dwellers. The total annual replenishable recharge (TARR) was estimated to be 48.6 Mm^3/year using the rainfall infiltration factor method and by using the groundwater table fluctuation method, TARR was calculated to be 62.2 Mm^3/year. The net annual groundwater availability was calculated to be 13.7 Mm^3 using rainfall infiltration techniques. The different

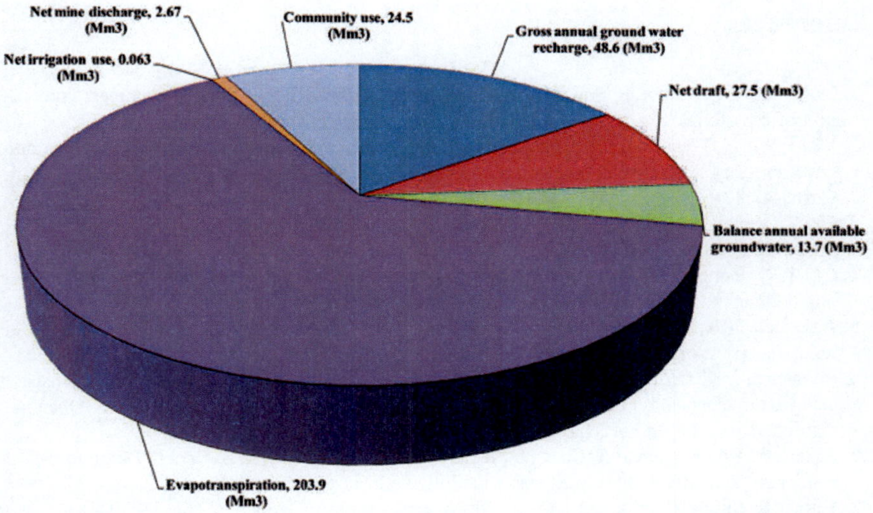

Fig. 16 Overall groundwater balance in the East Bokaro coalfield

stages of groundwater development was estimated which has been considered under the safe category. (*i.e.* 67%). Large aquifers of mining area modelling appear to be practically difficult task. Present study will be helpful for the local people in knowing the risks to the availability of water resources in the EB coalfield area.

6 Recommendations

• Evaluation of specific yield of the wells by pumping test studies at suitable locations.
• Evaluation of the Transmissivity & Storativity of the underlying aquifer.
• Contaminant Transport Modeling can be elaborated by visual MODFLOW using data of few years.
• Similar studies may be taken up for mining areas other than coal mining like Copper and Uranium mining.

Acknowledgments The authors gratefully acknowledge the Ministry of Human Resource Development, GoI, New Delhi (for the ISM/JRF fellowship) and the University Grants Commission (for Dr. D. S. Kothari's postdoctoral fellowship-OT/15-16/0017) for their financial sponsor. Thanks to the editors and reviewers for their insightful comments and advice.

References

1. Vorosmarty CJ, Green P, Salisbury J, Lammers RB (2006) Global Water Resources: vulnerability from climate change and population growth. Science 289:284–288
2. Wei J, Wei Y, Western A (2017) Evolution of the societal value of water resources for economic development versus environmental sustainability in Australia from 1843 to 2011. Glob Environ Chang 42:82–92
3. Oki T, Kanae S (2006) Global hydrological cycles and world water resources. Science 313: 1068–1072
4. Oki T, Quiocho RE (2020) Economically challenged and water scarce: identification of global populations most vulnerable to water crises. Int J Water Resourc Dev 36(2–3):416–428
5. Abhishek TRK, Sinha SK (2006) Status of Surface and Groundwater quality in Coal mining and industrial areas of Jharia Coalfield. Int J Environ Prot 26(10):905–910
6. Srivastava VK, Giri DN, Bharadwaj P (2012) Study and mapping of ground water prospect using remote sensing, GIS and geoelectrical resistivity techniques – a case study of Dhanbad district, Jharkhand, India. J Ind Geophys Union 16(2):55–63
7. Saxena S, Ramakrishnan S, Soni D, Meena KK, Arora S (2020) Water quality monitoring of residential area affected by solid waste landfill site in Jaipur city
8. Kumar R, Singh RD, Sharma KD (2005) Water resources of India. Curr Sci 89:794–811
9. Tiwari AK, Singh AK (2014) Hydrogeochemical investigation and groundwater quality assessment of Pratapgarh district, Uttar Pradesh. J Geol Soc India 83(3):329–343
10. Singh PK, Bhakat D, Singh G (2010) Qualitative and quantitative assessment of groundwater of coal mining areas, India. In: Fifth int. conf. on environmental science and technology, Houston, Texas, USA, July 12–16
11. Kumar CP (2012) Assessment of groundwater potential. Int J Eng Sci 1(1):64–79
12. Mahato MK, Singh G, Giri S, Mishra LP, Tiwari AK (2018) Quantitative assessment of groundwater resource potential in a coalfield of Damodar River Basin India. Sustain Water Resourc Manag 4:509–517
13. Mahala, A (2021) Delineating the status of groundwater in a plateau fringe region using Multi-influencing Factor (MIF) and GIS: a study of Bankura District, West Bengal, India. Geostatistics and Geospatial Technologies for Groundwater Resources in India. 215–237
14. Tripathi S, Srinivas VV, Ravi S, Nanjundiah (2006) Down scaling of precipitation for climate change scenarios: a support vector machine approach. J Hydrol 330:621–640
15. Hem JD (1985) Study and interpretation of the chemical characteristics of natural water (3rd ed.). US Geol Survey Water Supply 2254:263–264
16. Sophocleous MA, Perkins SP (2000) Methodology and application of combined watershed and groundwater models in Kansas. J Hydrol 236:185
17. Chandra S, Singh PK, Tiwari AK, Panigrahya BP, Kumar A (2015) Evaluation of hydrogeological factors and their relationship with seasonal water table fluctuation in Dhanbad district, Jharkhand, India. J Hydraulic Eng 21(2):193–206
18. Raju KCB (1998) Importance of recharging depleted aquifers: state of the art of artificial recharge in India. J Geol Soc India 51:429–454
19. Raja Rao CS (1987) Bulletin of geology survey of India, Sr A; No.45; IV(I):8–60
20. Sastry MVA, Acharya SK, Shah SC, Satsangi PP, Ghosh SC, Raha PK, Singh G, Ghosh RN (1977) Stratigraphic lexicon of Gondwana formations of India. Geol Surv India Misc Publ 36: 1–170
21. Tripathi C, Satsangi PP (1963) Lystrosaurus Fauna of the panchet series of the Raniganj Coalfield. Pal Ind Geol Surv India 37(3):1–53
22. Maheswari HK, Banerji J (1975) Lower triassic palynomorphs from the Maitur formation, West Bengal, India. Palaeonogr B 152:149–190
23. Kumar P, Singh AK (2020) Hydrogeochemistry and quality assessment of surface and sub-surface water resources in Raniganj coalfield area, Damodar Valley, India. Int J Environ Anal Chem, 1–24

24. Sharma PJ, Patel PL, Jothiprakash V (2019) Impact of rainfall variability and anthropogenic activities on streamflow changes and water stress conditions across Tapi Basin in India. Sci Total Environ 687:885–897
25. Kumar V, Kaushal RK, Taloor AK, Jain V (2021) Incorporation of slope and rainfall variability in channel network extraction from DEM data at basin scale. Geocarto Int, 1–16
26. Wang YH, Tseng YH (2015) Raster mapping of topographic parameters derived from high resolution Digital Elevation Models. In 36th Asian conference on remote sensing: fostering resilient growth in Asia, ACRS
27. Singh TBN (2009) Ground water information booklet, Central Groundwater Board, Patna
28. CGWB (Central Ground Water Board) (2009) "Ground Water Information Booklet" Bokaro District, Jharkhand State. http//cgwb.gov.in/NEW/Annual-Reports/Annual%20Report%202009-10.pdf
29. GEC (Groundwater Resource Estimation Methodology) (1997) Report of the Groundwater Resource Estimation Committee, Ministry of Water Resources, Government of India, New Delhi
30. Villholth KG (2006) Groundwater assessment and management, implications and opportunities of globalization. Hydrogeol J 14:330–339
31. Jan CD, Chen TH, Lo WC (2007) Effect of rainfall intensity and distribution and distribution groundwater level fluctuations. J Hydrol 332:348–360
32. Kumar G, Singh SK, Sinha BK, Prasad BM (2010) Groundwater recharge potential of the Tasra Block of Chasnala—a case study. Proceedings of the XI International Seminar on Mineral Processing Technology (MPT-2010), 1111–1119
33. Sophocleous M, Perkins SP (2000) Methodology and application of combined watershed and ground-water models in Kansas. J Hydrol 236(3–4):185–201. https://doi.org/10.1016/S0022-1694(00)00293-6

Part VI
Control of Water Quality

Distribution and Statistical Source Identification of Heavy Metals and Radionuclides in Soil: A Case Study from a Proposed Uranium Mining Site, Jharkhand, India

Soma Giri, Gurdeep Singh, and V. N. Jha

Abstract Singhbhum Thrust Belt is well recognized for its uranium and copper deposits and thus radionuclides as well as metals can be anticipated in its environs inclusive of soil. The enrichment of metals and radionclides in the vicinity of the uranium mining areas can be attributed to their release due to mining activities and waste disposal. As a part of the baseline study of the Bagjata mining area, 10 soil samples were collected from the villages in proximity of the mining area and analyzed for radionuclides (U(nat), ^{226}Ra, ^{230}Th, ^{210}Po) and metals (Fe, Mn, Zn, Pb, Cu, and Ni). The geometric mean activity of ^{226}Ra, ^{230}Th and ^{210}Po was 23.24, 65.54 and 55.15 Bqkg^{-1}, respectively while geometric mean concentration of U (nat), Fe, Mn, Zn, Pb, Cu and Ni was calculated to be 2.37, 26,106, 267.6, 53.2, 37, 159.6 and 141.4 mgkg^{-1}, respectively. Cu and Ni surpassed the Maximum Allowable limits (MAL) in the majority of locations. Owing to uneven and sporadic mineralization of the region, widespread spatial variation in the heavy metals and radionuclides has been observed. The Principal Component Analysis (PCA) depicted the geogenic contribution as the main source of radionuclides and metals. The secondary sources of metals were illustrated to be anthropogenic sources like mining and related vehicular load. The association of the radionuclides with the Fe is also depicted from the study.

Keywords Radionuclides · Heavy metals · Soil · Principal component analysis · Uranium mining

S. Giri (✉)
Natural Resource and Environmental Management Research Group, CSIR-Central Institute of Mining and Fuel Research, Dhanbad, Jharkhand, India

G. Singh
IIT (Indian School of Mines), Dhanbad, Jharkhand, India

V. N. Jha
Environmental Assessment Division, Bhabha Atomic Research Centre, Mumbai, India

S. Yadav et al. (eds.), *Water Quality, Assessment and Management in India*, Earth and Environmental Sciences Library, https://doi.org/10.1007/978-3-030-95687-5_18

1 Introduction

In recent years, owing to rapid industrial growth, one of the major environmental problems that have surfaced is land degradation. This is further aggravated by the immense increase in the mining activities. As in case of other mining activities such as metal mining or coal mining, uranium mining can also be apprehended to pose risk to the adjoining environment due to the release of enormous amount of waste rich in metals and radioactive elements. However, the extent of the influence is dependent on a number of factors such as the method and quantity of uranium excavation, infrastructure for the treatment of uranium, facility of waste disposal etc. [1, 2].

In the contaminant transport system, soil-plants-food-man, the soil forms the first line of contamination and thus it is important to know activity of the soil. The background radiation in soil is extensively contributed by the gamma radiations from U and Th series elements and radioactive K [3]. A significant amount of radiation exposure to the human population is contributed by these natural radionuclides of the soil [4, 5].

The transport and buildup of the pollutants in the soil involves intricate with processes like absorption, adsorption, leaching, plant uptake, runoff, etc. [6, 7]. Small amount of U and Th with their daughter radionuclides are naturally present in soil however, their levels are dependent on the lithology of the area and the type of the rocks which forms the soil [8].

The multivariate statistical tools such as principal component analysis (PCA) help in source apportionment of the contaminants with data reduction and identification of the factors which explains the most of the data variance. The factors are extracted in such a manner so as to protect the associations in the original data [9–11].

In several parts of eastern Singhbhum uranium mining have started subsequent to the discovery of low-grade uranium deposits in the region. With the increase in demand of nuclear energy, new mining sites have been excavated for uranium. One such underground uranium mine site was at Bagjata. Mining activities indiscriminately affect the environmental status of the mining and adjoining areas and thus baseline studies are important to ascertain the pre-existing environmental condition [12]. The study presented here was a part of baseline study conducted for the Bagjata mining area of which soil was also a component. The soils from the mining and surrounding areas were studied for their content of radionuclides and heavy metals.

2 Materials and Methods

2.1 *Study Area*

Singhbhum Thrust Belt (STB) is known worldwide for its rich deposits of copper, apatite, magnetite and kyanite. STB is an EW trending belt of approximately 160 km

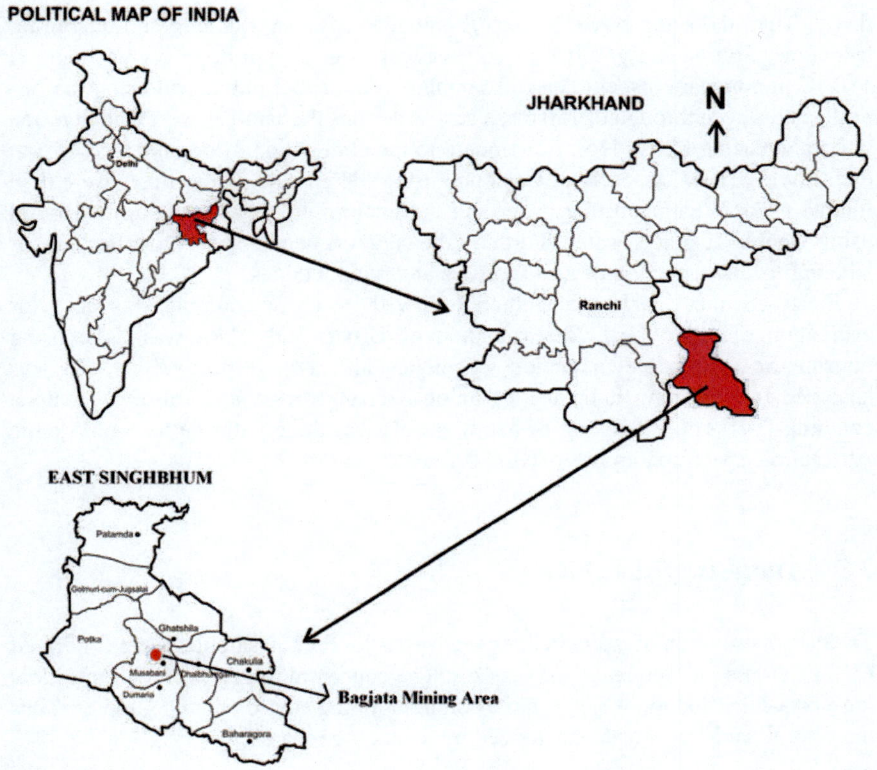

Fig. 1 Locations detail of the study area (adapted from [12])

length in which the mineralization occurred in 2 phases. The first phase was the deposition of apatite and magnetite and subsequently uranium. The second phase included the deposition of sulphides including chalcopyrites. U-Cu mineralization coexists in the area besides other minerals of Fe, Mn, etc. The study area under consideration i.e. Bagjata mining area is situated in the Bhalki-Kanyaluka deposit [13] and has latitude of 22°26′07″N to 22°28′34″N and longitude of 86°25′16″E to 86°31′29″E. The key minerals of U of the region consist of uraninite and pitchblende along with autonite as the secondary mineral. Sulphides of various metals like Cu, Ni, Co, Bi, As, etc. are also associated to the U minerals [14].

Figure 1 depicts the location map of the study area.

2.2 *Sample Collection and Laboratory Analysis*

The soil samples were collected from approximately 0–20 cm below the ground surface. The grassroots and other debris which were present in the soil were removed firstly. Then the samples were mixed homogenously using coning and quartering technique. Subsequently 100 g of soil was taken and oven dried for 24 hours at 110 °C in porcelain dishes. The soil samples were further pulverized using mortar-pestle and sieved through of 100 mesh size. After this the samples were acid digested in hotplate using HNO_3 [15]. Subsequent to the elimination of organic matter from the samples, repeated leaching was done using 8N HNO_3. The samples were then filtered using Whatmann filer paper and the final volume was made up to 100 ml using double distilled water. Further, the aliquots were preserved after proper labeling for the detection of heavy metals and radionuclides.

Electrochemical exchange method followed by alpha counting was used for evaluation of ^{210}Po [16]. Determination of U(nat) and ^{226}Ra was done using fluorimetric and radon emanation techniques [17, 18], respectively. ^{230}Th was analyzed by screening it by use of anion exchange resin and subsequent alpha counting [19]. The evaluation of heavy metals was done with the help of Atomic Absorption Spectrophotometer (GBC Avanta).

2.3 Analysis of the Data

The data distribution of radionuclides and metals for the soil samples were confirmed by the cumulative frequency curves [20]. The concentrations were found to follow log-normal distribution which is quite common for the environmental samples. Thus the central tendency was characterized by geometric mean [21].

2.4 Principal Component Analysis (PCA)

Principal component analysis (PCA) is a statistical source apportionment tool which replaces a larger number of interrelated variables with smaller number of independent variables called principal components (PCs). In this method, the covariance matrix of original variables is used to obtain the eigenvalues and eigenvectors [22]. The PCs are weighted linear groupings of original variables that give idea about the most dominant factors and in the process there is least loss of original information [23–26]. Further reduction of the effect of the less important variables is achieved by factor analysis which is done by extracting the varifactors obtained by rotation of axis as defined by PCA [27, 28].

2.5 *Enrichment Factor (EF) and Geo-accumulation Index (I_{geo})*

For the assessment of anthropogenic influence on the metal concentration of the soil of the study area enrichment factors (EF) were evaluated by the equation:

$$EF = (C_X/C_{Fe})_s/(C_X/C_{Fe})_c$$

Where, C_X and C_{Fe} refer to the concentration of metal X and Fe in the soil (s) and earth's crust (c), respectively. Iron is taken as the reference metal.

The degree of metal pollution in the soil of the study area was quantitatively evaluated using the Index of Geoaccumulation (I_{geo}) [29] as:

$$I_{geo} = \log_2 (C_n/1.5\, B_n)$$

Where, C_n is the measured concentration of the heavy metal 'n' in the soil, and B_n is the geochemical background value in average shale [30].

3 Results and Discussion

3.1 *Radionuclides*

Table 1 depicts the descriptive statistics of the analytical results of radionuclides and heavy metal in the soil samples of the study area and Table 2 provides the average concentration of the radionuclides in the soil with respect to sampling locations. It can be observed from the Tables that Uranium varied from 0.46 to 12.73 mg kg^{-1}, the highest being at Bagjata village. The activities of ^{226}Ra, ^{230}Th, and ^{210}Po vary from 4.1 to 145, 14.3 to 407.1 and 7.8 to 172.2 $Bqkg^{-1}$, respectively. The geometric

Table 1 Descriptive statistics of the activity of radionuclides and concentration of heavy metals in soil samples (unit: ^{226}Ra, ^{230}Th, ^{210}Po in $Bqkg^{-1}$; all other parameters in $mgkg^{-1}$)

Parameters	N	Minimum	Maximum	Geometric mean
U(nat)	40	0.46	12.73	2.37
^{226}Ra	40	4.1	145.0	23.24
^{230}Th	40	14.29	407.14	65.64
^{210}Po	40	7.78	172.22	55.15
Fe	40	13,132	64,798	26,106
Mn	40	104	586	267.6
Zn	40	26	115.1	53.2
Pb	40	16.6	81.0	37.0
Cu	40	70.1	392.2	159.6
Ni	40	64.0	296.0	141.4

Table 2 Average activity/concentration of radionuclides in soil samples with respect to different locations (unit: U (nat) in mgkg^{-1}; other parameters in Bqkg^{-1})

Code	Location	U(nat)	^{226}Ra	^{230}Th	^{210}Po
1	Bagjata	11.0	74.2	204.2	104.8
2	Bhaduya	2.3	20.6	110.0	100.5
3	Phuljhari	3.9	16.0	54.0	89.5
4	Manajhari	0.9	9.0	66.3	48.8
5	Balidungri	1.3	16.0	52.8	25.2
6	Bakra	3.3	69.5	43.1	50.6
7	Katsakra	1.9	4.7	74.8	31.8
8	Gohala	1.5	23.4	62.7	59.4
9	Nimdih	7.4	66.4	52.7	128.7
10	Surda	1.8	16.1	61.0	39.6

mean concentration of U(nat) was calculated to be 2.37 mgkg^{-1} while the geometric mean activity of ^{226}Ra, ^{230}Th and ^{210}Po were analyzed to be 23.24, 65.54 and 55.15 Bqkg^{-1}, respectively. The highest activity of all the considered four radionuclides was found in the same location i.e. Bagjata village. This may be attributed to the location being in close proximity to the mining site.

An uneven distribution of the specific activity of the considered radionuclides in the soil is observed in the study area. There is widespread uranium mineralization in this area, and accordingly wide variation in the concentration of radionuclides is expected [31]. This may be due to the intense weathering of the regolith following precipitation. The chemical characteristics of various radionuclides are different and so the extent of leaching. Apart from this, organic complexation of uranium and adsorption of polonium over ferric hydroxides may also lead to disequilibrium between the radionuclides of uranium series.

3.2 *Heavy Metals*

Owing to uneven mineralization, widespread spatial variation in the heavy metals has been observed. Table 1 depicts the descriptive statistics of heavy metal concentrations in the soil samples of the study area while the comprehensive results are presented in Tables 3. Taking into consideration all the four sampling episodes, the concentration of Fe was found in the range of 13,132 to 64,798 mgkg^{-1}; with the geometric mean of 26,106 mgkg^{-1}. The concentration of Mn and Zn varied from 104 to 586 and 26 to 115.1 mgkg^{-1}, respectively. The geometric mean concentration taking into account all the locations and all the sampling periods for Zn was found to be 53.2 mg kg^{-1}. The Pb concentration was evaluated to be in between 16.6 to 81.0 mgkg^{-1}. None of the locations exceeded the Maximum Allowable limits for Zn and Pb of 300 and 100 mgkg^{-1}, respectively as stated by Kloke [32].

However, both Cu and Ni surpassed the Maximum Allowable limits given by Kloke [32] i.e. 100 and 50 mgkg^{-1} for Cu and Ni, respectively. The area falls in the uranium-copper-nickel deposits of Singhbhum Thrust Belt, with extensive mining of

Table 3 Average concentration of heavy metals in soil samples for different locations (Unit: $mgkg^{-1}$)

Location	Fe	Mn	Zn	Pb	Cu	Ni
Bagjata	52,531	482.8	50.2	64.5	323.5	256.0
Bhaduya	22,467	274.2	64.9	41.8	91.3	131.8
Phuljhari	31,947	393.5	38.0	49.4	122.3	100.0
Manajhari	18,267	171.9	57.5	25.4	145.0	131.1
Balidungri	30,177	309.0	95.9	45.4	194.0	91.3
Bakra	22,818	311.0	85.0	46.8	198.3	176.0
Katsakra	21,478	251.0	39.4	28.7	150.1	135.4
Gohala	21,281	393.8	40.0	26.8	231.2	141.3
Nimdih	42,356	189.7	55.8	33.8	112.6	177.7
Surda	20,784	145.8	47.8	45.6	188.7	162.5

Table 4 Principal Component Analysis (PCA) for the elements in the soil of the study area (n = 40)

Elements	PC1	PC2	PC3
U(nat)	**0.935**	−0.117	−0.002
^{226}Ra	**0.691**	0.376	0.002
^{230}Th	**0.573**	−0.416	−0.008
^{210}Po	**0.695**	0.002	−0.488
Fe	**0.855**	−0.006	0.276
Mn	**0.588**	−0.009	0.048
Zn	0.005	**0.831**	0.237
Pb	0.191	**0.675**	0.437
Cu	−0.114	0.108	**0.615**
Ni	−0.126	−0.006	**0.877**
Eigen values	**2.99**	**2.36**	**1.48**
%Total variance	**29.7**	**20.0**	**18.6**
Cumulative variance	**29.7**	**49.7**	**68.3**

Cu. This may be attributed for the widespread spatial variation and elevated concentration of Cu and Ni in the soil samples of the study area. The concentration of Cu and Ni varied between 70.1 to 392.2 and 64 to 296 mg kg^{-1}, respectively. The level of Cu and Ni surpassed the Indian limits for soil [33] also at some of the locations. The highest concentration of Cu and Ni were found in the Bagjata village.

This is an obvious observation that the maximum concentration of heavy metals and radionuclides was found in the same location (Bagjata village) which is in very close proximity to Bagjata mining site. Though the concentration of metals and radionuclides were found to be highest in the same location no definite trend could be established in the spatial variation with respect to distance from the site.

3.3 *Principal Component Analysis (PCA)*

The results of PCA which is a statistical source apportionment tool based on inter-element correlations are presented I Table 4. The PCs obtained corresponds to cluster of metals that are linked together and so they may be associated with a common source and similar behavior [34]. The PCA of soil of Bagjata area depicts that 3 PCs obtained after varimax rotation explain the 68.3% of data variance. The number of principal components having eigenvalue greater than 1 are selected which is based on the Kaiser criterion [35, 36].

The first factor explaining 29.7% of variance and having high loading for U(nat), ^{226}Ra, Th(α), ^{210}Po, Fe and Mn appears to be linked to the geological setting of the region. The factor may also be related to the linkage of Fe with other radionuclides since it is a well known fact that iron oxides and hydroxides adsorb radionuclides inclusive of Uranium, Radium and Thorium [37–40]. The second component (PC2) depicted a strong association of Pb and Zn justifying 20% of variance and may be linked to vehicular pollution particularly from the mining related vehicles. Pb is known to be associated to vehicle related emissions [41]. Zn is used in the tires of the vehicles as vulcanization agent [42] which has strong probability of getting weared out due to high temperature of the region and thus contribute in increase in Zn concentration of the soil. The third factor (PC3) explains 18.6% of variance and is emphasized by the metals Cu and Ni. This factor can be allied to the widespread Copper mining activities of the region.

3.4 Enrichment Factor and Geo-accumulation Index

The Enrichment factor calculated for the metals of soil of the study area depicted moderate contamination (EF = 2 to 5) [43, 44] for Zn, Pb and Ni and significant contamination (EF = 5 to 20) [43, 44] with respect to Cu (Table 5). This may be attributed to the anthropogenic activities i.e. copper mining and vehicular pollution of the area which is also depicted from the PCA results.

The Geoaccumulation index (I_{geo}) values were calculated and classified as per Muller [29] (Table 6). The I_{geo} values suggested an unpolluted status (Class 0) of the soil with respect to Fe, Mn, and Zn. Ni and Pb falls in Class 1 indicating an unpolluted to moderately polluted status in the soil. However, the I_{geo} values for Cu were found to be in between 1 and 2 (Class 2), thus suggesting a moderate pollution with respect to Cu in the soil samples. This is in accordance to the calculated enrichment factors.

Table 5 Enrichment Factor of elements in the study area

	Fe	Mn	Zn	Pb	Cu	Ni
Average background shale [30]	47,200	850	95	20	45	68
Geometric mean	26,106	267.7	53.2	37	159.6	141.4
Enrichment Factor	1	0.569	**1.012**	**3.345**	**6.412**	**3.760**

Table 6 Geo-accumulation Index of Elements in the study area

I_{geo} Value	I_{geo} Class [29]	Sediment Quality (Muller [29])	Element
>5	6	Very highly polluted	
4–5	5	Highly polluted to very highly polluted	
3–4	4	Highly polluted	
2–3	3	Moderately polluted to highly polluted	
1–2	2	Moderately polluted	Cu
0–1	1	Unpolluted to moderately polluted	Pb, Ni
<0	0	Unpolluted	Fe, Mn, Zn

4 Conclusions

The present study which was a part of the baseline study of the Bagjata mining areas revealed a natural background concentration with respect to radionuclides in the soil. Amongst the studied metals, Cu and Ni surpassed the Maximum Allowable limits for the majority of the locations. The radionuclides and metals depicted an uneven distribution in the area which may be accredited to the widespread U-Cu mineralization in the region. The PCA suggested that the major source of radionuclides and metals in the soil is geogenic; however, some anthropogenic sources like mining and vehicular load also contribute to the metal content in the soil. The association of the radionuclides with Fe is also illustrated from the study. Moderate pollution with respect to Cu in the soil of the area is also depicted from the study. The study holds importance owing to the extensive industrial and mining activities in the region.

Acknowledgement The financial support for the baseline study of which the present study is a part was provided by Board of Research in Nuclear Sciences (BRNS), Department of Atomic Energy, GoI. The required laboratory facilities for carrying out the study were provided by Health Physics Unit (HPU), Jadugoda and Department of Environmental Science and Engineering, IIT-Indian School of Mines (IIT-ISM), Dhanbad. The authors extend their gratitude BRNS, HPU and IIT-ISM for supporting the study.

Recommendations

i. Regular monitoring of soil samples in the environs of the mining areas is recommended so as to know the post mining effect with respect to radionuclide and metals.
ii. The content of radionuclides and metals in the agricultural produces should be done since there is a strong probability for the pollutants to get into the food chain through the soil
iii. Human health risk assessment is suggested to be carried out owing to the exposure related to radionuclides and metals in the soil through the ingestion, dermal and inhalation pathways.

References

1. Lozano JC, Blanco RP, Vera Tome F (2002) Distribution of long-lived radionuclides of the ^{238}U series in the sediments of a small river in a uranium mineralized region of Spain. J Environ Radioact 63:153–171

2. Srivastava RR, Pathak P, Perween M (2020) Environmental and health impact due to uranium mining. In: Uranium in plants and the environment. Springer, Cham, pp 69–89
3. Noureddine A, Baggoura B, Larosa JJ, Vadja N (1997) Gamma and alpha emitting radionuclides in some Algerian soil samples. Appl Radiation Isotopes 48:1145–1148
4. Miah FK, Roy S, Touhiduzzamin M, Alam B (1998) Distribution of radionuclides in soil samples in and around Dhaka city. Appl Radiation Isotopes 49:133–137
5. Ajithra AK, Venkatraman B, Jose MT, Chandrasekar S, Shanthi G (2017) Assessment of natural radioactivity and associated radiation indices in soil samples from the high background radiation area, Kanyakumari district, Tamil Nadu, India. Radiat Protect Environ 40(1):27
6. Mortverdt JJ (1994) Plant and soil relationships of uranium and thorium decay series radionuclides – a review. J Environ Qual 23:643–650
7. Zhang X, Yang H, Cui Z (2018) Evaluation and analysis of soil migration and distribution characteristics of heavy metals in iron tailings. J Cleaner Prod 172:475–480
8. Zahid CS, Hasan MK, Aslam M, Iqbal S, Orfi SD (2001) Study of 137Cs contamination in soil and food samples of Jhanger valley, Pakistan. Nucleus 38:101–105
9. Ouyang Y, Nkedi-Kizza P, Wu QT, Shinde D, Huang CH (2006) Assessment of seasonal variations in surface water quality. Water Res 40:3800–3810
10. Shrestha S, Kazama F (2007) Assessment of surface water quality using multivariate statistical techniques: a case study of the Fuji river basin; Japan. Environ Model Softw 22:464–475
11. Ogunlaja A, Ogunlaja OO, Okewole DM, Morenikeji OA (2019) Risk assessment and source identification of heavy metal contamination by multivariate and hazard index analyses of a pipeline vandalised area in Lagos State, Nigeria. Sci Total Environ 651:2943–2952
12. Giri S, Singh G, Jha VN, Tripathi RM (2010) Ingestion of U(nat) , ^{226}Ra, ^{230}Th, and ^{210}Po in vegetables by adult inhabitants of Bagjata uranium mining area, Jharkhand, India. Radioprotection 45:183–199
13. Bhola KL, Dar KK, Ramarao YN, Suri Sastry C, Mehta NR (1964) A review of Uranium and Thorium deposits in India. In: 3rd International conference of the peaceful uses of Atomic Energy. Multilingual Edition United Nations 12: 750–756
14. Sarangi AK, Singh AS (2006) Vein type Uranium Mineralisation in Jaduguda Uranium Deposits, Singhbhum, India. In: International symposium on understanding the genesis of ore deposits to meet the demands of 21th century: association on the genesis of ore deposits, Moscow, 12: 54–61
15. Richards LA (1968) Diagnosis and improvement of saline and alkaline soils (1st ed.). Agri. Handbook No. 60, IBH Publications Company, New Delhi
16. Figgins PE (1961) Radiochemistry of polonium. National Academy of Science, NS Series, 3037
17. Kolthoff IM, Elving PJ (1962) Treatise on Analytical Chemistry, part II, Vol. 9
18. Ragjavayya M, Iyengar MAR, Markose PM (1980) Estimation of Radium–226 by emanometry. Bull Indian Assoc Radiat Protect 3:11–15
19. Hyde EK (1960) Radiochemistry of thorium. National Academy of Science, NS series, 3004
20. Miller JC, Miller JN (1989) Statistics for analytical chemistry, 2nd edn. Ellis Horwood Limited, New York
21. Wayne RO (1990) A physical explanation of the lognormality of pollutant concentrations. J Air Manage Assoc 40:1378–1383
22. Simeonov V, Stratis JA, Samara C, Zachariadis G, Voutsa D, Anthemidis A (2003) Assessment of the surface water quality in northern Greece. Water Res 37:4119–4124
23. Hair JF, Anderson RE, Tatham RL, Black WC (1995) Multivariate data analysis with readings, 4th edn. Prentice-Hall, London
24. Sharma S (1996) Applied multivariate techniques. Wiley, New York
25. Vega M, Pardo R, Barrato E, Deban L (1998) Assessment of seasonal and polluting effects on the quality of river water by exploratory data analysis. Water Res 32:3581–3592
26. Yang Y, Yang X, He M, Christakos G (2020) Beyond mere pollution source identification: Determination of land covers emitting soil heavy metals by combining PCA/APCS. GeoDetector and GIS Anal Catena 185:104297

27. Wunderlin DA, Días MP, Amémaría V, Pesce SF, Hued AC, Bistoni MÁ (2001) Pattern recognition techniques for the evaluation of spatial and temporal variations in water quality. A case study: Suquia river basin (Cordoba–Argentina). Water Res 35:2881–2894
28. Salim I, Sajjad RU, Paule-Mercado MC, Memon SA, Lee BY, Sukhbaatar C, Lee CH (2019) Comparison of two receptor models PCA-MLR and PMF for source identification and apportionment of pollution carried by runoff from catchment and sub-watershed areas with mixed land cover in South Korea. Sci Tot Environ 663:764–775
29. Muller G (1979) Schwermetalle in den Sedimenten des Rheins-Veranderungen seit 1971. Umschau 79:778–783
30. Turekian KK, Wedepohl KH (1961) Distribution of the elements in some major units of the earth's crust. Bull Geol Soc Am 72:175–192
31. Jha VN, Sethy NK, Sahoo SK, Shukla AK, Tripathi RM, Khan AH (2005) A comparison of radioactivity level in discharge waste and natural sources in uranium mineralized areas of Singhbhum, Jharkhand. In: Proceedings of 27th IARP National conference of Occupational and Environmental Radiation Protection, BARC, Mumbai, 284–286
32. Kloke A (1980) Richwerte '80, Orientierungsdaten für tolerierbare Gesamtgehalte einiger Elemente in Kulturböden, Mitt. VDLUFA H 2:9–11
33. Awasthi SK (2000) Prevention of Food Adulteration Act No. 37 of 1954, Central and State rules as amended for 1999. Ashoka Law House, New Delhi, India
34. Tahri M, Benyaich F, Bounakhla M, Bilal E, Gruffat JJ, Moutte J (2005) Multivariate Analysis of heavy metal contents in soils, sediments and water in the region of Meknes (Central Morocco). Environ Monit Assess 102:405–417
35. Kaiser HF (1960) The application of electronic computers to factor analysis. Educat Psychol Meas 20:141–151
36. Vitale R, Westerhuis JA, Næs T, Smilde AK, De Noord OE, Ferrer A (2017) Selecting the number of factors in principal component analysis by permutation testing—numerical and practical aspects. J Chemometrics 12:e2937
37. Musić S, Ristić M (1988) Adsorption of trace elements or radionuclides on hydrous iron oxides. J Radioanalyt Nuclear Chem 122:289–304
38. Wanty RB, Johnson SL, Briggs PH (1991) Radon-222 and its parent nuclides in groundwater from two study areas in New Jersey and Maryland, USA. Appl Geochem 6:305–318
39. Ek J, Ek BM (1996) Radium and uranium concentrations in two eskers with enhanced radon emission. Environ Int 22:495–498
40. Navas A, Machín J, Soto J (2005) Mobility of natural radionuclides and selected major and trace elements along a soil toposequence in the central Spanish Pyrenees. Soil Sci 170:743–757
41. Howard JL, Sova JE (1993) Sequential extraction analysis of lead in michigan roadside soils: mobilization in the vadose zone by deicing salts. J Soil Contam 2:1–18
42. Alloway BJ (1990) Soil processes and the behaviour of metals. In: Alloway BJ (ed) Heavy metals in soils. Blackie, London, pp 11–37
43. Kartal S, Aydin Z, Tokalioglu S (2006) Fractionation of metals in street sediment samples by using the BCR sequential extraction procedure and multivariate statistical elucidation of the data. J Hazard Mater 132:80–89
44. Giri S, Singh AK, Mahato MK (2017) Metal contamination of agricultural soils in the copper mining areas of Singhbhum shear zone in India. J Earth Syst Sci 126:49

Part VII
Conclusions and Recommendations

Update, Conclusions and Recommendations for "Water Quality, Assessment and Management in India"

Abdelazim M. Negm, El-Sayed E. Omran, Shalini Yadav, and Ram Narayan Yadava

Abstract This chapter sheds light on the key conclusions and recommendations of the 19 chapters that were provided in the book titled "Water quality, assessment, and management in India". Also, some observations from a few recently published research works relevant to water quality. Therefore, this chapter provides details on (a) status of water resources in India, (b) water harvesting and water qualities, (c) water quality assessment and management and groundwater quality assessment and control. Besides, a collection of guidelines for the future research activity is pointed out to guide future research towards environmental protection, which is a main subject of strategic significance under Indian circumstances.

Keywords Water resources · Groundwater · Harvesting · Environment · Management · Quality · Quantity · India · Sustainability

A. M. Negm (✉)
Water and Water structures Engineering Dept., Faculty of Engineering, Zagazig University, Zagazig, Egypt
e-mail: amnegm@zu.edu.eg

E.-S. E. Omran
Soil and Water Department, Faculty of Agriculture, Suez Canal University, Ismailia, Egypt

Institute of African Research and Studies and Nile Basin Countries. Aswan University, Aswan, Egypt

S. Yadav
Centre of Excellence in Advanced Water and Environmental Research, Rabindranath Tagore University, Bhopal, Madhya Pradesh, India

R. N. Yadava
Research and International Affairs Department, Madhyanchal Professional University, Bhopal, Madhya Pradesh, India

S. Yadav et al. (eds.), *Water Quality, Assessment and Management in India*, Earth and Environmental Sciences Library, https://doi.org/10.1007/978-3-030-95687-5_19

1 Background

India is rich in water supplies, being equipped with a network of rivers and gifted with snow cover in the Himalayan range that can fulfill a variety of water requirements of the region. Throughout India, the quality of surface water is higher than groundwater. However, due to the decentralized nature of groundwater, it is readily usable and forms the main part of India's irrigation and drinking water supply. However, with the rapid rise in the population of the country and the desire to meet the growing need of irrigation, human and industrial use, the available water supplies in many areas of the world are getting depleted and the water quality has deteriorated. Indian rivers are contaminated due to the discharge of untreated waste and industrial effluents. Water contamination not only impacts water quality but also threatens human health, economic growth and social stability. To maintain that the water quality is being preserved or restored at the desired level, it is vital that it is monitored on a regular basis. Water quality monitoring helps in assessing the nature and degree of pollution control required and the efficacy of pollution control measures already in operation. This also assists in identifying water quality patterns and prioritizing pollution reduction measures.

Therefore, this chapter provides general findings of the water quality, assessment and management in India. In planning sustainable management of water supplies, it is important to give due concern to different resources used, which make the resultant production system unsustainable. So, the purpose of the book is to strengthen and discuss the following key theme:

- Status of Water Resources in India
- Water Harvesting and Water Qualities
- Water Quality Assessment and Management
- Groundwater Quality Assessment and Control

The next section presents a brief of the essential findings of some of the recent (updated) published studies on water quality, assessment and management, then the main conclusions of the book chapters in addition to the main recommendations for researchers and decision-makers. The update, conclusions, and recommendations presented in this chapter come from the data presented in this book.

2 Update

The following are the major update for the book project based on the main book theme:

2.1 Status of Water Resources in India

Two overviews were identified to discuss the status of water resources in India. The first overview is "the interlinking of Indian Rivers for sustainable use of water resources In India: An Overview." The interlinking of rivers (ILR) project is based on the National Perspective Plan (NPP), developed by the Ministry of Water Resources for water development in August 1980, ensuring the best possible use of water resources. India is known as a better-endowed nation in terms of its share in global water supplies with annual precipitation of 4000 km^3 and a per capita water supply of 1820 m^3 in 2001. India gets the majority of its rainfall during the monsoon. But, there is both a spatial as well as a temporal difference in the precipitation received all over the world. The plan to interlink rivers aims to exploit the available water resources in a fair and productive manner. The idea of moving water from water surplus areas to water shortage areas received attention by many water management planner and professionals. The concept of connecting rivers of India stems from the views of the (Indian Engineer Bharat Ratna Mokshagundam Visvesvaraya).

The second synopsis is "An overview—water resource management aspects in India." India has tremendous potential in terms of water resources; the total utilizable water supply of the country is estimated as 1123 billion cubic meters (BCM). The production and management of water resources are essential to meet rising domestic, agriculture, industrial, hydropower demands. Considering the current rate of increase in population by the year 2025, the per capita supply is likely to drop to below 1000 m^3, which could be classified as a situation of water scarcity [41]. Sustainable water management in India poses numerous challenges: bridging the increasing gap between demand and supply, providing enough water for production of food, balancing the uses between competing demands, meeting the growing demands of big cities, treatment of wastewater, sharing of water with the neighboring countries and among the co-basin states, etc. [1]. In India, climate changes can influence long term rainfall patterns impacting water availability, along with the threat of increasing occurrences of droughts and floods. Changes in climate over in India, especially the South-West monsoon, would have a major impact on agricultural production, water resources management, overall economy and livelihood sector of the country. Deterioration of water quality due to water contamination is a major environmental problem in India. Many of the rivers, lakes and surface water in India are contaminated. Water pollution in India has now reached a critical point. Almost every river system in India is now polluted to a considerable extent. Indian agriculture contributes about 46 percent of the gross national product and is also the main occupation of the people and the preoccupation of the government having the responsibility to provide adequate food for a population that makes up about 16 percent of the world and holds a potential agricultural which only 14 percent of the world's total. The agriculture sector has been successful in keeping pace with rising demand for food. For food and ecological sustainability of Indian agriculture, the state's interventions must be on the basis of the conservation agriculture approach [2].

2.2 *Water Harvesting and Water Qualities*

Seven potential approaches were identified for water harvesting and water quality in India: The first possible approach is "Drought characterization during monsoon months based on standardized precipitation index (SPI) in Nuapada District, Odisha, India." Droughts attracted the attention of scientific experts from backgrounds in the fields of environment, ecology, hydrology, meteorology, geology, and agriculture because it is recognized as an environmental catastrophe. Drought understanding and experience of past droughts events and their effects assist in planning as well as water resource management. Awareness of basic theories of drought would, therefore, be useful in the creation of models for exploring the different properties of drought. Information on the extent, duration, and severity of the drought is needed to address the adverse effects of drought [42]. Indices of drought can be used to achieve drought forecasting and tracking. Few studies attempted in monsoon months for the assessment of drought in case of drought-prone areas like Odisha, India [3].

A second possible approach is "Forest-water interactions at catchment scales for pollution management." Forests are essential constituents to the water cycle: they regulate stream flow, care groundwater recharge, and by evapotranspiration bestow to cloud generation and precipitation. Forests currently cover only about one-third of Earth's surfaces [4]. Forest uses water to rise, and therefore, fast-growing species will use water more quickly [5]. Trees often release water into the environment by evapotranspiration, which also returns locally as precipitation [6]. Thus, forest management may have both negative and positive impacts on water quantity & quality, plants, ranges, tree densities and other managerial aspects. Hydrological studies [7] found that water is very rarely considered first in forest management possibly because forest and water co-occurrence is so common. However, with global climate air temperatures and climate instability continuing to increase, the relationship between forests and water flow has begun to change drastically.

A third possible approach is "Assessment of groundwater quality in and around Nemawar, Madhya Pradesh, India." There are so many variables, such as disposal of various forms of pollution, land use activities, rock formation; the weather patterns and permeability are responsible for influencing groundwater quality. The quality of groundwater within an area is controlled by both natural processes such as precipitation intensity, weathering activities, and soil erosion and anthropogenic effects such as urban, commercial, and agricultural activities, as well as human exploitation of water resources. Groundwater movement is influenced by the physical and geochemical characteristics of contaminant, groundwater and the geological structure through which the polluted groundwater flows. Under those limitations, the presence of multiple ionic contents can make it unacceptable for irrigation, domestic or industrial usage. Our climate has been contaminated, impure, toxic and, therefore, dangerous to human health.

A fourth possible approach is "Hydrological response of factors affecting rainfall water discharge and water balance—A case study of Tons watershed." Hydrology is one of the working disciplines within the larger framework of hydrogeological sciences, itself a subcategory of earth sciences study. The hydrological cycle and its components, in which water circulates through various pathways above and below the ground, are of particular importance to hydrology. However, the hydrological cycle is only one example of the role that water plays in the operation of earth systems. The overflowing of the rivers is causing great harm to the people and their property around them. The most serious effect of the flood is to wash the topsoil cover away. A number of attempts have been made to analyze streamflow and sediment transfer data from the Himalayan region.

A fifth possible approach is "Water quality status of the Narmada river across its basin: predicting water quality using artificial neural network." A strong water quality management system plays a key role in pollution control and river basin planning. Proper management requires data on water quality for planning and decision-making. Narmada is Central India's largest perennial river receiving huge amounts of waste from domestic and municipal sewage discharges, industrial effluents, and agricultural run-offs. The rapid urbanization and industrialization will only increase the pollution rate of Narmada [8]. That makes it increasingly important to model and forecast the river's water quality. However, water quality control in data-constraints countries such as India is a challenge due to the lack of water quality data for many rivers and their tributaries [9]. Therefore, new mechanisms need to be developed that might aid in predicting the water quality status based on the limited data setting. Machine learning is a well-developed method which is known to play a role in predicting parameters of water quality [10]. Machine learning incorporates the use of various techniques such as tree-based approaches (decision tree, random forest), perceptron based approaches (Artificial Neural Network), statistical-based approaches (Bayesian Networks), instance-based learning (k-Nearest Neighbours) and support vector machines [11].

A sixth possible approach is "Rain water harvesting methods in Rajasthan." Earth when watched from the outer space appears to be a unique and splendid planet, "a blue globe" [12] with red areas from dessert and land and white mass of cloud. Water is important to life and it plays a significant role in the atmosphere on earth. This is an invaluable universal resource but not an absolute constant. Water is a scarce natural resource and plays an essential role in sustaining a safe climate. However, industrialization has resulted in the degradation and pollution of water resources by industrial effluents. Industrial pollution affects water, air, and soil. According to the report, the water quality issue affects about 44 million people worldwide [13]. Rain water collection is one of the environmentally sustainable methods that could be done individually to meet the demands of dropping groundwater levels. Rainwater is the primary fresh water source. Roof top rainwater could be made fit for drinking taking advantage of the available technologies. The water from the first rain is usually discarded.

A seventh possible approach is "Groundwater exploration using remote sensing and GIS techniques coupled with vertical electrical soundings from hard rock terrain: A Case study in Salem District, Southern India." Because of the broad and inconsistent variation of vital parameters, which characterize the groundwater system, production of groundwater resources in hard rock is fraught with many difficulties. The situation becomes even more precarious in the Indian context due to negligible primary porosity and low permeability of host rocks that restrict the scope of groundwater storage and movement, and generally low rainfall and high percentage of surface run-off resulting in limited recharge to the groundwater regime. Therefore it is important to perform thorough investigations to distinguish further possible zones from hard rock regions. Geophysical, Remote Sensing (RS), and GIS are some of the main approaches used to delineate possible groundwater zones. A systematic integration of these data with follow up of hydrogeological investigation provides rapid and cost-effective delineation of groundwater potential.

2.3 Water Quality Assessment and Management

Four methods were identified for water quality assessment and management as a potential way for sustainability. First is "Management of coal fly ash leachates generated from disposal sites near thermal power plants." Fly ash (FA) is the left-over product for generating electricity after coal is combusted in thermal power plant (TPP) furnaces. Upon combustion, small FA particles are stored in electrostatic precipitators along with flue gas and then either used or disposed of [14]. The dry type of FA is used for the manufacture of cement, for repair, for the building of roads or embankments, for the filling of mines, for the manufacture of bricks or tiles and as an agricultural amendment [15]. FA utilization for biomass production in forestry with economic trees has also been reported with high reclamation costs [16]. Up to now, 100% FA use is a difficult problem and the majority of unused FA is disposed of in terrestrial soil, causing tremendous pollution in the local area [16]. In addition, the projected demands for coal in the power sector indicate that more FA will be generated in the near future which will continue to surpass its consumption [17].

Second is "Assessment of water quality by evaluating water quality index over Kolkata using statistical approach—A step towards water pollution management." Kolkata is ecologically blessed with the Ganga beside its western end, historically big groundwater reserve and large wetlands region in its eastern periphery which naturally treats its wastewater and turns that as raw water for fishery and agriculture. Rapid population growth, industrialization, and urbanization contributed to a major increase in wastewater production. Indian rivers are contaminated by the disposal of untreated wastewater and industrial effluents. The huge problem of our river ecosystem's depletion has necessitated monitoring the water quality of numerous rivers across the country to assess their production capacity, utility potential and plan restorative actions. Heavy metals like Cu, Fe, Zn, Ni, Mn are essential micronutrients for plants and microorganisms while metals such as Pb, Cr, and Cd have no physiological activity and are toxic even if present in trace quantities. In recent years, various studies have been carried out in India to evaluate the water quality in aquatic ecosystems.

The third is "Assessment of water quality of the Ramganga River in Moradabad Region." The major causes of pollution of River water are sewage discharge, run-off from agriculture, wastewater from households and cattle bath, industrial effluents, etc. [18, 19]. The contaminated water from the river often contaminates the underground water belts, rendering it unsafe for human, livestock, industrial and irrigation use. As per the Indian Standard IS: 2296, the water is divided into five classes from A to E. Class A water is appropriate for consumption (drinking) without any treatment and class B water is appropriate for cleaning. But, class C is appropriate for consumption and domestic use only after treatment and disinfection. Furthermore, class D is appropriate for aquatic life and class E is recommended for cooling, waste disposal and dilution, industrial and irrigation applications [20]. Moradabad is a significant city on the western Uttar Pradesh (UP) frontier. As one moves from the national capital New Delhi to Lucknow, UP's capital. Moradabad is one of the largest cities on the way. It may be termed as the gateway of eastern UP. The city was established in 1625 AD and named after MuradBaksh, the son of Emperor Shah Jahan. Due to its brass industries, it is also called as PeetalNagri alias Brass City. About 40% of the handicraft export from India originate from Moradabad [21]. Ramganga River is one of the main tributaries of the Ganges, which originates from Dudhatoli-PauriGarhwal (Uttrakhand) and merges in the Ganges in Farukhabad (UP) in North India. The pollution of Ramganga water adds to the pollution of the Ganges to a large extent.

Added to the above, the "Assessment of surface water quality of Indian Rivers in terms of water quality index (WQI)" is discussed. Surface water is one of the most important inland resources required to sustain life, agriculture, industry and recreation [22]. At any level, the quality of a river reflects significant factors including basin lithology, atmospheric inputs, climatic conditions and anthropogenic inputs [22, 23]. Rivers play a significant role in the assimilation or transport of many contaminants from urban and industrial wastewater and runoff from agricultural land [24]. Study [25] has been performed to access water quality of Gomti River which has known to be polluted heavily through anthropogenic activities. Loktak Lake is situated in North East India, and the local people use the water for drinking purposes. Study [26] has been carried out in upper and middle regions of the Damodar river basin (DRB). The river Damodar is a source of water for millions of people around and within it. Mahananda River water quality has been evaluated [27] to test its suitability for domestic, agricultural, and industrial uses.

2.4 *Groundwater Quality Assessment and Control*

Five potential methodologies were identified for groundwater quality assessment and control. The first approach is "Groundwater quality assessment in the semi-arid blocks of Rajasthan, India: A combined approach of fuzzy aggregation technique with GIS." Groundwater has been a crucial resource for a society's growth, it serves many consuming water demands and provides freshwater with a direct potable source. Groundwater quality has a major impact on human health, agriculture and the availability of industrial water. It is extremely important to assess the quality and quantity of groundwater for long-term use and to ensure its environmental sustainability [28]. Contamination of groundwater occurs by different mechanisms, such as contact of groundwater with weathering of mineral rocks and degradation of soil salt concentrations. Other mechanisms include sub-surface pollutant intrusion which results in the accumulation of dissolved salts in groundwater. Agricultural practices also impact the quality of groundwater in aquifers by intruding contamination from non-point sources. Major contributors to non-point source emissions are fertilizers, unnecessary pesticides and enhancers dependent on nitrogen and phosphorous in agricultural fields [29]. In some arid blocks of Rajasthan, India, groundwater levels were observed to drop to 30 m along with constant water quality degradation [30]. The drop in groundwater levels increases the interaction between water and minerals and results in an increase in salinity and nitrate, iron and fluoride concentrations [28]. There are currently different methods for evaluating groundwater quality and researchers studied groundwater quality through geochemical interactions, numerical analysis and correlation analysis [31, 32].

The second methodology is "Assessment of groundwater quality parameters for drinking purpose using IDW, GIS and statistical analysis methods: A case study of basaltic hard rock area in Mahesh River Basin, Akola and Buldhana Districts (MS), India." Throughout India, for irrigation and domestic purposes, about 80 percent of the rural population and 50% of the metropolitan population use groundwater. About 33 percent of the land water wealth area of the country is not suitable for use in drinking and irrigation. Groundwater is an essential natural resource in India for the development of agriculture, human health, sustainable development and environment capability. Further, its procedures are one of the key natural resources for the production of groundwater activities. Groundwater quality is one of the most significant features of assessing groundwater quality. Recently there has been wonderful demand for fresh groundwater due to declines in rainfall, population growth and intensive farming activities. According to [33] India day by day has included more toxic anthropogenic activities such as industrial waste, open defecation, and unregulated disposal of sewerage in the standard of fresh groundwater and surface water. The composition of the recharged groundwater, hydrological cycle and human factors in any country influence its effect of considerable variations in groundwater quality parameters. Groundwater quality plays an important role in upholding human health and agricultural production standards, since it is directly linked to human welfare [34]. In India, due to unsafe water, exposure to potable water from groundwater supplies has increased, leading to significant adverse effects on human health. The lack of safe and potable water has emerged in recent years as one of the most serious issues in many parts of India one such example is the state of Telangana. Globally, around 748 million people still do not have access to an improved source of drinking water, and its demand for manufacturing is expected to increase by 400 percent by 2050 [35]. Groundwater quality inconsistency is a function of physical and chemical parameters that have a significant effect due to either natural processes or/and anthropogenic behavior. Improper management of industrial, municipal solid waste, agricultural runoff, urbanization and over-exploitation induced groundwater quality degradation [36, 37]. Evaluation of groundwater quality and establishment of a database is necessary in order to develop more strategies for the growth of civilization and water supplies [38].

The third methodology is "Evaluation of hydro-meteorological conditions and water resource potential for a Coalfield Area of Damodar Valley, India." Groundwater is like an essential commodity to meet the water demands of the Indian agricultural, industrial and domestic sectors. Groundwater is a natural water resource that is abundant and replenishable but its availability is limited in hard rock terrain. Access to clean drinking water remains an essential requirement, as 30 percent of urban populations and 90 percent of rural Indians still depend entirely on untreated surface or groundwater supplies. Though in recent years the shortage of safe and drinking water has emerged as one of the most severe developmental problems in many parts of West Bengal, Jharkhand, Orissa, Western Uttar Pradesh, Andhra Pradesh, Rajasthan and Punjab [39]. Groundwater table rise and fall are closely related to the area's hydro- meteorological conditions. The rainfall is the principal source for replenishing groundwater by percolation by deep soil zones. A major part of the rainfall drains out as surface run-off into the local drainages, which in turn discharges its load into the master drainage of that area. Establishment of coal-related industries and coal mining activities has marginally impacted the geology, geomorphology and soil structure with the recharge of aquifers related to their unique hydrogeological properties [40]. Various techniques are available to assess recharge-potential, and their capability in estimating recharge is also variable.

The fourth methodology is "Distribution and statistical source identification of heavy metals and radionuclides in soil: A case study from a proposed uranium mining site, Jharkhand, India." Natural soil radionuclides produce a significant component of the population's exposure to background radiation. Migration and accumulation of pollutants in soils is complex and includes processes such as leaching, capillary rise, drainage, sorption, root absorption and re-suspension into the atmosphere. Soil, along with its progeny, contains small amounts of radioactive elements U and Th. The actual terrestrial radiation rates are related to the composition of each lithologically separate region and the nature of the rock from which the soils derive. Following the discovery of deposits of low-grade uranium in several parts of eastern Singhbhum, uranium ore mining has begun in several places in the area. However, with the increasing demand for nuclear energy, it was suggested that new mining sites be excavated for uranium. One such proposed site was at Bagjata, which is an underground mine, in Singhbhum. Mining at Bagjata will affect the area's pre-existing environmental status.

3 Conclusions

Several findings from this book were taken by the editorial teams in the process of the current book project. The chapter draws valuable lessons from the cases in the book, in particular the positive features of water quality, assessment, and management in India, in addition to methodological observations. Such conclusions are significant for India's environmental management. Those are addressed in no specific order in the following.

3.1 Status of Water Resources in India

Linking of Indian rivers has been proposed as a solution to the looming water crisis. Interlinking of Indian rivers is a grand project aimed at a number of proposed benefits that would solve a number of water-related problems in the country, by diverting the extra flood water to the areas where the water quantity is limited to meet the demand of various sectors. The project aims at overcoming the regional imbalances in water availability in India. However right since its inception, the project has been shrouded in controversy. The paper analyses the policy, a summary of the benefits and a review of the major concerns. It also questions the viability of such a project to achieve the proposed goals and also provides certain alternatives to solve water issues in India. In this work the effort is made to suggest the different sustainable options keeping in mind the economic, technological and environmental feasibility analysis. It will also suggest the different possible alternatives before the project passes through the final stage of government approval.

Water resources management has always been an issue of great concern in India due to increase in water demand with the increase in population, development of agriculture, commencement of new industries, growth in other sectors and deteriorating water quality. Despite having enormous water resources, there has always been a shortage of safe drinking water and irrigation water. India is an agrarian country, where 70% of the population lives in rural areas and depends on agriculture for their livelihood. This sector has a very high water demand. This calls for sustainable water resources development in the country to bridge the increasing gap between demand and supply, providing enough water for domestic, agriculture and industries. The rainfall in the major part of the country is mainly governed by the southwest monsoon contributing enormous water to rivers, lakes, reservoirs, and groundwater. However the high temporal and spatial variation is the main cause for floods and droughts in the country. India has a network of rivers for the development of surface water resources however groundwater is the most preferred source of water.

3.2 Water Harvesting and Water Qualities

Thirty five years of monthly precipitation data utilized during monsoon months for the computation of SPI in Nuapada District's five blocks. Estimated SPI value were used for the identification of drought characteristic to suggest the strategy for water resources management. Maximum total (mod+sev + ext) drought events occurred in Bodan and Sinapali blocks whereas Komna block experienced the lowest number of total drought events as per SPI 3 Aug and Sep value. Therefore, the blocks experiencing frequent droughts like Bodan and Sinapali blocks, runoff water should be harvested through water harvesting structures for utilizing the limited water resources. Drought resistant varieties of crop should also be grown in the drought areas. Maximum drought duration of 23 months observed in case of Komna block. Different threshold limits of agricultural drought value were identified among the five blocks of Nuapada District as per the computed value of SPI 3 Aug and Sep. Bodan block experienced minimum agricultural threshold limit value in the case of both SPI 3 Aug and SPI 3 Sep.

With the increasing demand for agricultural and urban land (owing to population explosion & more affluent lifestyles) the majority of forests are put under tremendous pressure (further worsened by climate change). Water and land use policies in tropical countries, like India are often influenced to a big extent by many perceived effects from hydrological functioning of forested catchments towards soil erosion control and sediment-reduction benefits. Therefore, hydrological processes become indispensable for any informed discussion of forest-water interactions. Clean & abundant water is not only an extraordinary ecosystem service offered by forests; rather it serves as a base input to govern climatic as well as socio-economic concerns of the region & habitats.

For the water quality studies of Nemawar and adjacent areas of Madhya Pradesh, 15 groundwater samples were collected. It is concluded that the agricultural water quality of the study area was determined on the basis of Bicarbonate hazard, Salinity hazard and Sodium hazard. When the natural water of the study area classifies with respect to SAR and EC, the most of the water samples of the study area belongs to medium to high salinity class and Low sodium water. Kelly's ratio shows that 90% of natural water in pre and post monsoon belongs to a suitable class. Classification based on RSC clearly indicates that that all the water of the study area belongs to Safe class and thus they are suitable for agricultural purposes. As per the magnesium ratio classification, all the water of the study area belongs to suitable class and there are no magnesium hazards in natural water of the study area. The natural water of the study area is quite suitable for irrigation purposes.

A sediment budget and rates of denudation for the middle order streams and fourth order streams of the Tons watershed of Lesser Himalaya shows that the average sediment transported is about 113231.12 ton every year. The average rate of erosion is 1.02 mm/y. Out of the total sediment budget, 32.62% bed load was measured, which was the most contribution in denudation for load type and39.31% suspended load and only 28.07% was dissolved load generates the denudation rate in Tons watershed. The generation and flow of sediment load in the Himalaya region may be the result of rainfall variability, land use, slopes, natural hazards, neotectonics and anthropogenic activities within the catchment. Degradation may be the result of enhanced sediment delivery to the streams in the Himalaya.

The prediction model developed using Artificial Neural Network (ANN) was developed for the Narmada River which could predict the Chemical Oxygen Demand level category in the sample based on other water quality parameters. The performance of the model is evaluated using the accuracy metric. The study provides an approach that could be used to develop a database of all limited data and develop a model, which could be used to predict different water quality parameters using other water quality parameters. This could solve the very critical problem in the water quality management for countries suffering from data constraints. These countries with limited data set could estimate the overall water quality status of their river based on the sequential and/or parallel prediction models.

In order to meet the demands for water we need to start practicing rainwater harvesting. The amount of water harvested varies from place to place. It depends on the amount of annual rainfall data. Rainwater is found to be the most pure form of water as it is free from impurities and contaminants. Rainwater has much potential to increase the ground water level. Rainwater harvesting is the need of the hour. A useful technique should be developed in a country like India.

The application of Remote Sensing (RS) and Geographical Information System (GIS) techniques coupled with geophysical techniques in the exploration and assessment of groundwater especially in hard rock terrain, it is crucial for successful implementation. From the potential map output the panchayat union wise areal extent have been accessed and most prospective panchayat union falls under the category of excellent, good, and good to moderate that is Konganapuram (68.1%) followed by Verapandy (64.4%) and Mc Donalds Choulery (62.4%). Based on the results of 21 VES, 13 deep wells were drilled. The total depth of the drilled wells ranges from 60–140 m. The yield of wells range from 1.5 to 755 lpm (liters per minute) with one well did not give any yield. It indicates that high yielding wells are located in Excellent and good groundwater potential zones. The yield ranging from 250 to 755 lpm on excellent category, 109 to 276 lpm on good category, 49 lpm on good to moderate, and Dry to 8.6 lpm on moderate to poor class. This indicates that the integration of Remote sensing and GIS techniques coupled with geophysical data reduces the time, cost and economy while identification suitable sites for drilling wells in the drought affected regions where the success rate are very low due to the decision taken only on conventional data.

3.3 *Water Quality Assessment and Management*

In developing countries like India, water is crucial for industrial development. Water treatment strategies are the most studies topics across the globe. Various sophisticated techniques have been found till date which can remove impurities in every drop of water. However, prevention of water pollution should be the primary concern which can be more useful for the economic development of a country and reduce treatment costs. In this context, manipulating the fly ash leachate quality before its origination should be the primary concern of the thermal power authorities. Research on demineralization of coal can be a suitable alternative in this context. Additionally the idea of switching over to alternative sources of energy yielding fuels (viz., biomass to biofuel, algae to biofuel, biohydrogen, etc.) is of tremendous importance. It is a highly demanding area of research in the era of global warming and climate change caused by fossil fuels. However, to conclude in the present context leaching tests are an utmost parameter which should be followed to indicate possible future groundwater contamination. The leaching test method should be wisely chosen keeping in view the climate, geological strata, groundwater table, surface water bodies in the area where FA is to be disposed. Incorporation of attenuation barriers underneath fly ash ponds and establishment of a proper leachate collection system are of the utmost requirement before FA disposal. These strategies can be followed up with a suitable treatment technique as per leachate quality, so as to further discharge or recycle the water as per industry requirements and minimize freshwater consumption.

The study aims the assessment of Water Quality by evaluating water quality index over Kolkata. For this purpose, the analysis of different physicochemical parameters is considered. From the analysis of the parameters a distinct seasonal variability has been observed. In addition to that water quality has been monitored through calculation of Water Quality index. It is observed that during the wet season the water quality is found to be rather bad condition in terms of WQI. Thus, it can be stated that the endeavor of the present study is to identify the changes in water quality so that the stakeholders can take necessary actions to deal with.

Ramganga River gets polluted severely while it flows from the Moradabad city to Farukhabad. It adds significant pollution to the River Ganga. The industries such as metal surface, food processing, slaughterhouses, meat processing, pulp and paper, and milk processing are major contributors to water pollution of Ramganga. The sewage, agriculture runoff, solid wastes, domestic wastewater, and cattle bath discharge are the other main contributors. The efforts of UPPCB and CPCB are insufficient due to a large number of unorganized industry units or subsidiaries, which are not registered. Such small units are operated in the households in Moradabad city and have no treatment facilities. A comprehensive remedial method is proposed with a large emphasis on the monitoring of the River flow through satellite imagery and an assessment of its water quality on a real-time basis. The assessment results should be put in the public domain to make people aware and responsible. Awareness and training of the public are required to a large extent. To control the population of the Ramganga River and to meet the further pressure on it due to the increase in population, rainwater harvesting and development of local water bodies is of utmost importance. Pollution control measures need further improvement.

The study clearly demonstrated how the fuzzy approach is useful in defining the sustainability indicators under uncertain environment. The fuzzy set theory has helped in handling linguistic variables with variation in GW quantity and quality attributes. It is very difficult to compare and incorporate non-comparable attributes in a mathematical model, but the fuzzy set theory gave a flexible and robust computational framework for the study. Integrating GIS in the fuzzy model was helpful in standardizing the input data to perform calculations accurately for selected regions. The final integrated model was able to downscale the wide array of data and obtain priority maps of the study area that can be readily used by decision-makers for sustainable groundwater management. The dual approach of quality and quantity for solving the sustainability assessment problem for GW in Rajasthan resulted an in-depth insight into the problem at regional as well as at the state level.

This chapter shows the simplicity and ease of water quality index to access any water bodies. WQI is a mathematical tool which has the ability to provide a single number for the large quantities of water quality data in a comprehensive manner. Most of the reviewed Indian Rivers show a wide range of quality varying from bad to excellent category. Downstream ends of most of the rivers have been fallen under bad category owing to anthropogenic pollution and illegal sewage and effluent discharges. Such bad quality water is not suitable to domestic, agricultural and industrial purposes.

3.4 Groundwater Quality Assessment and Control

Groundwater quality analysis of Mahesh river basin for the drinking purpose has shown the slightly Saline nature and the high ranges of alkalinity in the basin area. The pH and TDS profiles of both pre-post monsoons alongside the river in the study area presented a progressive rise of TDS concentration in the lower drainage way. Some parts of the groundwater are moderate affected by the nitrate and Chloride concentration. The groundwater of the study area is good to permissible limits in the study area. The groundwater quality maps have shown the EC value ranges from 345 to 1587 and 348 and 1598 for pre-post monsoons respectively. The various indices and Statistical analysis and various groundwater quality plotting's derived in the study indicates that the most of groundwater of the study area is suitable for drinking and irrigation needs absence of rainwater and that time groundwater playing important role in drinking and irrigation purposes.

Rainfall is the principal recharge source to groundwater. The EBC area experiences an average annual rainfall of about 1366 mm and the highest annual rainfall was recorded as 2543.8 mm in 2011. Besides rainfall, the mine water discharge from the local mining areas and existing water bodies including waterlogged in abundant mine quarries are also contributed to the groundwater recharge as return flow. However, the marginal declining trend of water level fluctuation might be attributed due to the increasing overdraft by the local population. The Total annual replenishable recharge (TARR) was calculated to be 48.56 Mm^3/year and 62.21 Mm^3/year by rainfall infiltration factor method and groundwater table fluctuation method. The net annual groundwater availability was estimated to be 13.7 Mm^3 by rainfall infiltration method. Stage of groundwater development was calculated under the safe category. Modeling of large aquifers of mining area seems an almost impossible assignment. However, the essential information for water availability is the information on fluxes in and out of the storage.

The concentration of the discussed radionuclides and heavy metals in the soil from various locations of Bagjata mining areas assumes importance due to large-scale industrial and mining activity in the region. The concentration levels of the radionuclides depict the natural background concentration. Cu and Ni exceeded the Maximum Allowable limits (MAL) given by Kloke, 1980 in most of the locations. The radionuclides and the heavy metals in the soil are distributed unevenly which may be attributed to the widespread uranium mineralization in this area. The Principal Component Analysis (PCA) suggests that the main source of heavy metals is the lithogenic origin. The association of the radionuclides with the Fe is also depicted from the analysis. The other source of heavy metals in the area is vehicular load and Copper mining in the area. The concentration of Cu depicts moderate pollution with respect to I_{geo} values.

4 Recommendations

Throughout the course of this book project, the editorial teams acknowledged other areas that could be explored in order to develop further. Based on the authors' observations and conclusions, this section provides a set of recommendations that provide suggestions for potential researchers to go beyond the scope of this book. The following suggestions are derived primarily from the chapters set out in this volume:

- The improper distribution and differential consumption of water, not ignoring the aspects of consumer wastages and leakages etc., is one of main culprits behind water shortage in India. Rainwater harvesting at local level has produced many desirable results. When water is allowed to percolate naturally or artificially into the ground, even dried seasonal rivers have been rejuvenated (as demonstrated in Rajasthan by Rajendra Singh of Tarun Bharat Sangh). The IRL is believed to be crucial for food security in India. But improvement in farming technologies and cultivation of high yielding varieties has produced good results. The proponents of IRL need to examine the Indian political system, which is a huge nexus between politicians, beaurocrats, and contractors that doesn't work with public interest in their minds. For them the prime motive is profit; the project is estimated at 5 lakh crores and is a virtual goldmine for them. Even if we assume that the project has zero technical error and execution, political interference may completely defeat the purpose of this project.
- Sustainable water resources development is important in India to bridge the increasing gap between demand and supply, providing enough water for domestic, agriculture and industries. Identification of potential areas for surface water resources development through a vast network of rivers and implementing programs for augmentation of groundwater recharge through natural and artificial means will be the key to tackle this issue. Systematic scientific approach for assessment of water scarcity, droughts, formulate mitigation plan, assessment of extent and magnitude of floods, flood management, development of policies under climate change scenarios are key issues to tackle flood and droughts. Agriculture is the main occupation of the people in the country. Policies, provision, plans have to be made for assured water availability to meet irrigation demand, thereby increasing food production to meet increasing food demand. Deterioration of surface and groundwater quality is a major immediate action, which is needed to implement policy options for reducing industrial water,

industrial effluent treatment and urban waste-water treatment for better handling of organic wastes.

- Existing cropping pattern should be modified in case of drought prone areas by growing the crops having low water requirement. Categories of drought severity were not uniform across the different blocks of Nuapada District. Therefore, management of water resources for coping the drought should be different for different blocks.
- Managing forests could be an easier & effective task once a comprehensive 'forest & water agenda' be thought of keeping sustainable water productivity and effectual conservation aspects as two big targets or goals. Integrated advanced models & modelling approaches must be given high priorities for forests and forested catchments by accommodating their prevailing hydrological, climatic, meteorological, geo-physical, socio-economic, and environmental considerations. Forests and forest-based catchments or a landscape happens to be historical creatures with a variety of abandoned bits of traditional knowledge in regards to effectual natural resource management. It will certainly be a prolific exertion to plan or incorporate such elements in forest policy or forest & water-based development planning.
- In Tons watershed, the study recommends that the results will be worthwhile in planning to reduce the inflow of silt into the reservoir, conserve soil and minimize run-off in the extreme rainfall years. Results suggest to planners and engineers for formulating strategies to control erosion and landslides for watershed management on precursory basis. The outcome of this study has proved to recommend that useful in understanding the geographic conditions and hydrological behavior (discharge and sediment flux) of the basin in the context of setting up small hydropower plants in the region. The study also recommends that an agency should act as a valuable reference for hydropower station management and governmental oversight. It is also necessary to understand natural processes within the regional ecological systems, the human impacts on these processes, and ecological responses. There is an urgent need to discover the consequences for future planning towards sustainable development.
- This study provides certain recommendations. Firstly, the study recommends the use of supervised machine learning algorithms to develop models for estimating and predicting the water quality status of rivers in limited data settings. Secondly, the study recommends use of the ANN algorithm-based model to predict the Narmada water quality status.
- Following recommendations have been listed for preventing water pollution from fly ash waste disposal sites: FA disposal by the Thermal Power Authorities should include prior involvement of a series of leaching tests which will indicate future possibilities of groundwater contamination. The leaching test method should be chosen on the basis of climate and geological strata. Also before disposal of the fly ash Thermal Power Authorities should note the groundwater table and position of surface water bodies from the disposal site. This will determine the strategy of pollution prevention to be used like attenuation barriers underneath fly ash ponds and the establishment of a proper leachate collection system. After disposal of fly

ash, regular manipulation of the originating FA leachates should be done before discharging the leachates. This can be done through a suitable treatment technique as per the leachate quality, so as to further discharge or recycle the water as per industry requirements and minimize freshwater consumption. Regular monitoring of water quality near the disposal site and planning of remediation measures accordingly are recommended.

- Establishment of Water Quality Monitoring and Assessment Collaboration across Jurisdictions is necessary in this regard. To determine the socio-economic impact setting up of Research and Monitoring program can be recommended. The availability of data on daily basis would give more accuracy to evaluate the water quality parameters over a specific location.
- The future scope of the study is to map the region of Rajasthan at regional level and at a much higher resolution. This can only be achieved if the governing agencies and privately funded research projects collaborate to develop an updated unified database from the region. Decision-makers should prioritize the critical regions identified in terms of GW Scarcity, GW Quality, and GW Sustainability for developing water infrastructures in the near future. Although the mapping and classification technique used in the study use limited number of parameters, the indicators can be modified and adapted for any number of indicators, provided a similar approach is used.
- The selected groundwater quality parameters of Maharashtra are so important for drinking and irrigation purposes in the basaltic hard rock area. The regular monitoring of groundwater quality analysis is needed for the development, planning and management of human, animal and sustainable agriculture. Results can be used for development, planning and management of policy in study area. These results may utilization to sustainable water resource management and agriculture development in the basaltic hard-rock regions.
- Evaluation of specific yield of the wells by pumping test studies at suitable locations. Evaluation of the Transmissivity & Storativity of the underlying aquifer. Contaminant Transport Modeling can be elaborated by visual MODFLOW using data of few years. Similar studies may be taken up for mining areas other than coal mining like Copper and Uranium mining.
- Regular monitoring of soil samples in the vicinity of the mining areas is recommended so as to know the post mining effect with respect to radionuclide and metals. The content of radionuclides and metals in the agricultural produces should be done since there is a strong probability for the pollutants to get into the food chain through the soil. Human health risk assessment is suggested due to the exposure of the radionuclides and metals in the soil through the ingestion, dermal and inhalation pathways.
- Monitoring: Regular and stringent monitoring of water bodies is required to safeguard them from pollution. Improving the sewerage system: Cities located near rivers should be completely sewered and all the wastewater even in low-lying areas near the river should be sent (through pumps if necessary) for treatment and disposal insuring ‘zero’ discharge in the river. It can be done by upgrading the existing STPs, which do not meet the required disposal standards.

Flow augmentation to increase capacity: To increase the flow capacity of the rivers, water stored during monsoon period should be used and during dry seasons allow water to be released in rivers. Efficient aeration should be done for streams and various open drains carrying the waste water. Risk: Health risk assessment is required to be evaluated for all parameters having concentration higher than the permissible range to limit the water borne diseases.

References

1. Jain SK (2012) Sustainable Water Management in India considering likely climate and other changes. Curr Sci 102(2)
2. Gupta R, Mehra M, Sahoo R, Abrol I (2019) Indian agriculture redefining strategies and priorities. Econ Politic Weekly LIII:84–91
3. Sudarsan Rao A, Padhi J, Das B (2018) Assessment of drought in Balangir District of Odisha, India using drought Indices. In: Singh V, Yadav S, Yadava R (eds) Climate change impacts. Water Science and Technology Library, vol 82. Springer, Singapore
4. FAO (2016) State of the world's forests 2016. Available at http://www.fao.org/publications/sofo/2016/en/ Accessed on 23.12.2018
5. Filoso S, Bezerra MO, Weiss KCB, Palmer MA (2017) Impacts of forest restoration on water yield: a systematic review. PLoS One 12(8):e0183210
6. Ellison D et al (2017) Trees, forests and water: cool insights for a hot world. Glob Environ Chang 43:51–61
7. USDA (2018). Management option for dealing with changing forest-water relations. Available at https://www.srs.fs.usda.gov/pubs/chap/chap_6_2018_mcnulty.pdf. Accessed on 07.09.2019
8. Katakwar M (2016) Narmada river water: pollution and its impact on the human health. Int J Chem Stud IJCS 4:66–70
9. Srikanth R (2009) Challenges of sustainable water quality management in rural India. Curr Sci 97:317–325
10. Chau K (2006) A review on integration of artificial intelligence into water quality modelling. Mar Pollut Bull 52:726–733
11. Kotsiantis SB (2007) Supervised machine learning: a review of classification techniques. Informatica 31:249–268
12. Ishaku HT, Majid MR, Johar F (2012) Rainwater harvesting: an alternative to safe water supply in Nigerian rural communities. Water Resour Manag 26(2):295–305. https://doi.org/10.1007/s11269-011-9918-7
13. Fletcher TD, Andrieu H, Hamel P (2013) Understanding, management and modelling of urban hydrology and its consequences for receiving waters: a state of the art. Adv Water Resour 51:261–279. https://doi.org/10.1016/j.advwatres.2012.09.001
14. Belviso C, Cavalcante F, Di Gennaro S, Palma A, Ragone P, Fiore S (2015) Mobility of trace elements in fly ash and in zeolitised coal fly ash. Fuel 144:369–379
15. CEA (Central Electricity Authority) (2017) Annual report on fly-ash utilization, report on fly ash generation at coal/lignite based thermal power stations and its utilization in the country for the year 2016–17, New Delhi
16. Tiwari MK, Bajpai S, Dewangan UK (2016) Fly ash utilization: a brief review in Indian context. Int Res J Eng Technol 3(4):949–956
17. Wang P, Wang J, Qin Q, Wang H (2017) Life cycle assessment of magnetized fly-ash compound fertilizer production: a case study in China. Renew Sust Ener Rev 73:706–713
18. Singh P (2018a) Water quality assessment of River Ramganga at Bareilly. Res Rev J Ecol Environ Sci 6(4):33–40

19. Singh P (2018b) Hydrobiological study on summer season of River Ramganga at Bareilly. Environ Pollut Clim Change 2(3):1–4. https://doi.org/10.4172/2573-458x.1000162
20. UPPCB (2018) Action plan for restoration of polluted stretch of River Ramganga from Moradabad to Kannauj. Uttar Pradesh Pollution Control Board, Vibhuti Khand, Gomtinagar, Lucknow: 1–131
21. India Population (2019) Population of Moradabad 2019. Available on https://indiapopulation2019.com/population-of-moradabad-2019.html. Accessed on 23.01.20
22. Biglin A, Konanc MU (2016) Evaluation of surface water quality and heavy metal pollution of Coruh river basin (Turkey) by multivariate statistical methods. Environ Earth Sci 75:1029
23. Jung KY, Lee KL, Im TH, Lee IJ, Kim S, Han KY, Ahn JM (2016) Evaluation of water quality for the Nakdong River watershed using multivariate analysis. Environ Technol Innov 5:67–82
24. Malik RN, Hashmi MZ (2017) Multivariate statistical techniques for the evaluation of surface water quality of the Himalayan foothills streams, Pakistan. Appl Water Sci. https://doi.org/10.1007/s13201-017-0532-6
25. Iqbal K, Ahmad S, Dutta V (2019) Pollution mapping in a urban segment of a tropical river: is water quality. J Appl Water Sci 9:197
26. Verma RK, Murthy S, Tiwary RK, Verma S (2019) Development of simplified WQIs for assessment of spatial and temporal variations of surface water quality in upper Damodar river basin, eastern India. J Appl Water Sci 9:21
27. Shil S, Singh UK, Mehta P (2019) Water quality assessment of a tropical river using water quality index (WQI), multivariate statistical techniques and GIS. J Appl Water Sci 9:168
28. Singh AP, Khakoliya A, Tavanshetti S, Yadav J (2019) Groundwater quality assessment using GIS and fuzzy logic – a case study of Jhunjhunu District. Pollut Res 38(3):140–147
29. Ahada CPS, Suthar S (2017) Hydrochemistry of groundwater in North Rajasthan, India: chemical and multivariate analysis. Environ Earth Sci 76(5):203
30. Chintalapudi P, Pujari P, Khadse G, Sanam R, Labhasetwar P (2017) Groundwater quality assessment in emerging industrial cluster of alluvial aquifer near Jaipur, India. Environ Earth Sci 76(1):1–14
31. Bhakar P, Singh AP (2018b) Groundwater quality assessment in a hyper-arid region of Rajasthan, India. Nat Resour Res 28(2):505–522
32. Bhakar P, Singh AP (2018a) Life cycle assessment of groundwater supply system in a hyper-arid region of India. Procedia CIRP 69(May):603–608
33. Mukate S, Panaskar D, Wagh V, Muley A, Jangam C, Pawar RS (2018) Impact of anthropogenic inputs on water quality in Chincholi industrial area of Solapur, Maharashtra, India. Groundw Sustain Dev 7:359–371
34. FAO (2017) Water pollution from agriculture: a global review published by the UN Food and Agriculture Organization (FAO), and the International Water Management Institute (IWMI) on behalf of the Water Land and Ecosystems research program Colombo, 2017 (www.fao.org/publications)
35. WWAP (United Nations World Water Assessment Programme) (2015) The United Nations World Water Development Report 2015: Water for a Sustainable World. Paris, UNESCO
36. Aradhi KK, Kurakalva RM, Dararam B (2019) Assessment of groundwater quality, toxicity and health risk in an industrial area using multivariate statistical methods. Environ Syst Res 8:26–34
37. Zhang X, Miao J, Hu BX, Liu H, Zhang H, Ma Z (2017) Hydrogeochemical characterization and groundwater quality assessment in intruded coastal brine aquifers (Laizhou Bay, China). Environ Sci Pollut Res Int 26:21073–21090
38. Rawat KS, Singh SK (2018) Water quality indices and GIS based evaluation of a decadal groundwater quality. Geol Ecol Landsc 2(4):240–255

39. Tiwari AK, Singh AK (2014) Hydrogeochemical investigation and groundwater quality assessment of Pratapgarh district, Uttar Pradesh. J Geol Soc India 83(3):329–343
40. Chandra S, Singh PK, Tiwari AK, Panigrahya BP, Kumar A (2015) Evaluation of hydrogeological factors and their relationship with seasonal water table fluctuation in Dhanbad district, Jharkhand, India. J Hydraulic Eng 21(2):193–206
41. Government of India (2006) Report of sub-group of minor irrigation, CAD and private sector and beneficiaries participation for the XI Five Year Plan (2007–12), Ministry of Water Resources, GOI
42. Jain VK, Pandey RP, Jain MK, Byun HR (2015) Comparison of drought indices for appraisal of drought characteristics in the Ken River Basin. Weath Clim Ext 8:1–11